THE

WORLD ATLAS

A SIGNET BOOK
NEW AMERICAN LIBRARY
TIMES MIRROR

ISBN Number: 0-451-08150-1

Library of Congress Catalog Card Number: 78-57590

SIGNET TRADEMARK REG. U.S. PAT. OFF. AND FOREIGN COUNTRIES
REGISTERED TRADEMARK—MARCA REGISTRADA
HECHO EN CHICAGO, U.S.A.

SIGNET, SIGNET CLASSICS, MENTOR, PLUME and MERIDIAN BOOKS are published by The New American Library, Inc., 1301 Avenue of the Americas, New York, New York 10019

First Signet Printing, July, 1978

1 2 3 4 5 6 7 8 9

PRINTED IN THE UNITED STATES OF AMERICA

CONTENTS

GAZETTEER-INDEX OF THE WORLD

This alphabetical list of grand divisions, countries, states, colonial possessions, etc., gives area, population, capital, seat of government or chief town, and index references and numbers of plates on which they are shown on the largest scale. The index reference shows the square on the respective map in which the name of the country, state or colonial possession is located. *Indicates members of the United Nations.

Country	Area (Sq. Miles)	Population	Capital or Chief Town	Index Ref.	Plate No.
*Afghanistan	250,775	20,000,000	Kabul	B 1	64
Africa	11,707,000	451,200,000			76-77
Alabama, U.S.A.	51,609	3,444,165	Montgomery		114-115
Alaska, U.S.A.	586,412	302,173	Juneau		116-117
*Albania	11,100	2,482,000	Tiranë	B 3	49
Alberta, Canada	255,285	1,627,874	Edmonton		230-231
*Algeria	919,591	16,776,000	Algiers	E 3	78
American Samoa	83	30,000	Pago Pago	K 7	85
Andorra	188	26,558	Andorra la Vella	G 1	41
*Angola	481,351	6,761,000	Luanda	D 4	82
Antarctica	5,500,000				90-91
Antigua	171	73,000	St. John's	G 3	107
*Argentina	1,068,296	26,100,000	Buenos Aires		98-99
Arizona, U.S.A.	113,909	1,772,482	Phoenix		118-119
Arkansas, U.S.A.	53,104	1,923,295	Little Rock		120-121
Armenian S.S.R., U.S.S.R.	11,506	2,491,900	Erivan	F 6	51
Asia	17,128,500	2,535,333,000			54-55
*Australia	2,967,909	14,000,000	Canberra		86-89
*Austria	32,375	7,540,000	Vienna		46
Azerbaidzhan S.S.R., U.S.S.R.	33,436	5,117,100	Baku	G 6	51
*Bahamas	5,382	220,000	Nassau	C 1	106
*Bahrain	240	300,000	Manama	G 4	57
*Bangladesh	55,126	82,900,000	Dacca	E 2	65
*Barbados	166	253,620	Bridgetown	G 4	107
*Belgium	11,781	9,813,000	Brussels		37
Belize	8,867	144,000	Belmopan	B 1	104
*Benin	43,483	3,200,000	Porto-Novo	E 7	79
Bermuda	21	57,000	Hamilton	G 2	107
*Bhutan	18,147	1,200,000	Thimphu	E 2	65
*Bolivia	424,163	4,804,000	La Paz, Sucre	D 7	95
*Botswana	224,764	700,000	Gaborone	E 6	82
*Brazil	3,286,470	116,000,000	Brasília		94-98
British Columbia, Can.	366,255	2,184,621	Victoria		232-233
British Indian Ocean Territory	29	1,400	(London, U.K.)	F 6	55
British Virgin Islands	59	10,484	Road Town	H 1	107
Brunei	2,226	155,000	Bandar Seri Begawan	E 5	74
*Bulgaria	42,823	8,800,000	Sofia	D 3	48
*Burma	261,789	31,240,000	Rangoon	A 2	66
*Burundi	10,747	4,100,000	Bujumbura	F 2	83
*Byelorussian (White Russian) S.S.R., U.S.S.R.	80,154	9,522,000	Minsk	C 4	51
California, U.S.A.	158,693	19,953,134	Sacramento		122-123
*Cambodia	69,898	8,110,000	Phnom Penh	D 4	67
*Cameroon	183,568	6,600,000	Yaoundé	G 7	79
*Canada	3,851,809	23,400,000	Ottawa		108-109
Canal Zone, U.S.A.	647	44,650	Balboa Heights	E 3	105

Country	Area (Sq. Miles)	Population	Capital or Chief Town	Index Ref.	Plate No.
*Cape Verde	1,557	302,000	Praia	G 9	77
Cayman Islands	100	10,652	Georgetown	B 3	106
*Central African Empire	236,293	1,800,000	Bangui	A 6	81
Central America	197,575	20,500,000	....		104-105
*Ceylon (Sri Lanka)	25,332	14,000,000	Colombo	D 4	64
*Chad	495,752	4,178,000	N'Djamena	A 4	81
Channel Islands	74	128,000	Saint Helier, St. Peter Port	E 6	31
*Chile	292,256	10,700,000	Santiago		98-99
*China (People's Rep.)	3,691,000	853,000,000	Peking		68-69
China (Taiwan)	13,971	16,426,386	Taipei	F 3	69
*Colombia	439,735	25,000,000	Bogotá	C 3	94
Colorado, U.S.A.	104,247	2,207,259	Denver		124-125
*Comoro Islands	719	266,000	Moroni	J 4	83
*Congo	132,046	1,400,000	Brazzaville	C 2	82
Connecticut, U.S.A.	5,009	3,032,217	Hartford		126-127
Cook Islands	91	17,046	Avarua	L 7	85
*Costa Rica	19,652	2,100,000	San José	C 3	104
*Cuba	42,827	9,500,000	Havana	B 2	106
*Cyprus	3,473	639,000	Nicosia	E 5	58
*Czechoslovakia	49,373	14,900,000	Prague		46-47
Delaware, U.S.A.	2,057	548,104	Dover		128-129
*Denmark	16,629	5,065,313	Copenhagen	A 3	33
District of Columbia, U.S.A.	67	756,510	Washington	A 3	152
*Djibouti	8,880	250,000	Djibouti	F 5	81
Dominica	290	70,302	Roseau	G 4	107
*Dominican Republic	18,816	5,000,000	Santo Domingo	D 3	106
*East Germany (German Democratic Rep.)	41,768	16,850,000	Berlin (East)		34-35
*Ecuador	109,483	7,400,000	Quito	B 4	94
*Egypt	386,659	37,900,000	Cairo	C 2	80
*El Salvador	8,260	4,300,000	San Salvador	B 2	104
England, U.K.	50,516	46,417,600	London		30-31
*Equatorial Guinea	10,831	320,000	Malabo	F 8	79
Estonian S.S.R., U.S.S.R.	17,413	1,356,100	Tallinn	C 3	50
*Ethiopia	471,776	30,000,000	Addis Ababa	E 5	81
Europe	4,057,000	666,116,000	....		28-29
Faerøe Islands, Den.	540	38,000	Tórshavn	D 2	28
Falkland Islands	6,198	1,905	Stanley	D 8	99
*Fiji	7,055	569,468	Suva	H 7	84
*Finland	130,128	4,729,000	Helsinki		32-33
Florida, U.S.A.	58,560	6,789,443	Tallahassee		130-131
*France	210,038	53,300,000	Paris		38-39
French Guiana	35,135	62,000	Cayenne	B 2	96
French Polynesia	1,544	135,000	Papeete	M 7	85
*Gabon	103,346	526,000	Libreville	C 2	82
*Gambia	4,127	524,000	Banjul	A 6	79
Georgia, U.S.A.	58,876	4,589,575	Atlanta		132-133
Georgian S.S.R., U.S.S.R.	26,911	4,686,000	Tbilisi	F 6	51
*Germany, East (German Democratic Rep.)	41,768	16,850,000	Berlin (East)		34-35
*Germany, West (Federal Republic)	95,985	61,644,600	Bonn		34-35
*Ghana	92,099	10,200,000	Accra	D 7	79
Gibraltar	2.28	30,000	Gibraltar	D 4	40
Gilbert Islands	354	47,711	Bairiki	H 5	84
*Great Britain & Northern Ireland (U.K.)	94,399	56,076,000	London		30-31
*Greece	50,944	9,046,000	Athens		49
Greenland	840,000	54,000	Godthåb	P 2	100

Country	Area (Sq. Miles)	Population	Capital or Chief Town	Index Ref.	Plate No.
* Grenada	133	96,000	Saint George's	F 4	107
Guadeloupe	687	324,000	Basse-Terre	F 3	107
Guam	212	111,000	Agaña	E 4	84
* Guatemala	42,042	6,200,000	Guatemala	B 2	104
* Guinea	94,925	4,500,000	Conakry	B 6	79
* Guinea-Bissau	13,948	517,000	Bissau	A 6	79
* Guyana	83,000	800,000	Georgetown	A 1	96
* Haiti	10,694	4,867,190	Port-au-Prince	D 3	106
Hawaii, U.S.A.	6,450	769,913	Honolulu		134-135
* Holland (Netherlands)	15,892	13,800,000	The Hague, Amsterdam		36-37
* Honduras	43,277	3,000,000	Tegucigalpa	C 2	104
Hong Kong	403	4,400,000	Victoria	E 3	69
* Hungary	35,919	10,590,000	Budapest		47
* Iceland	39,768	220,000	Reykjavík	C 2	28
Idaho, U.S.A.	83,557	713,008	Boise		136-137
Illinois, U.S.A.	56,400	11,113,976	Springfield		138-139
* India	1,269,339	605,614,000	New Delhi		64-65
Indiana, U.S.A.	36,291	5,193,669	Indianapolis		140-141
* Indonesia	788,430	131,255,000	Djakarta		74-75
Iowa, U.S.A.	56,290	2,825,041	Des Moines		142-143
* Iran	636,293	32,900,000	Tehran		62-63
* Iraq	172,476	11,400,000	Baghdad		62
* Ireland	27,136	3,109,000	Dublin		30-31
Ireland, Northern, U.K.	5,452	1,537,200	Belfast	C 3	30-31
* Israel	7,847	3,459,000	Jerusalem		60-61
* Italy	116,303	56,110,000	Rome		42-43
* Ivory Coast	127,520	7,000,000	Abidjan	C 7	79
* Jamaica	4,244	2,200,000	Kingston	C 3	106
* Japan	145,730	112,200,000	Tokyo		70-71
* Jordan	37,737	2,700,000	Amman		60-61
Kansas, U.S.A.	82,264	2,249,071	Topeka		144-145
Kazakh S.S.R., U.S.S.R.	1,048,300	14,185,000	Alma-Ata	C 5	52
Kentucky, U.S.A.	40,395	3,219,311	Frankfort		146-147
* Kenya	224,960	14,000,000	Nairobi	E 7	87
Kirgiz S.S.R., U.S.S.R.	76,641	2,932,800	Frunze	D 5	62
Korea, North	46,540	17,000,000	P'yŏngyang		70
Korea, South	38,175	36,000,000	Seoul		70
* Kuwait	6,532	1,100,000	Al Kuwait	F 4	56
* Laos	91,428	3,500,000	Vientiane	D 3	66
Latvian S.S.R., U.S.S.R.	24,595	2,364,100	Riga	C 3	50
* Lebanon	4,015	3,207,000	Beirut	F 6	59
* Lesotho	11,720	1,100,000	Maseru	M 7	83
* Liberia	43,000	1,600,000	Monrovia	C 7	79
* Libya	679,358	2,500,000	Tripoli G 3 78,	A 2	80
Liechtenstein	61	25,000	Vaduz	B 1	42
Lithuanian S.S.R., U.S.S.R.	25,174	3,128,000	Vilna	B 3	50
Louisiana, U.S.A.	48,523	3,643,180	Baton Rouge		148-149
* Luxembourg	999	358,000	Luxembourg	H 8	37
Macao	6.2	300,000	Macao	E 3	69
* Madagascar	226,657	8,200,000	Tananarive	K 6	83
Maine, U.S.A.	33,215	993,663	Augusta		150-151
* Malawi	45,747	5,100,000	Lilongwe	G 4	83
* Malaysia	128,308	12,368,000	Kuala Lumpur	D 5	74
* Maldives	115	136,000	Male	C 4	64
* Mali	464,873	6,100,000	Bamako	C 6	79
* Malta	122	319,000	Valletta	E 7	43
Man, Isle of	227	59,000	Douglas	D 3	31
Manitoba, Canada	251,000	988,247	Winnipeg		226-227
Martinique	425	360,000	Fort-de-France	G 4	107

Country	Area (Sq. Miles)	Population	Capital or Chief Town	Index Ref.	Plate No.
Maryland, U.S.A.	10,577	3,922,399	Annapolis		152-153
Massachusetts, U.S.A.	8,257	5,689,170	Boston		154-155
* Mauritania	452,702	1,318,000	Nouakchott	B 5	79
* Mauritius	790	899,000	Port Louis	H 7	77
Mayotte	144	40,000	Mamoutzou	J 4	83
* Mexico	761,600	62,500,000	Mexico City		102-103
Michigan, U.S.A.	58,216	8,875,083	Lansing		156-157
Midway Islands	1.9	2,220	....	J 3	84
Minnesota, U.S.A.	84,068	3,805,069	St. Paul		158-159
Mississippi, U.S.A.	47,716	2,216,912	Jackson		160-161
Missouri, U.S.A.	69,686	4,677,399	Jefferson City		162-163
Moldavian S.S.R., U.S.S.R.	13,012	3,823,000	Kishinev	C 5	51
Monaco	368 acres	23,035	Monaco	G 6	39
* Mongolia	606,163	1,500,000	Ulan Bator		68-69
Montana, U.S.A.	147,138	694,409	Helena		164-165
Montserrat	40	12,300	Plymouth	G 3	107
* Morocco	241,224	18,000,000	Rabat	C 2	78
* Mozambique	308,641	9,300,000	Maputo	G 6	83
Namibia (South-West Africa)	317,827	883,000	Windhoek	D 6	82
Nauru	7.7	7,100	Yaren (dist.)	G 6	84
Nebraska, U.S.A.	77,227	1,483,791	Lincoln		166-167
* Nepal	54,663	12,900,000	Kathmandu	D 2	64
* Netherlands	15,892	13,800,000	The Hague, Amsterdam		36-37
Netherlands Antilles	390	242,000	Willemstad	E 4	107
Nevada, U.S.A.	110,540	488,738	Carson City		168-169
New Brunswick, Canada	28,354	634,557	Fredericton		218-219
New Caledonia	7,335	136,000	Nouméa	G 8	84
Newfoundland, Canada	156,185	522,104	St. John's		214-215
New Hampshire, U.S.A.	9,304	737,681	Concord		170-171
New Hebrides	5,700	97,468	Vila	G 7	84
New Jersey, U.S.A.	7,836	7,168,164	Trenton		172-173
New Mexico, U.S.A.	121,666	1,016,000	Santa Fe		174-175
New South Wales, Australia	309,433	4,847,800	Sydney	H 6	89
New York, U.S.A.	49,576	18,241,266	Albany		176-177
* New Zealand	103,736	3,121,904	Wellington	L 6	89
* Nicaragua	45,698	2,300,000	Managua	C 2	104
* Niger	489,189	4,700,000	Niamey	F 5	79
* Nigeria	379,628	83,800,000	Lagos	F 7	79
Niue	100	3,843	Alofi	K 7	85
North America	9,363,000	358,400,000	....		100-101
North Carolina, U.S.A.	52,586	5,082,059	Raleigh		178-179
North Dakota, U.S.A.	70,665	617,761	Bismarck		180-181
Northern Ireland, U.K.	5,452	1,537,200	Belfast	C 3	30-31
Northern Terr., Aust.	520,280	98,400	Darwin	E 3	86
North Korea	46,540	17,000,000	P'yŏngyang		70
Northwest Territories, Canada	1,304,903	34,807	Yellowknife		236-237
* Norway	125,053	4,027,000	Oslo		32-33
Nova Scotia, Canada	21,425	788,960	Halifax		216-217
Oceania	3,292,000	21,500,000	....		84-85
Ohio, U.S.A.	41,222	10,652,017	Columbus		182-183
Oklahoma, U.S.A.	69,919	2,559,253	Oklahoma City		184-185
* Oman	120,000	800,000	Muscat	H 5	57
Ontario, Canada	412,582	7,703,106	Toronto		224-225
Oregon, U.S.A.	96,981	2,091,385	Salem		186-187
Pacific Islands, Territory of the	707	120,000	Kolonia (Ponape)	F 5	84

Country	Area (Sq. Miles)	Population	Capital or Chief Town	Index Ref.	Plate No.
* Pakistan	310,403	72,370,000	Islamabad	B 2	64
* Panama	29,209	1,800,000	Panamá	D 3	105
Panama Canal Zone, U.S.A.	647	44,650	Balboa Heights	E 3	105
* Papua New Guinea	183,540	2,800,000	Port Moresby	E 6	84
* Paraguay	157,047	2,750,000	Asunción	E 2	98
Pennsylvania, U.S.A.	45,333	11,793,909	Harrisburg		188-189
* Persia (Iran)	636,293	32,900,000	Tehran		62-63
* Peru	496,222	16,300,000	Lima	B 5	95
* Philippines	115,707	43,751,000	Manila		72-73
Pitcairn Islands	18	67	Adamstown	O 8	85
* Poland	120,725	34,364,000	Warsaw		44-45
* Portugal	35,549	9,800,000	Lisbon		40
Prince Edward I., Can.	2,184	111,641	Charlottetown		220-221
Puerto Rico	3,435	3,300,000	San Juan	G 1	107
* Qatar	4,247	150,000	Doha	G 4	57
Québec, Canada	594,860	6,027,764	Québec		222-223
Queensland, Aust.	666,991	2,015,300	Brisbane		88-89
Réunion	969	475,700	Saint-Denis	H 7	77
Rhode Island, U.S.A.	1,214	949,723	Providence		190-191
Rhodesia	150,803	6,600,000	Salisbury	F 5	83
* Rumania	91,699	21,500,000	Bucharest		48
Russian S.F.S.R., U.S.S.R.	6,592,812	133,913,000	Moscow		50-53
* Rwanda	10,169	4,241,000	Kigali	C 2	83
Saint Christopher-Nevis-Anguilla	138	71,500	Basseterre	F 3	107
Saint Helena	162	6,438	Jamestown	B 6	77
Saint Lucia	238	110,000	Castries	G 4	107
Saint Pierre & Miquelon	93.5	6,000	Saint-Pierre	B 4	215
Saint Vincent	150	89,129	Kingstown	G 4	107
San Marino	23.4	20,000	San Marino	D 3	42
* São Tomé e Príncipe	372	80,000	São Tomé	A 1	82
Saskatchewan, Canada	251,700	926,242	Regina		228-229
* Saudi Arabia	829,995	7,800,000	Riyadh, Mecca		56-57
Scotland, U.K.	30,414	5,261,000	Edinburgh		30
* Senegal	76,124	5,085,388	Dakar	B 6	79
* Seychelles	145	60,000	Victoria	E 6	55
* Sierra Leone	27,925	3,100,000	Freetown	B 7	79
* Singapore	226	2,300,000	Singapore	E 6	67
Solomon Islands	11,500	196,708	Honiara	F 6	84
* Somalia	246,200	3,170,000	Mogadishu	G 6	81
* South Africa	458,179	26,200,000	Cape Town, Pretoria		82-83
South America	6,875,000	230,700,000	....		92-93
South Australia, Aust.	380,070	1,247,100	Adelaide		87
South Carolina, U.S.A.	31,055	2,590,516	Columbia		192-193
South Dakota, U.S.A.	77,047	666,257	Pierre		194-195
South Korea	38,175	36,000,000	Seoul		70
South-West Africa	317,827	883,000	Windhoek	D 6	82
* Spain	194,881	36,000,000	Madrid		40-41
* Sri Lanka	25,332	14,000,000	Colombo	D 4	64
* Sudan	967,494	18,347,000	Khartoum	C 5	81
* Surinam	55,144	430,000	Paramaribo	A 2	96
* Swaziland	6,705	500,000	Mbabane	G 7	83
* Sweden	173,665	8,236,461	Stockholm		32-33
Switzerland	15,943	6,489,000	Bern	B 1	42
* Syria	71,498	7,585,000	Damascus	G 5	59
Tadzhik S.S.R., U.S.S.R.	55,251	2,900,000	Dushanbe	C 6	52
* Tanzania	363,708	15,506,000	Dar es Salaam	G 3	83
Tasmania, Australia	26,383	410,800	Hobart	H 8	89

Country	Area (Sq. Miles)	Population	Capital or Chief Town	Index Ref.	Plate No.
Tennessee, U.S.A.	42,244	3,924,164	Nashville		196-197
Texas, U.S.A.	267,339	11,196,730	Austin		198-199
* Thailand	198,455	42,700,000	Bangkok		66-67
Tibet, China	471,660	1,270,000	Lhasa	B 2	68
* Togo	21,622	2,300,000	Lomé	E 7	79
Tokelau	3.9	1,603	Fenuafala (Fakaofo)	J 6	85
Tonga	270	102,000	Nuku'alofa	J 7	85
* Trinidad and Tobago	1,980	1,040,000	Port-of-Spain	G 5	107
* Tunisia	63,170	5,776,000	Tunis	F 2	78
* Turkey	300,946	40,284,000	Ankara		58-59
Turkmen S.S.R., U.S.S.R.	188,455	2,158,880	Ashkhabad	B 6	52
Turks and Caicos Is.	166	6,000	Cockburn Town (Grand Turk)	D 2	106
Tuvalu	9.78	5,887	Fongafale (Funafuti)	H 6	84
Uganda	91,076	12,000,000	Kampala	D 7	81
* Ukrainian S.S.R., U.S.S.R.	233,089	49,438,000	Kiev	D 5	51
* Union of Soviet Socialist Republics	8,649,490	258,402,000	Moscow		50-53
* United Arab Emirates	32,278	650,000	Abu Dhabi	G 5	57
* United Kingdom	94,399	56,076,000	London		30-31
* United States of America	3,615,123	224,000,000	Washington		110-111
* Upper Volta	105,869	6,144,013	Ouagadougou	D 6	79
* Uruguay	72,172	2,900,000	Montevideo	E 4	98
Utah, U.S.A.	84,916	1,059,273	Salt Lake City		200-201
Uzbek S.S.R., U.S.S.R.	173,591	11,960,000	Tashkent	C 5	52
Vatican City	116 acres	726	....	D 4	42
* Venezuela	352,143	12,750,000	Caracas	D 2	94
Vermont, U.S.A.	9,609	444,732	Montpelier		202-203
Victoria, Australia	87,884	3,713,200	Melbourne	G 7	89
* Vietnam	128,405	46,600,000	Hanoi		66-67
Virginia, U.S.A.	40,817	4,648,494	Richmond		204-205
Virgin Islands, British	59	10,484	Road Town	H 1	107
Virgin Islands, U.S.A.	133	110,000	Charlotte Amalie	H 1	107
- Wake Island	2.5	150	Wake Islet	G 4	84
Wales, U.K.	8,017	2,778,000	Cardiff	E 4	31
Wallis and Futuna	106	9,000	Matautu	J 7	85
Washington, U.S.A.	68,192	3,409,169	Olympia		206-207
Western Australia, Australia	975,920	1,148,100	Perth		86-87
* Western Samoa	1,133	159,000	Apia	J 7	85
* West Germany (Federal Republic)	95,985	61,644,600	Bonn		34-35
West Virginia, U.S.A.	24,181	1,744,237	Charleston		208-209
* White Russian (Byelo-russian) S.S.R., U.S.S.R.	80,154	9,522,000	Minsk	C 4	51
Wisconsin, U.S.A.	56,154	4,417,933	Madison		210-211
World	57,970,000	4,240,700,000	....		26-27
Wyoming, U.S.A.	97,914	332,416	Cheyenne		212-213
* Yemen Arab Republic	77,220	5,600,000	San'a	E 7	56
* Yemen, Peoples Dem. Rep. of	111,101	1,700,000	Aden	F 7	56
* Yugoslavia	98,766	21,520,000	Belgrade		48
Yukon Territory, Can.	207,076	18,388	Whitehorse		234-235
* Zaire	918,962	25,600,000	Kinshasa	B 8 81, E 2	82
* Zambia	290,586	5,250,000	Lusaka		82-83

GLOSSARY OF GEOGRAPHICAL TERMS

A. = Arabic Camb. = Cambodian Ch. = Chinese Dan. = Danish Du. = Dutch
Finn. = Finnish Fr. = French Ger. = German Ice. = Icelandic It. = Italian
Jap. = Japanese Mong. = Mongol Nor. = Norwegian Per. = Persian
Port.=Portuguese Russ.=Russian Sp.=Spanish Sw.=Swedish Turk. =Turkish

Å........................ Nor., Sw. ......Stream
Abajo.................. Sp. ....Lower
Ada, Adasi.......... Turk. ....Island
Altiplano............. Sp. ....Plateau
Älv, Alf, Elf......... Sw. ....River
Arrecife............. Sp. ....Reef
Baai.................... Du. ....Bay
Bahía.................. Sp. ....Bay
Bahr.................. Arabic...Marsh, Lake, Sea, River
Baia.................... Port. ....Bay
Baie..................... Fr. ....Bay, Gulf
Bañados............. Sp. ....Marshes
Barra.................. Sp. ....Reef
Belt..................... Ger. ....Strait
Ben...................... Gaelic ....Mountain
Berg.................. Ger., Du. ...Mountain
Bir....................... Aarbic ....Well
Boca.................. Sp. ....Gulf, Inlet
Bolshoi, Bolshaya.......... Russ. ....Big
Bolsón................ Sp. ....Depression
Bong.................. Korean ......Mountain
Bucht.................. Ger. ....Bay
Bugt.................... Dan. ....Bay
Bukhta................ Russ. ....Bay
Burnu, Burun...... Turk. .....Cape, Point
By....................... Dan., Nor., Sw. Town
Cabo.................. Port., Sp. ....Cape
Campos.............. Port. ....Plains
Canal.................. Port., Sp. ....Channel
Cap, Capo........... Fr., It. ....Cape
Catarátas............ Sp. ....Falls
Central, Centrale........... Fr., It. ....Middle
Cerrito, Cerro...... Sp. ....Hill
Ciénaga.............. Sp. ....Swamp
Ciudad................ Sp. ....City
Col...................... Fr. ....Pass
Cordillera............ Sp. ....Mt. Range
Côte.................... Fr. ....Coast
Cuchilla.............. Sp. ....Mt. Range
Dağ, Dagh.......... Turk. ....Mountain
Dağlari................ Turk. ......Mt. Range
Dal...................... Nor., Sw. ....Valley
Darya.................. Per. ....Salt Lake
Dasht.................. Per. .....Desert, Plain
Deniz, Denizi...... Turk. ....Sea, Lake
Desierto............. Sp. ....Desert
Eiland................. Du. ....Island
Elv...................... Dan., Nor. ....River
Emi..................... Berber ......Mountain
Erg..................... Arabic...Dune, Desert
Est, Este............ Fr., Port., Sp. ...East
Estrecho, Estreito........... Sp., Port. ......Strait

Étang................. Fr. ....Pond, Lagoon, Lake
Fjørd.................. Dan., Nor. ....Fiord
Fleuve................. Fr. ....River
Gebel.................. Arabic ......Mountain
Gebirge.............. Ger. ....Mt. Range
Gobi.................... Mongol ....Desert
Gol...................... Mongol, Turk. Lake, Stream
Golf..................... Ger., Du. ....Gulf
Golfe................... Fr. ....Gulf
Golfo.................. Sp., It., Port. ...Gulf
Gölü.................... Turk. ....Lake
Gora................... Russ. ....Mountain
Grand, Grande.... Fr., Sp. ....Big
Groot.................. Du. ....Big
Gross.................. Ger. ....Big
Grosso................ It., Port. ....Big
Guba.................. Russ. ....Bay, Gulf
Gunto.................. Jap. ......Archipelago
Gunung.............. Malay ....Mountain
Higashi, Higasi.... Jap. ....East
Ho....................... Ch. ....River
Hoek................... Du. ....Cape
Holm.................. Dan., Nor., Sw. Island
Hu....................... Ch. ....Lake
Hwang................ Ch. ....Yellow
Île........................ Fr. ....Island
Insel.................... Ger. ....Island
Irmak.................. Turk. ....River
Isla...................... Sp. ....Island
Isola.................... Sp. ....Island
Jabal, Jebel........ Arabic .....Mountains
Järvi.................... Finn. ....Lake
Jaure.................. Sw. ....Lake
Jezira.................. Arabic ....Island
Jima.................... Jap. ....Island
Joki..................... Finn. ....River
Kaap................... Du. ....Cape
Kabir, Kebir........ Arabic ....Big
Kanal.................. Russ., Ger. ....Canal, Channel
Kap, Kapp........... Nor., Sw., Ice. ..Cape
Kawa................... Jap. ....River
Khrebet.............. Russ. ....Mt. Range
Kiang.................. Ch. ....River
Kita..................... Jap. ....North
Klein................... Du., Ger. ....Small
Kô....................... Jap. ....Lake
Ko....................... Thai. ....Island
Koh..................... Camb., Khmer Island
Köping................ Sw. ....Borough
Körfez, Körfezi... Turk. ....Gulf
Kuh..................... Per. ....Mountain

Kul ... Sinkiang Turki ... Lake

Kum ... Turk. ... Desert

Lac ... Fr. ... Lake

Lago ... Port., Sp., It. ... Lake

Lagôa ... Port. ... Lagoon

Laguna ... Sp. ... Lagoon

Lagune ... Fr. ... Lagoon

Llanos ... Sp. ... Plains

Mar ... Sp., Port. ... Sea

Mare ... It. ... Sea

Meer ... Du. ... Lake

Meer ... Ger. ... Sea

Mer ... Fr. ... Sea

Meseta ... Sp. ... Plateau

Minami ... Jap. ... Southern

Misaki ... Jap. ... Cape

Mittel ... Ger. ... Middle

Mont ... Fr. ... Mountain

Montagne ... Fr. ... Mountain

Montaña ... Sp. ... Mountains

Monte ... Sp., It., Port. ... Mountain

More ... Russ. ... Sea

Muong ... Siamese ... Town

Mys ... Russ. ... Cape

Nam ... Burm., Lao ... River

Nevado ... Sp. ... Snow covered peak

Nieder ... Ger. ... Lower

Nishi, Nisi ... Jap. ... West

Nizhnl, Nizhnyaya ... Russ. ... Lower

Nor ... Mong. ... Lake

Nord ... Fr., Ger. ... North

Norte ... Sp., It., Port. North

Nos ... Russ. ... Cape

Novi, Novaya ... Russ. ... New

Nusa ... Malay ... Island

O ... Jap. ... Big

ö ... Nor., Sw ... Island

Ober ... Ger. ... Upper

Occidental, Occidentale ... Sp., It. ... Western

Oeste ... Port. ... West

Oriental ... Sp., Fr. ... Eastern

Orientale ... It. ... Eastern

Ost ... Ger. ... East

Ostrov ... Russ. ... Island

Ouest ... Fr. ... West

öy ... Nor. ... Island

Ozero ... Russ. ... Lake

Pampa ... Sp. ... Plain

Paso ... Sp. ... Pass

Passo ... It., Port. ... Pass

Pequeño ... Sp. ... Small

Peski ... Russ. ... Desert

Petit ... Fr. ... Small

Pic ... Fr. ... Mountain

Pico ... Port., Sp. ... Mountain, Peak

Pik ... Russ. ... Peak

Pointe ... Fr. ... Point

Poluostrov ... Russ. ... Peninsula

Ponta ... Port. ... Point

Presa ... Sp. ... Reservoir

Proliv ... Russ. ... Strait

Pulou, Pulo ... Malay ... Island

Punta ... Sp., It., Port. Point

Ras ... Arabic ... Cape

Ría ... Sp. ... Estuary

Río ... Sp. ... River

Rivier, Rivière ... Du., Fr. ... River

Rud ... Per. ... River

Saki ... Jap. ... Cape

Salto ... Sp., Port. ... Falls

San ... Ch., Jap., Korean Hill

See ... Ger. ... Sea, Lake

Selvas ... Sp., Port. ... Forest

Serra ... Port. ... Mts.

Serranía ... Sp. ... Mts.

Severni, Servernaya ... Russ. ... North

Shan ... Ch., Jap. ... Hill, Mts.

Shima ... Jap. ... Island

Shoto ... Jap. ... Islands

Sierra ... Sp. ... Mountains

Sjö ... Nor., Sw. Lake, Sea

Spitze ... Ger. ... Mt. Peak

Srednl, Srednyaya ... Russ. ... Middle

Stad ... Dan., Nor., Sw. City

Stari, Staraya ... Russ. ... Old

Su ... Turk. ... River

Sud, Süd ... Sp., Fr., Ger. ... South

Sul ... Port. ... South

Sungei ... Malay ... River

Sur ... Sp. ... South

Tagh ... Turk. ... Mt. Range

Tal ... Ger. ... Valley

Tandjong, Tanjung ... Malay ... Cape, Point

Tso ... Tibetan ... Lake

Val ... Fr. ... Valley

Velho ... Port. ... Old

Verkhni ... Russ. ... Upper

Vesi ... Finn. ... Lake

Vishni, Vishnyaya ... Russ. ... High

Vostochni, Vostochnaya ... Russ. ... East, Eastern

Wadi ... Arabic ... Dry River

Wald ... Ger. ... Forest

Wan ... Jap. ... Bay

Yama ... Jap. ... Mountain

Yug, Yuzhni, Yuzhnaya ... Russ. South, Southern

Zaliv ... Russ. ... Bay, Gulf

Zapadni, Zapadnaya ... Russ. ... Western

Zee ... Du. ... Sea

Zemlya ... Russ. ... Land

Zuid ... Du. ... South

WORLD STATISTICAL TABLES

OCEANS AND SEAS

	AREA IN SQ. MILES	GREATEST DEPTH IN FEET	VOLUME IN CUBIC MILES
Pacific Ocean	64,186,000	36,198	167,025,000
Atlantic Ocean	31,862,000	28,374	77,580,000
Indian Ocean	28,350,000	25,344	68,213,000
Arctic Ocean	5,427,000	17,880	3,026,000
Caribbean Sea	970,000	24,720	2,298,400
Mediterranean Sea	969,000	16,896	1,019,400
South China Sea	895,000	15,000	....
Bering Sea	875,000	15,800	788,500
Gulf of Mexico	600,000	12,300	....
Sea of Okhotsk	590,000	11,070	454,700
East China Sea	482,000	9,500	52,700
Japan Sea	389,000	12,280	383,200
Hudson Bay	317,500	846	37,590
North Sea	222,000	2,200	12,890
Black Sea	185,000	7,365	....
Red Sea	169,000	7,200	53,700
Baltic Sea	163,000	1,506	5,360

PRINCIPAL MOUNTAINS

	FEET
Everest, Nepal-China	29,028
K2 (Godwin Austen), India	28,250
Kanchenjunga, Nepal-India	28,208
Lhotse, Nepal-China	27,923
Makalu, Nepal-China	27,824
Dhaulagiri, Nepal	26,810
Nanga Parbat, India	26,660
Annapurna, Nepal	26,504
Nanda Devi, India	25,645
Kamet, India	25,447
Tirich Mir, Pakistan	25,230
Minya Konka, China	24,902
Muztagh Ata, China	24,757
Communism Peak, U.S.S.R.	24,599
Pobeda Peak, U.S.S.R.	24,406
Chomo Lhari, Bhutan-China	23,997
Muztagh, China	23,891
Aconcagua, Argentina	22,831
Ojos del Salado, Chile-Arg.	22,572
Tupungato, Chile-Arg.	22,310
Mercedario, Argentina	22,211
Huascarán, Peru	22,205
Llullaillaco, Chile-Arg.	22,057
Ancohuma, Bolivia	21,489
Illampu, Bolivia	21,276
Chimborazo, Ecuador	20,561
McKinley, Alaska	20,320

	FEET
Logan, Yukon	19,850
Cotopaxi, Ecuador	19,347
Kilimanjaro, Tanzania	19,340
El Misti, Peru	19,101
Huila, Colombia	18,865
Citlaltépetl (Orizaba), Mexico	18,855
El'brus, U.S.S.R.	18,510
Demavend, Iran	18,376
St. Elias, Alaska-Yukon	18,008
Popocatépetl, Mexico	17,887
Dykh-Tau, U.S.S.R.	17,070
Kenya, Kenya	17,058
Ararat, Turkey	16,946
Vinson Massif, Antarc.	16,864
Margherita (Ruwenzori), Africa	16,795
Kazbek, U.S.S.R.	16,512
Djaja, Indonesia	16,503
Blanc, France	15,771
Klyuchevskaya Sopka, U.S.S.R.	15,584
Rosa (Dufourspitze), Italy-Switzerland	15,203
Ras Dashan, Ethiopia	15,157
Matterhorn, Switzerland	14,688
Whitney, California	14,494

WORLD STATISTICAL TABLES

LAKES AND INLAND SEAS

	AREA IN SQ. MILES		AREA IN SQ. MILES
Caspian Sea	143,243	Lake Chad	5,300
Lake Superior	31,700	Lake Onega	3,710
Lake Victoria	26,724	Lake Titicaca	3,200
Aral Sea	25,676	Lake Nicaragua	3,100
Lake Huron	23,010	Lake Athabasca	3,064
Lake Michigan	22,300	Reindeer Lake	2,568
Lake Tanganyika	12,650	Lake Rudolf	2,463
Lake Baykal	12,162	Issyk-Kul'	2,425
Great Bear Lake	12,096	Vänern	2,156
Lake Nyasa	11,555	Lake Winnipegosis	2,075
Great Slave Lake	11,269	Lake Albert	2,075
Lake Erie	9,910	Kariba Lake	2,050
Lake Winnipeg	9,417	Lake Urmia	1,815
Lake Ontario	7,340	Lake of the Woods	1,679
Lake Ladoga	7,104	Lake Peipus	1,400
Lake Balkhash	7,027	Great Salt Lake	1,100

LONGEST RIVERS

	LENGTH IN MILES		LENGTH IN MILES
Nile, Africa	4,145	Zambezi, Africa	1,600
Amazon, S.A.	3,915	Nelson, Canada	1,600
Mississippi-Missouri, U.S.A.	3,710	Orinoco, S.A.	1,600
Yangtze, China	3,434	Paraguay, S.A.	1,584
Ob-Irtysh, U.S.S.R.	3,362	Kolyma, U.S.S.R.	1,562
Yenisey-Angara, U.S.S.R.	3,100	Ganges, Asia	1,550
Hwang (Yellow), China	2,877	Ural, U.S.S.R.	1,509
Amur, Asia	2,744	Japurá, S.A.	1,500
Lena, U.S.S.R.	2,734	Arkansas, U.S.A.	1,450
Congo, Africa	2,718	Colorado, U.S.A.-Mexico	1,450
Mackenzie-Peace, Canada	2,635	Negro, S.A.	1,400
Mekong, Asia	2,610	Dnieper, U.S.S.R.	1,368
Niger, Africa	2,548	Orange, Africa	1,350
Paraná, S.A.	2,450	Irrwaddy, Burma	1,325
Murray-Darling, Australia	2,310	Ohio-Allegheny, U.S.A.	1,306
Volga, U.S.S.R.	2,194	Kama, U.S.S.R.	1,262
Madeira, S.A.	2,013	Columbia, U.S.A.-Canada	1,243
Purus, S.A.	1,995	Red, U.S.A.	1,222
Yukon, Alaska-Canada	1,979	Don, U.S.S.R.	1,222
St. Lawrence, Canada-U.S.	1,900	Brazos, U.S.A.	1,210
Rio Grande, U.S.-Mexico	1,885	Saskatchewan, Canada	1,205
Syr-Dar'ya, U.S.S.R.	1,859	Peace-Finlay, Canada	1,195
São Francisco, Brazil	1,811	Tigris, Asia	1,181
Indus, Asia	1,800	Darling, Australia	1,160
Danube, Europe	1,775	Angara, U.S.S.R.	1,135
Salween, Asia	1,770	Sungari, Asia	1,130
Brahmaputra, Asia	1,700	Pechora, U.S.S.R.	1,124
Euphrates, Asia	1,700	Snake, U.S.A.	1,038
Tocantins, Brazil	1,677	Churchill, Canada	1,000
Si, China	1,650	Pilcomayo, S.A.	1,000
Amu-Dar'ya, Asia	1,616	Magdalena, Colombia	1,000

AIR DISTANCES BETWEEN MAJOR WORLD CITIES (See also, U.S. AIR DISTANCES, p. 199)

SOURCE: USAF Aeronautical Chart and Information Center (in statute miles)

	Bangkok	Berlin	Cairo	Cape Town	Caracas	Chicago	Hong Kong	Honolulu	Istanbul	Lima	London	Madrid	Melbourne
Accra	6,850	3,330	2,672	2,974	4,576	5,837	7,615	10,052	3,039	5,421	3,169	2,412	9,325
Amsterdam	5,707	360	2,015	5,997	4,883	4,118	5,772	7,254	1,372	6,538	222	921	10,286
Anchorage	6,022	4,545	6,116	10,478	5,353	2,858	5,073	2,778	5,388	6,385	4,491	5,181	7,729
Athens	4,930	1,121	671	4,957	5,815	5,447	5,316	8,353	352	7,312	1,488	1,474	9,297
Auckland	4,645	9,995	8,825	6,574	9,620	9,507	4,625	5,346	9,203	7,989	10,570	10,884	160
Baghdad	3,756	2,029	798	4,924	7,020	6,430	4,260	8,399	1,006	8,487	2,547	2,675	8,105
Bangkok	—	5,351	4,521	6,301	10,558	8,569	1,076	6,610	4,648	12,241	5,929	6,334	4,579
Beirut	4,272	1,689	341	4,794	6,520	6,097	4,756	8,536	614	7,972	2,151	2,190	8,579
Belgrade	5,073	623	1,147	5,419	5,587	5,000	5,327	7,882	500	7,169	1,053	1,263	9,578
Berlin	5,351	—	1,768	5,958	5,242	4,415	5,443	7,323	1,075	6,893	580	1,162	9,929
Bombay	1,870	3,915	2,717	5,103	9,034	8,066	2,679	8,036	3,000	10,389	4,478	4,689	6,101
Buenos Aires	10,490	7,395	7,360	4,285	3,155	5,582	11,478	7,554	7,608	1,945	6,907	6,236	7,219
Cairo	4,521	1,768	—	4,510	6,337	6,116	5,057	8,818	741	7,725	2,158	2,069	8,700
Cape Town	6,301	5,958	4,510	—	6,361	8,489	7,377	11,534	5,204	6,074	5,988	5,306	6,428
Caracas	10,558	5,242	6,337	6,361	—	2,500	10,171	6,024	6,050	1,699	4,662	4,351	9,703
Chicago	8,569	4,415	6,116	8,489	2,500	—	7,797	4,256	5,485	3,772	3,960	4,192	9,667
Copenhagen	5,361	222	1,964	6,179	5,215	4,263	5,392	7,101	1,252	6,886	595	1,289	9,936
Denver	8,409	5,092	6,846	9,331	3,078	920	7,476	3,346	6,164	3,986	4,701	5,028	8,755
Frankfurt	5,305	50	1,730	5,944	5,290	4,460	5,403	7,341	1,032	6,940	628	1,193	9,882
Helsinki	4,903	689	2,069	6,490	5,658	4,442	4,867	6,818	1,330	7,349	1,135	1,835	9,448
Hong Kong	1,076	5,443	5,057	7,377	10,171	7,797	—	5,557	4,989	11,415	5,986	6,556	4,605
Honolulu	6,610	7,323	8,818	11,534	6,024	4,256	5,557	—	8,118	5,944	7,241	7,874	5,501
Houston	9,261	5,337	7,005	8,608	2,262	942	8,349	3,902	6,400	3,123	4,860	5,014	8,979
Istanbul	4,648	1,075	741	5,204	6,050	5,485	4,989	8,118	—	7,593	1,551	1,701	9,100
Karachi	2,305	3,365	2,222	5,153	8,502	7,564	2,977	8,059	2,457	9,943	3,928	4,152	6,646
Keflavik	6,300	1,505	3,267	7,107	4,269	2,942	6,044	6,085	2,578	5,965	1,188	1,802	10,552
Kinshasa	5,974	3,916	2,618	2,047	5,752	7,085	6,904	11,178	3,241	6,322	3,951	3,305	8,112
Leningrad	4,718	826	2,034	6,500	5,843	4,589	4,687	6,816	1,306	7,534	1,307	1,985	9,263
Lima	12,241	6,893	7,725	6,074	1,699	3,772	11,415	5,944	7,593	—	6,316	5,907	8,052
Lisbon	6,651	1,442	2,352	5,301	4,040	4,001	6,862	7,835	2,015	5,591	989	317	11,049
London	5,929	580	2,158	5,988	4,662	3,960	5,986	7,241	1,551	6,316	—	786	10,508
Madrid	6,334	1,162	2,069	5,306	4,351	4,192	6,556	7,874	1,701	5,907	786	—	10,766
Melbourne	4,579	9,929	8,700	6,428	9,703	9,667	4,605	5,501	9,100	8,052	10,508	10,766	—
Mexico City	9,793	6,054	7,677	8,516	2,234	1,688	8,789	3,791	7,106	2,635	5,558	5,642	8,420
Montreal	8,337	3,740	5,403	7,920	2,443	746	7,736	4,919	4,798	3,967	3,256	3,449	10,390
Moscow	4,394	1,001	1,770	6,277	6,176	4,984	4,443	7,049	1,087	7,855	1,556	2,140	8,965
Nairobi	4,481	3,947	2,217	2,543	7,179	8,012	5,447	10,740	2,957	7,821	4,229	3,840	7,159
New Delhi	1,812	3,598	2,752	5,769	8,837	7,486	2,339	7,413	2,837	10,430	4,178	4,528	6,340
New York City	8,669	3,980	5,598	7,801	2,124	714	8,061	4,969	5,022	3,635	3,473	3,596	10,352
Oslo	5,395	523	2,243	6,477	5,167	4,050	5,342	6,801	1,518	6,857	718	1,485	9,934
Panama City	10,871	5,856	7,118	7,021	867	2,321	10,089	5,254	6,756	1,454	5,285	5,081	9,027
Paris	5,877	549	1,973	5,782	4,735	4,145	5,992	7,452	1,400	6,367	215	652	10,442
Peking	2,027	4,600	4,687	8,034	8,978	6,625	1,195	5,084	4,407	10,365	5,089	5,759	5,632
Rabat	6,652	1,623	2,230	4,954	4,111	4,282	6,954	8,177	2,008	5,590	1,254	474	10,856
Rio de Janeiro	9,987	6,207	6,153	3,773	2,805	5,288	11,002	8,295	6,378	2,351	5,751	5,045	8,218
Rome	5,493	735	1,305	5,231	5,198	4,823	5,773	8,040	853	6,748	892	849	9,940
Saigon (Ho Chi Minh City)	467	5,771	4,987	6,534	10,905	8,695	938	6,302	5,102	12,180	6,345	6,779	4,168
San Francisco	7,930	5,673	7,436	10,248	3,908	1,860	6,904	2,397	6,711	4,516	5,369	5,806	7,850
Santiago	10,967	7,772	7,967	4,947	3,033	5,295	11,615	6,861	8,135	1,528	7,241	6,639	7,017
Seattle	7,455	5,060	6,809	10,205	4,096	1,737	6,481	2,681	6,077	4,961	4,799	5,303	8,176
Shanghai	1,797	5,231	5,188	8,062	9,508	7,071	760	4,947	4,975	10,665	5,728	6,386	4,991
Shannon	6,256	940	2,534	6,188	4,320	3,583	6,246	7,006	1,938	5,992	387	884	10,826
Singapore	887	6,167	5,143	6,007	11,408	9,376	1,608	6,728	5,379	11,689	6,747	7,079	3,767
St. Louis	8,763	4,676	6,370	8,549	2,414	265	7,949	4,134	5,744	3,589	4,215	4,426	9,476
Stockholm	5,141	505	2,084	6,422	5,422	4,288	5,115	6,873	1,347	7,109	892	1,613	9,693
Teheran	3,392	2,184	1,220	5,240	7,322	6,502	3,844	8,072	1,274	8,850	2,739	2,974	7,838
Tokyo	2,865	5,557	5,937	9,155	8,813	6,313	1,792	3,860	5,574	9,628	5,956	6,704	5,070
Vienna	5,252	323	1,455	5,656	5,374	4,696	5,432	7,632	791	6,990	767	1,124	9,802
Warsaw	5,032	322	1,588	5,934	5,563	4,679	5,147	7,368	858	7,212	901	1,425	9,609
Washington D.C.	8,807	4,182	5,800	7,892	2,051	2,598	8,157	4,839	5,225	3,504	3,676	3,794	10,174

Mexico City	Montreal	Moscow	Nairobi	New Delhi	New York	Paris	Peking	Rio de Janeiro	Rome	San Francisco	Singapore	Stockholm	Teheran	Tokyo	Vienna	Warsaw
6,677	5,146	4,038	2,603	5,279	5,126	2,988	7,359	3,501	2,624	7,688	7,183	3,835	3,874	8,594	3,100	3,440
5,735	3,426	1,337	4,136	3,958	3,654	271	4,890	5,938	807	5,465	6,526	701	2,533	5,788	581	681
3,776	3,133	4,364	8,287	5,709	3,373	4,697	3,997	8,145	5,263	2,005	6,678	4,102	5,654	3,463	4,856	4,601
7,021	4,737	1,387	2,827	3,120	4,938	1,305	4,757	6,030	654	6,792	5,629	1,498	1,539	5,924	801	996
8,274	10,231	9,018	7,315	6,420	10,194	10,519	5,626	8,259	10,048	7,692	3,848	9,732	7,935	5,017	9,886	9,676
8,082	5,768	1,583	2,431	1,966	6,007	2,405	3,925	6,938	1,836	7,466	4,427	2,164	431	5,199	1,781	1,752
9,793	8,337	4,394	4,481	1,812	8,669	5,877	2,027	9,987	5,493	7,930	887	5,141	3,392	2,865	5,252	5,032
7,707	5,405	1,514	2,420	2,479	5,622	1,987	4,352	6,478	1,368	7,302	4,935	1,931	913	5,598	1,401	1,459
6,610	4,305	1,066	3,328	3,270	4,526	902	4,634	6,145	449	6,296	5,833	1,010	1,741	5,720	309	516
6,054	3,740	1,001	3,947	3,598	3,980	549	4,600	6,207	735	5,673	6,167	505	2,184	5,557	323	322
9,739	7,524	3,132	2,811	722	7,811	4,367	2,953	8,334	3,846	8,406	2,427	3,880	1,743	4,196	3,725	3,601
4,580	5,597	8,369	6,479	9,823	5,279	6,857	11,994	1,231	6,925	6,455	9,870	7,799	8,565	11,411	7,334	7,656
7,677	5,403	1,770	2,217	2,752	5,598	1,973	4,687	6,153	1,305	7,436	5,143	2,084	1,220	5,937	1,455	1,588
8,516	7,920	6,277	2,543	5,769	7,801	5,782	8,034	3,773	5,231	10,248	6,007	6,422	5,240	9,155	5,656	5,934
2,234	2,443	6,176	7,179	8,837	2,124	4,735	8,978	2,805	5,198	3,908	11,408	5,422	7,322	8,813	5,374	5,563
1,688	746	4,984	8,012	7,486	714	4,145	6,625	5,288	4,823	1,860	9,376	4,288	6,502	6,313	4,696	4,679
5,918	3,605	971	4,156	3,640	3,857	642	4,503	6,321	953	5,473	6,195	325	2,287	5,415	540	417
1,438	1,639	5,501	8,867	7,730	1,631	4,900	6,385	5,866	5,887	953	9,079	4,879	7,033	5,815	5,395	5,322
6,127	3,787	961	3,915	3,550	4,028	589	4,567	6,237	729	5,709	6,119	502	2,135	5,533	296	274
6,101	3,845	554	4,282	3,247	4,126	1,192	3,956	6,872	1,370	5,435	5,759	248	2,062	4,872	895	569
8,789	7,736	4,443	5,447	2,339	8,061	5,992	1,195	11,002	5,773	6,904	1,608	5,115	3,844	1,792	5,432	5,147
3,791	4,919	7,049	10,740	7,413	4,969	7,452	5,084	8,295	8,040	2,397	6,728	6,873	8,072	3,860	7,832	7,368
749	1,606	5,925	8,746	8,388	1,419	5,035	7,244	5,015	5,702	1,648	9,954	5,227	7,442	6,685	5,609	5,609
7,106	4,798	1,087	2,957	2,837	5,022	1,400	4,407	6,378	853	6,711	5,379	1,347	1,274	5,574	791	858
9,249	6,997	2,600	2,708	678	7,277	3,817	3,020	8,082	3,306	8,078	2,942	3,340	1,194	4,313	3,175	3,052
4,614	2,317	2,083	5,404	4,749	2,597	1,402	4,951	6,090	2,068	4,196	7,181	1,352	3,568	5,497	1,813	1,745
7,915	6,378	4,328	4,234	4,692	6,378	3,742	7,002	4,105	3,186	8,920	6,132	4,388	3,612	8,307	3,619	3,910
6,276	4,005	396	1,505	3,069	4,291	1,350	3,789	7,028	1,460	5,523	5,575	431	1,926	4,733	986	642
2,635	3,967	7,855	7,821	10,430	3,635	6,367	10,365	2,351	6,748	4,516	11,689	7,109	8,850	9,628	6,990	7,212
5,396	3,255	2,433	4,013	4,844	3,377	904	6,040	4,777	1,163	5,679	7,393	1,862	3,288	6,943	1,432	1,720
5,558	3,256	1,556	4,229	4,178	3,473	215	5,089	5,751	892	5,369	6,747	892	2,739	5,956	767	901
5,642	3,449	2,140	3,840	4,528	3,596	652	5,759	5,045	849	5,806	7,079	1,613	2,974	6,704	1,124	1,425
8,420	10,390	8,965	7,159	6,340	10,352	10,442	5,632	8,218	9,940	7,850	3,767	9,693	7,838	5,070	9,802	9,609
—	2,315	6,671	9,218	9,119	2,086	5,723	7,772	4,769	6,374	1,889	10,331	5,965	8,182	7,036	6,316	6,335
2,315	—	4,397	7,267	7,012	333	3,432	6,541	5,082	4,102	2,544	9,207	3,667	5,879	6,470	4,007	4,021
6,671	4,397	—	3,928	2,703	4,680	1,550	3,627	7,162	1,477	5,884	5,236	764	1,534	4,663	1,039	716
9,218	7,267	3,928	—	3,371	7,365	4,020	5,720	5,556	3,340	9,598	4,636	4,299	2,709	6,996	3,625	3,800
9,119	7,012	2,703	3,317	—	7,319	4,103	2,350	8,747	3,684	7,691	2,574	3,466	1,584	3,638	3,467	3,277
2,086	333	4,680	7,365	7,319	—	3,638	6,867	4,805	4,293	2,574	9,539	3,939	6,141	6,757	4,233	4,271
5,722	3,418	1,024	4,446	3,726	3,686	838	4,395	6,462	1,248	5,196	6,249	260	2,462	5,238	839	661
1,496	2,542	6,720	8,043	9,422	2,213	5,388	8,939	3,296	5,916	3,326	11,692	5,956	8,011	8,441	6,031	6,175
5,723	3,432	1,550	4,020	4,103	3,638	—	5,138	5,681	688	5,579	6,676	964	2,624	6,054	643	853
7,772	6,541	3,627	5,720	2,350	6,867	5,138	—	10,778	5,076	5,934	2,754	4,197	3,496	1,305	4,664	4,340
5,612	3,537	2,579	3,733	4,841	3,636	1,125	6,206	4,589	1,184	5,995	7,348	2,084	3,263	7,174	1,546	1,866
4,769	5,082	7,162	5,556	8,747	4,805	5,681	10,778	—	5,704	6,621	9,776	6,638	7,368	11,535	6,124	6,453
6,374	4,102	1,477	3,340	3,684	4,293	688	5,076	5,704	—	6,259	6,231	1,229	2,126	6,140	476	819
9,718	8,558	4,798	4,874	2,268	8,889	6,303	2,072	10,290	5,943	7,829	682	5,534	3,851	2,689	5,687	5,454
1,889	2,544	5,884	9,598	7,691	2,574	5,579	5,934	6,621	6,259	—	8,449	5,372	7,362	5,148	5,992	5,854
4,094	5,436	8,770	7,180	10,518	5,106	7,224	11,859	1,820	7,391	5,926	10,190	8,120	9,185	10,711	7,760	8,059
2,340	2,289	5,217	9,006	7,046	2,409	5,012	5,432	6,890	5,680	679	8,074	4,731	6,686	4,793	5,381	5,222
8,033	7,067	4,248	5,951	2,646	7,384	5,772	645	11,339	5,679	6,150	2,363	4,837	3,974	1,097	5,281	4,963
5,172	2,873	1,863	4,563	4,529	3,086	563	5,288	5,597	1,247	5,040	7,089	1,135	3,117	6,064	1,153	1,258
10,331	9,207	5,236	4,636	2,574	9,539	6,676	2,754	9,776	6,231	8,449	—	5,993	4,106	3,304	6,039	5,846
1,425	978	5,248	8,231	7,736	878	4,398	6,792	5,218	5,073	1,744	9,544	4,552	6,766	6,407	4,955	4,942
5,965	3,667	764	4,299	3,466	3,939	964	4,197	6,638	1,229	5,372	5,993	—	2,217	5,091	771	504
8,182	5,879	1,534	2,709	1,584	6,141	2,624	3,496	7,386	2,126	7,362	4,106	2,217	—	4,775	1,983	1,878
7,036	6,479	4,663	6,996	3,638	6,757	6,054	1,305	11,535	6,140	5,148	3,304	5,091	4,775	—	5,689	5,346
6,316	4,007	1,039	3,625	3,467	4,233	643	4,664	6,124	476	5,992	6,039	771	1,983	5,689	—	347
6,335	4,021	716	3,800	3,277	4,271	853	4,340	6,453	819	5,854	5,846	504	1,878	5,346	347	—
1,883	490	4,873	7,550	7,500	203	3,841	6,965	4,783	4,496	2,444	9,667	4,135	6,340	6,792	4,436	4,471

U.S. AIR DISTANCES

	Atlanta	Birmingham	Boston	Buffalo	Charleston, S.C.	Chicago	Cincinnati	Cleveland	Dallas	Denver	Des Moines	Detroit	Houston	Indianapolis	Jacksonville	Kansas City, Mo.	Los Angeles	Louisville	Miami	Minneapolis	Nashville	New Orleans	New York City	Omaha	Philadelphia	Phoenix	Pittsburgh	Portland, Ore.	St. Louis	Salt Lake City	San Francisco	Seattle	Tulsa	Washington, D.C.
Albuquerque	1272	1138	1972	1580	1539	1129	1251	1421	588	334	837	1364	754	1169	1488	720	664	1178	1698	983	1119	1029	1815	721	1753	330	1499	1107	942	484	896	1184	604	1653
Amarillo	999	866	1722	1338	1266	894	992	1173	334	358	626	1124	533	915	1219	481	937	915	1441	812	848	776	1560	526	1494	598	1244	1304	685	668	1157	1359	335	1391
Atlanta		140	937	697	267	587	369	554	721	1212	739	596	701	426	285	676	1936	319	604	907	214	424	748	817	666	1592	521	2172	467	1583	2139	2182	678	543
Billings	1519	1425	1861	1473	1761	1073	1304	1369	1092	453	798	1283	1315	1204	1796	846	959	1275	2085	742	1309	1479	1760	703	1727	872	1479	686	1057	387	904	668	930	1669
Birmingham	140		1052	776	402	578	406	618	581	1095	670	641	567	433	374	579	1802	331	665	862	182	312	864	732	783	1456	608	2066	400	1466	2013	2082	552	661
Boston	937	1052		400	820	851	740	551	1551	1769	1159	613	1605	807	1017	1251	2596	826	1255	1123	943	1359	188	1282	271	2300	483	2540	1038	2099	2699	2493	1398	393
Buffalo	697	776	400		699	454	393	173	1198	1370	760	216	1286	435	879	861	2198	483	1181	731	627	1086	292	883	279	1906	178	2156	662	1699	2300	2117	1023	292
Burlington, Vt.	951	1049	182	304	884	749	690	476	1501	1654	1049	516	1580	739	1079	1161	2485	780	1347	985	916	1361	260	1171	328	2202	445	2385	966	1969	2568	2333	1327	432
Charleston, S.C.	267	402	820	699		757	506	609	981	1474	967	681	936	594	197	928	2203	500	482	1104	455	630	641	1058	562	1857	528	2425	704	1845	2405	2428	945	453
Charlotte	227	361	721	538	177	587	335	435	930	1358	819	504	927	428	341	803	2119	343	652	939	340	649	533	918	451	1783	362	2290	568	1727	2301	2285	853	330
Cheyenne	1229	1119	1735	1335	1486	891	1082	1199	726	96	583	1125	947	986	1493	560	882	1033	1763	642	1032	1131	1604	463	1556	663	1298	947	795	371	967	973	588	1477
Chicago	587	578	851	454	757		252	308	803	920	309	238	940	165	863	414	1745	269	1188	355	397	833	713	432	666	1453	410	1758	262	1260	1858	1737	598	597
Cincinnati	369	406	740	393	506	252		222	814	1094	510	235	892	100	626	541	1897	90	952	605	238	706	570	622	503	1581	257	1985	309	1453	2043	1972	661	404
Cleveland	554	618	551	173	609	308	222		1025	1227	617	90	1114	263	770	700	2049	311	1087	630	459	924	405	739	360	1749	115	2055	492	1568	2166	2026	853	306
Dallas	721	581	1551	1198	981	803	814	1025		663	632	999	225	763	908	451	1240	726	1111	862	617	443	1374	586	1299	887	1070	1633	547	999	1483	1681	236	1185
Denver	1212	1095	1769	1370	1474	920	1094	1227	663		610	1156	879	1000	1467	558	831	1038	1726	700	1023	1082	1631	488	1579	586	1320	982	796	371	949	1021	550	1494
Des Moines	739	670	1159	760	967	309	510	617	632	610		546	821	411	1023	180	1438	476	1333	235	525	827	1022	123	973	1155	715	1475	273	953	1550	1467	396	896
Detroit	596	641	613	216	681	238	235	90	999	1156	546		1105	240	831	645	1983	316	1152	543	470	939	482	669	443	1690	205	1969	455	1492	2091	1938	813	396
El Paso	1291	1152	2072	1692	1552	1252	1335	1525	572	557	983	1479	676	1264	1473	839	701	1254	1643	1157	1169	983	1905	878	1836	346	1590	1286	1034	689	995	1376	674	1728
Fargo	1114	1060	1300	919	1317	569	820	835	972	642	397	745	1183	725	1399	549	1427	818	1716	214	902	1222	1210	390	1184	1225	949	1239	660	863	1446	1197	741	1140
Houston	701	567	1605	1286	936	940	892	1114	225	879	821	1105		865	821	644	1374	803	968	1056	665	318	1420	794	1341	1017	1137	1836	679	1200	1645	1891	442	1220
Indianapolis	426	433	807	435	594	165	100	263	763	1000	411	240	865		699	453	1809	107	1024	511	251	712	646	525	585	1499	330	1885	231	1356	1949	1872	591	494
Jacksonville	285	374	1017	879	197	863	626	770	908	1467	1023	831	821	699		950	2147	594	326	1191	499	504	838	1098	758	1794	703	2439	751	1837	2374	2455	921	647
Kansas City, Mo.	676	579	1251	861	928	414	541	700	451	558	180	645	644	453	950		1356	480	1241	413	473	680	1097	166	1038	1049	781	1497	238	925	1506	1506	216	945
Knoxville	155	235	818	548	316	454	219	400	767	1178	651	442	790	290	410	624	1941	188	736	792	161	547	632	745	552	1607	375	2115	392	1547	2121	2114	676	430
Little Rock	456	325	1259	913	723	552	524	740	293	780	478	723	388	483	690	325	1480	435	949	708	325	355	1081	492	1007	1137	779	1759	291	1148	1688	1785	231	892
Los Angeles	1936	1802	2596	2198	2203	1745	1897	2049	1240	831	1438	1983	1374	1809	2147	1356		1829	2339	1524	1780	1673	2451	1315	2394	357	2136	825	1589	579	347	959	1266	2300
Louisville	319	331	826	483	500	269	90	311	726	1038	476	316	803	107	594	480	1829		919	605	154	623	652	580	582	1508	344	1950	242	1402	1986	1943	582	476
Memphis	337	217	1137	803	604	482	410	630	420	879	485	623	484	384	590	369	1603	320	872	699	197	358	957	529	881	1263	660	1849	240	1250	1802	1867	341	765
Miami	604	665	1255	1181	482	1188	952	1087	1111	1726	1333	1152	968	1024	326	1241	2339	919		1511	815	669	1092	1397	1019	1982	1010	2708	1061	2089	2594	2734	1176	923
Minneapolis	907	862	1123	731	1104	355	605	630	862	700	235	543	1056	511	1191	413	1524	605	1511		697	1051	1018	290	985	1280	743	1427	466	987	1584	1395	626	934
Nashville	214	182	943	627	455	397	238	459	617	1023	525	470	665	251	499	473	1780	154	815	697		469	761	607	685	1446	472	1969	254	1393	1963	1975	515	569
New Orleans	424	312	1359	1086	630	833	706	924	443	1082	827	939	318	712	504	680	1673	623	669	1051	469		1171	847	1089	1316	919	2063	598	1434	1926	2101	548	966
New York City	748	864	188	292	641	713	570	405	1374	1631	1022	482	1420	646	838	1097	2451	652	1092	1018	761	1171		1144	83	2145	317	2445	875	1972	2571	2408	1231	205
Omaha	817	732	1282	883	1058	432	622	739	586	488	123	669	794	525	1098	166	1315	580	1397	290	607	847	1144		1094	1036	836	1371	354	833	1429	1369	352	1014
Philadelphia	666	783	271	279	562	666	503	360	1299	1579	973	443	1341	585	758	1038	2394	582	1019	985	685	1089	83	1094		2083	259	2412	811	1925	2523	2380	1163	123
Phoenix	1592	1456	2300	1906	1857	1453	1581	1749	887	586	1155	1690	1017	1499	1794	1049	357	1508	1982	1280	1446	1316	2145	1036	2083		1828	1005	1272	504	653	1114	932	1983
Pittsburgh	521	608	483	178	528	410	257	115	1070	1320	715	205	1137	330	703	781	2136	344	1010	743	472	919	317	836	259	1828		2165	559	1668	2264	2138	917	192
Portland, Ore.	2172	2066	2540	2156	2425	1758	1985	2055	1633	982	1475	1969	1836	1885	2439	1497	825	1950	2708	1427	1969	2063	2445	1371	2412	1005	2165		1723	636	534	145	1531	2354
Raleigh	356	491	609	490	220	642	396	428	1057	1463	902	510	1056	495	414	905	2237	429	695	996	457	776	426	1008	345	1903	330	2377	667	1829	2410	2367	972	233
St. Louis	467	400	1038	662	704	262	309	492	547	796	273	455	679	231	751	238	1589	242	1061	466	254	598	875	354	811	1272	559	1723		1162	1744	1724	361	712
Salt Lake City	1583	1466	2099	1699	1845	1260	1453	1568	999	371	953	1492	1200	1356	1837	925	579	1402	2089	987	1393	1434	1972	833	1925	504	1668	636	1162		600	701	917	1848
San Antonio	882	744	1766	1430	1122	1051	1039	1256	252	802	882	1238	189	999	1011	702	1204	949	1148	1110	823	507	1584	828	1507	849	1291	1720	792	1087	1490	1787	486	1388
San Francisco	2139	2013	2699	2300	2405	1858	2043	2166	1483	949	1550	2091	1645	1949	2374	1506	347	1986	2594	1584	1963	1926	2571	1429	2523	653	2264	534	1744	600		678	1461	2442
Seattle	2182	2082	2493	2117	2428	1737	1972	2026	1681	1021	1467	1938	1891	1872	2455	1506	959	1943	2734	1395	1975	2101	2408	1369	2380	1114	2138	145	1724	701	678		1560	2329
Spokane	1961	1865	2266	1888	2204	1508	1744	1796	1489	826	1240	1709	1704	1644	2237	1287	940	1717	2520	1166	1752	1898	2179	1146	2151	1019	1908	290	1500	550	727	229	1353	2100
Syracuse	781	875	264	138	738	592	514	303	1326	1508	898	354	1403	567	928	998	2336	603	1212	861	739	1187	194	1021	220	2044	268	2281	796	1835	2435	2238	1157	290
Tulsa	678	552	1398	1023	945	598	661	853	236	550	396	813	442	591	921	216	1266	582	1176	626	515	548	1231	352	1163	932	917	1531	361	917	1461	1560		1058
Washington, D.C.	543	661	393	292	453	597	404	306	1185	1494	896	396	1220	494	647	945	2300	476	923	934	569	966	205	1014	123	1983	192	2354	712	1848	2442	2329	1058	
Wichita	776	658	1424	1036	1039	591	702	873	340	437	334	821	559	620	1031	177	1197	633	1297	546	594	677	1266	257	1204	879	950	1411	394	808	1369	1437	130	1106

WORLD FLAGS and MAPS

AFGHANISTAN

ALBANIA

ALGERIA

ANDORRA

ANGOLA

ARGENTINA

AUSTRALIA

AUSTRIA

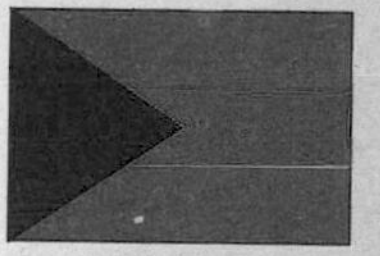
BAHAMAS

BAHRAIN

BANGLADESH

BARBADOS

BELGIUM

BELIZE

BENIN (DAHOMEY)

BHUTAN

BOLIVIA

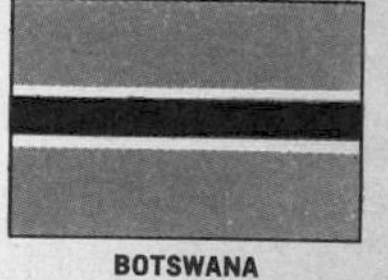
BOTSWANA

BRAZIL

BULGARIA

BURMA

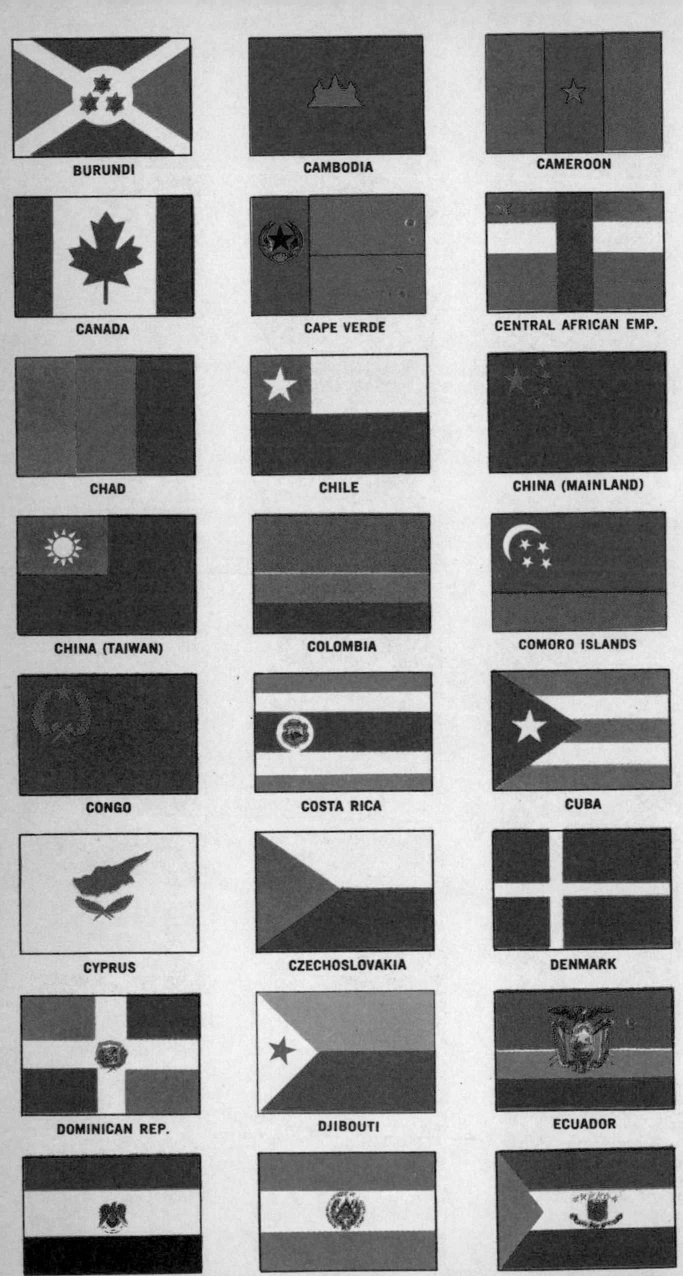
BURUNDI
CAMBODIA
CAMEROON
CANADA
CAPE VERDE
CENTRAL AFRICAN EMP.
CHAD
CHILE
CHINA (MAINLAND)
CHINA (TAIWAN)
COLOMBIA
COMORO ISLANDS
CONGO
COSTA RICA
CUBA
CYPRUS
CZECHOSLOVAKIA
DENMARK
DOMINICAN REP.
DJIBOUTI
ECUADOR
EGYPT
EL SALVADOR
EQUATORIAL GUINEA

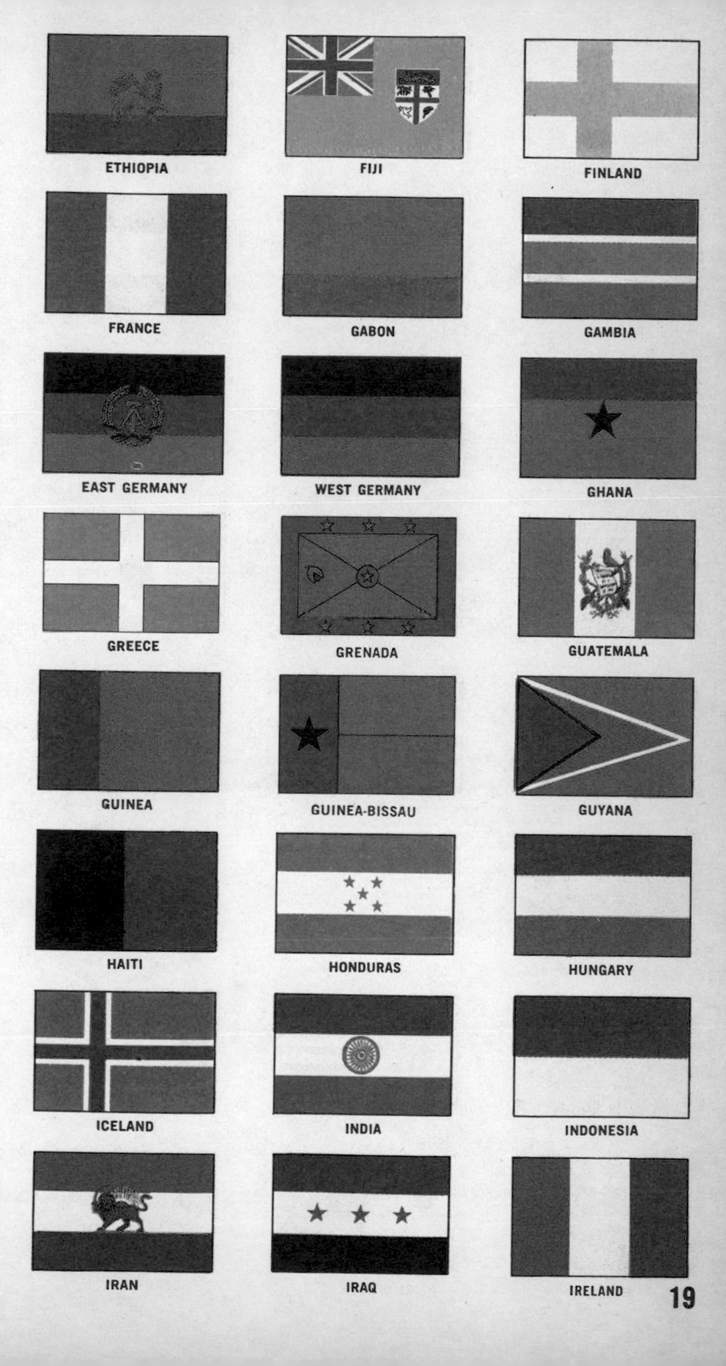
ETHIOPIA
FIJI
FINLAND
FRANCE
GABON
GAMBIA
EAST GERMANY
WEST GERMANY
GHANA
GREECE
GRENADA
GUATEMALA
GUINEA
GUINEA-BISSAU
GUYANA
HAITI
HONDURAS
HUNGARY
ICELAND
INDIA
INDONESIA
IRAN
IRAQ
IRELAND

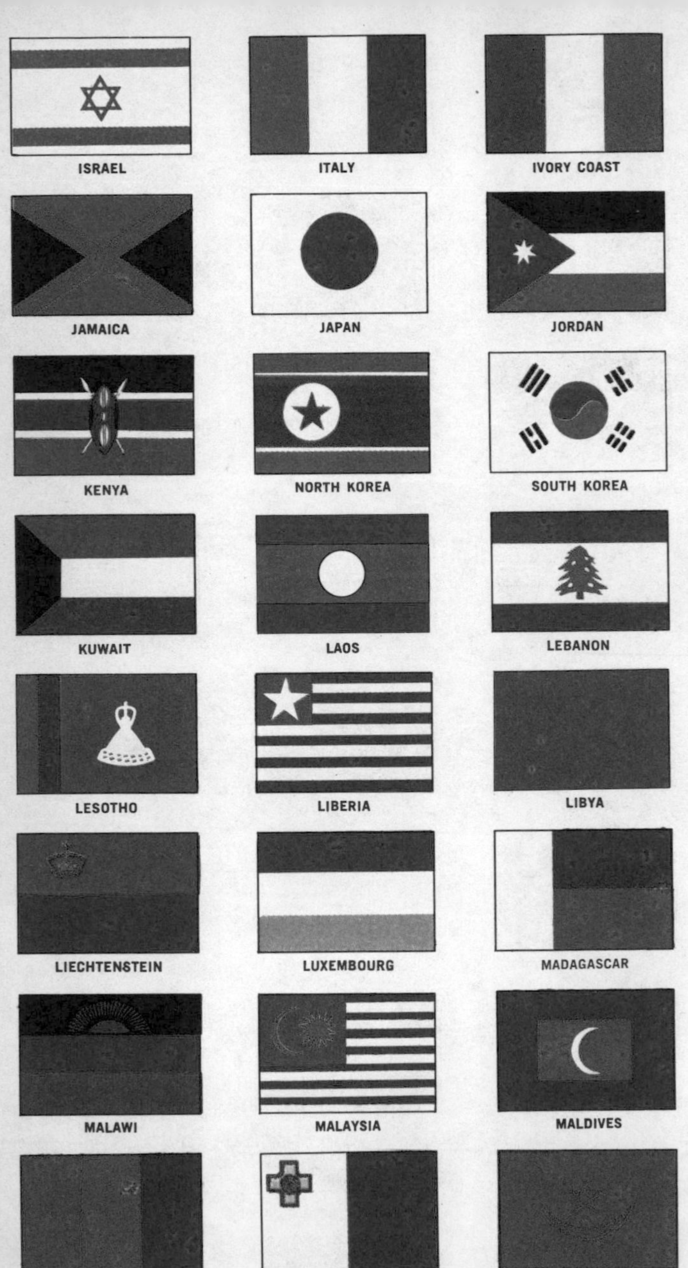
ISRAEL
ITALY
IVORY COAST
JAMAICA
JAPAN
JORDAN
KENYA
NORTH KOREA
SOUTH KOREA
KUWAIT
LAOS
LEBANON
LESOTHO
LIBERIA
LIBYA
LIECHTENSTEIN
LUXEMBOURG
MADAGASCAR
MALAWI
MALAYSIA
MALDIVES
MALI
MALTA
MAURITANIA

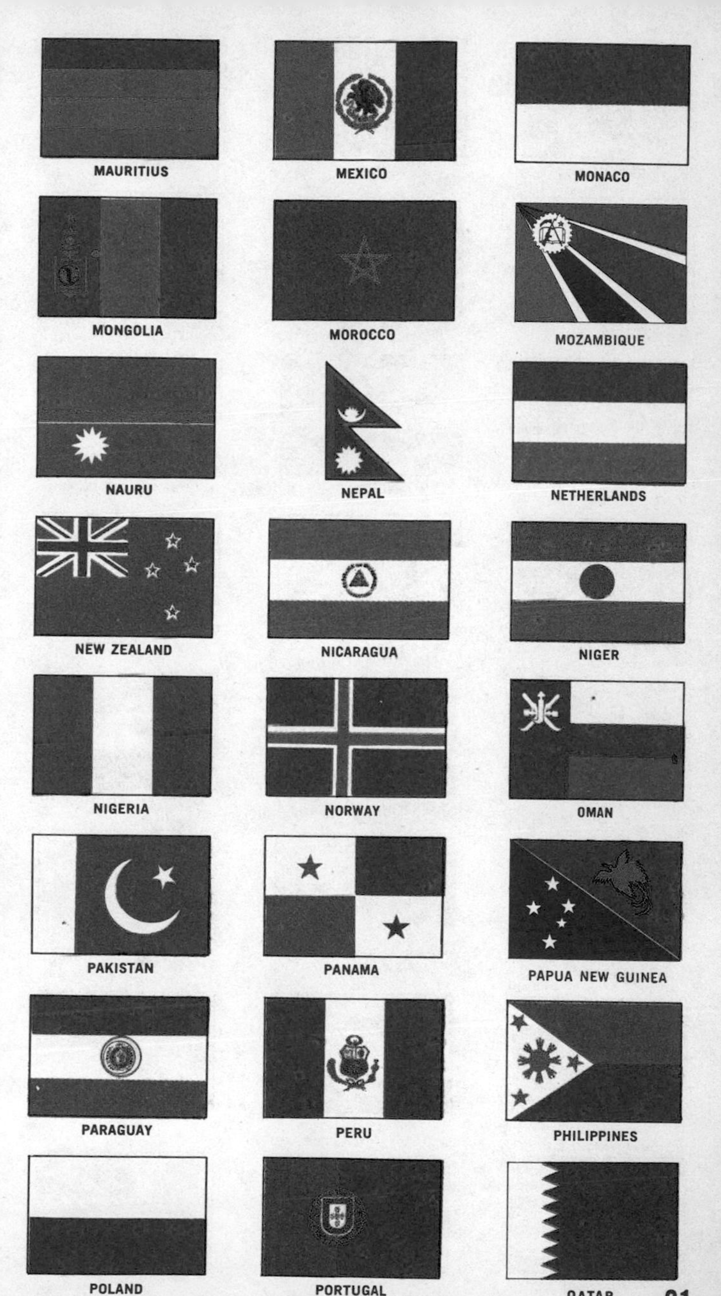
MAURITIUS
MEXICO
MONACO
MONGOLIA
MOROCCO
MOZAMBIQUE
NAURU
NEPAL
NETHERLANDS
NEW ZEALAND
NICARAGUA
NIGER
NIGERIA
NORWAY
OMAN
PAKISTAN
PANAMA
PAPUA NEW GUINEA
PARAGUAY
PERU
PHILIPPINES
POLAND
PORTUGAL
QATAR

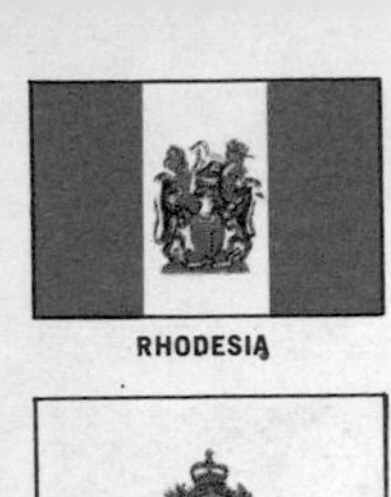
RHODESIA

RUMANIA

RWANDA

SAN MARINO

SÃO TOMÉ E PRÍNCIPE

SAUDI ARABIA

SENEGAL

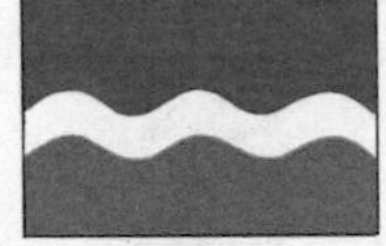
SEYCHELLES

SIERRA LEONE

SINGAPORE

SOMALIA

SOUTH AFRICA

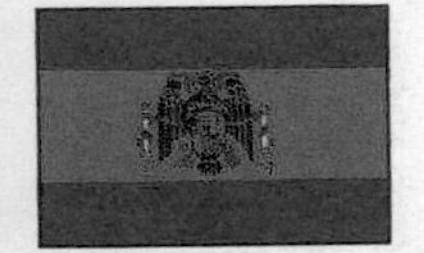
SPAIN

SRI LANKA (CEYLON)

SUDAN

SURINAM

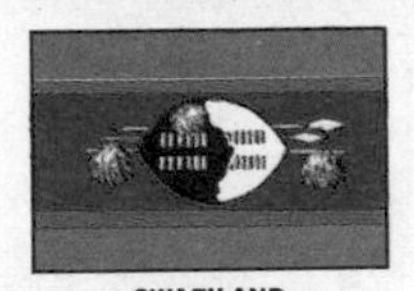
SWAZILAND

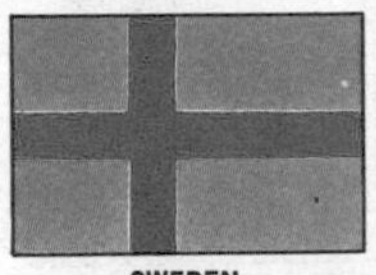
SWEDEN

SWITZERLAND

SYRIA

TANZANIA

THAILAND

TOGO

TONGA

TRINIDAD & TOBAGO

TUNISIA

TURKEY

UGANDA

U.S.S.R.

UNITED ARAB EMIRATES

UNITED KINGDOM

UNITED STATES

UPPER VOLTA

URUGUAY

VATICAN CITY

VENEZUELA

VIETNAM

WESTERN SAMOA

YEMEN (PEOPLES REP.)

YEMEN ARAB REP.

YUGOSLAVIA

ZAIRE

ZAMBIA

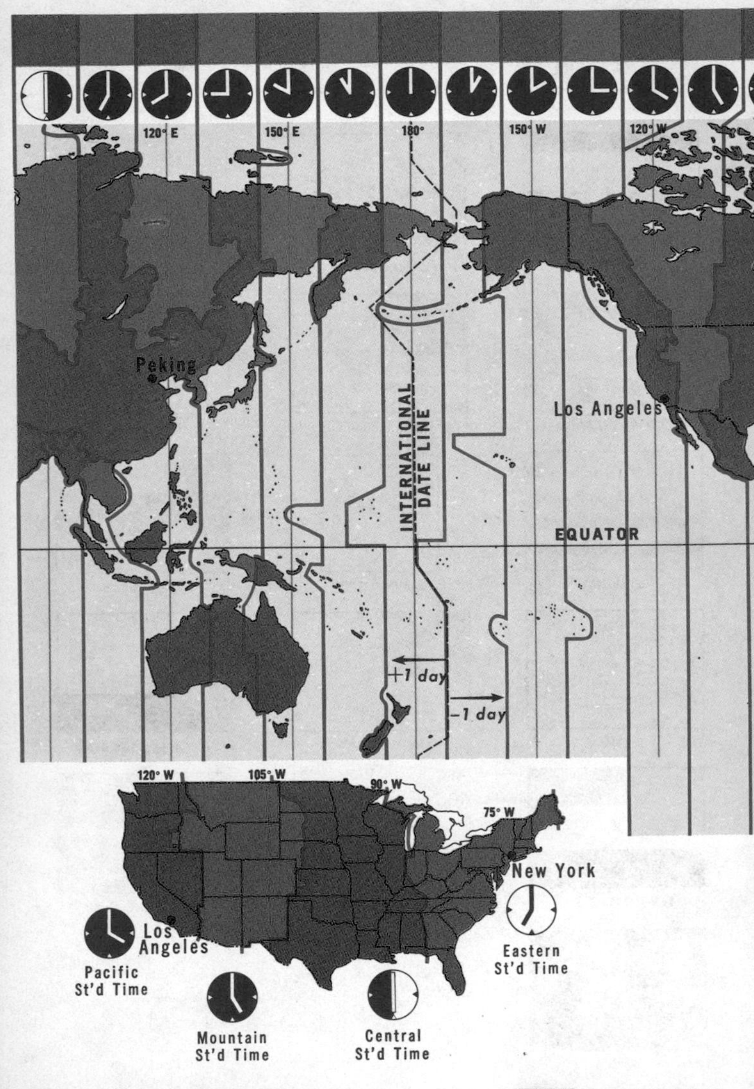
120° E
150° E
180°
150° W
120° W
Peking
INTERNATIONAL DATE LINE
Los Angeles
EQUATOR
+1 day
−1 day
120° W
105° W
90° W
75° W
New York
Los Angeles
Pacific St'd Time
Mountain St'd Time
Central St'd Time
Eastern St'd Time

TIME ZONES

Standard time zones

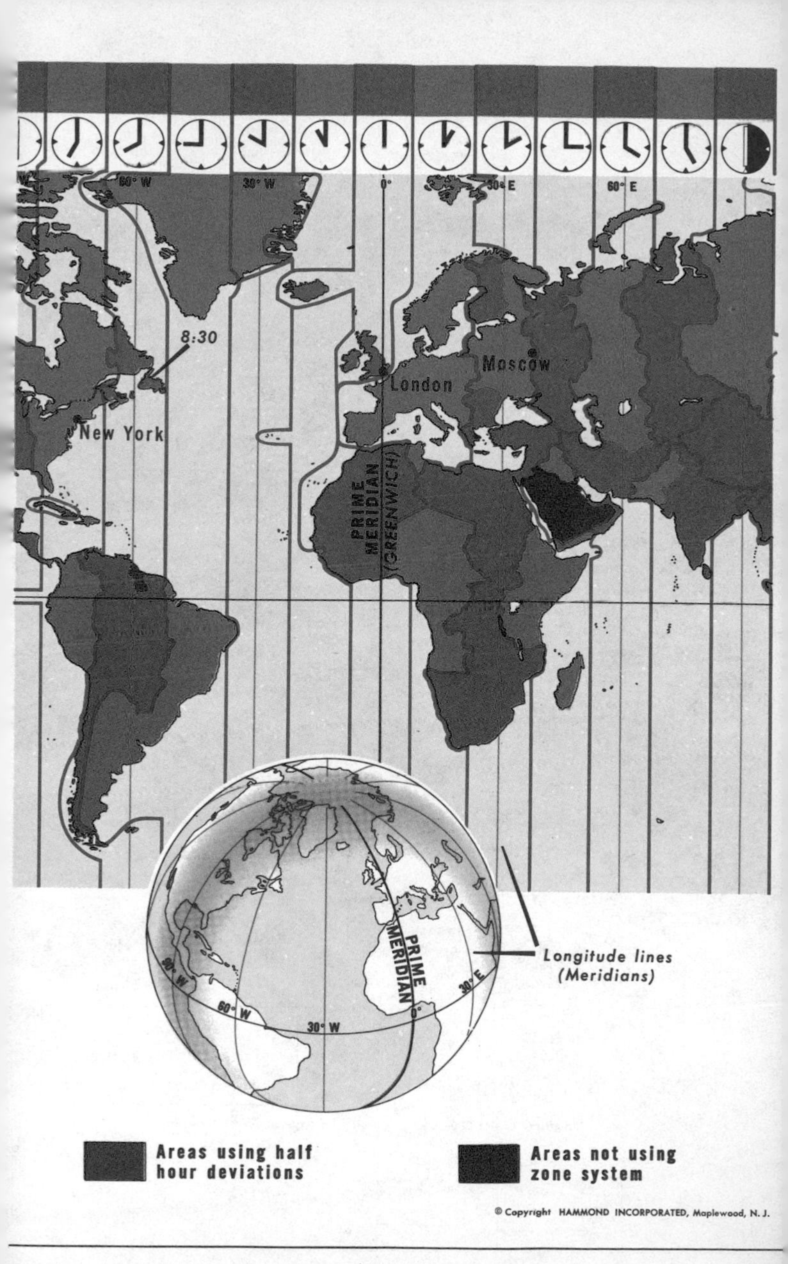
60° W
30° W
0°
30° E
60° E
8:30
Moscow
London
New York
PRIME
MERIDIAN
(GREENWICH)
PRIME
MERIDIAN
Longitude lines
(Meridians)
90° W
60° W
30° W
0°
30° E
Areas using half
hour deviations
Areas not using
zone system

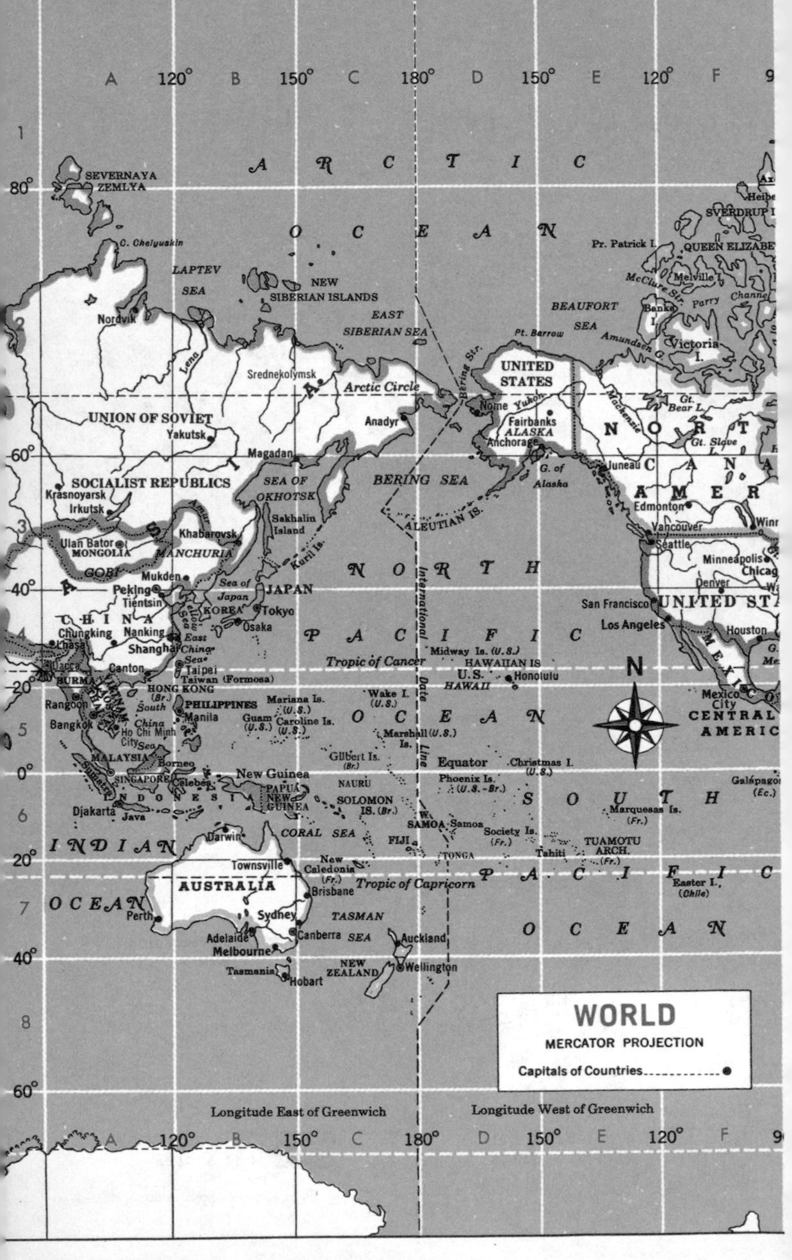
WORLD
MERCATOR PROJECTION
Capitals of Countries
ARCTIC OCEAN
NORTH PACIFIC OCEAN
SOUTH PACIFIC OCEAN
INDIAN OCEAN
UNION OF SOVIET SOCIALIST REPUBLICS
ASIA
CHINA
JAPAN
AUSTRALIA
NORTH AMERICA
UNITED STATES
CANADA
Longitude East of Greenwich
Longitude West of Greenwich

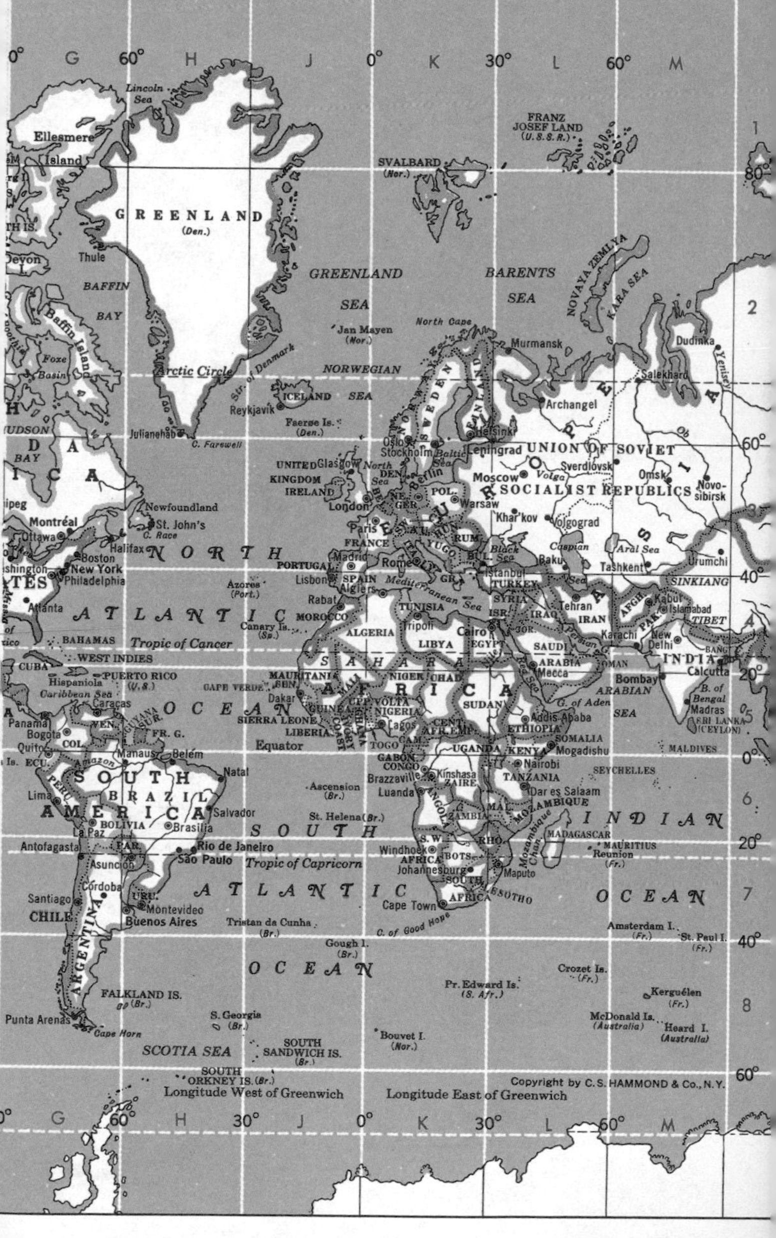

GREENLAND
(Den.)
Ellesmere Island
Lincoln Sea
Thule
BAFFIN BAY
Baffin Island
Foxe Basin
Arctic Circle
Str. of Denmark
Julianehåb
C. Farewell
GREENLAND SEA
Jan Mayen
(Nor.)
NORWEGIAN SEA
ICELAND
Reykjavík
Faerøe Is.
(Den.)
SVALBARD
(Nor.)
FRANZ JOSEF LAND
(U.S.S.R.)
BARENTS SEA
NOVAYA ZEMLYA
KARA SEA
North Cape
Murmansk
Archangel
Salekhard
Dudinka
Yenisey
Ob
UNION OF SOVIET SOCIALIST REPUBLICS
EUROPE
ASIA
Leningrad
Helsinki
Moscow
Volga
Sverdlovsk
Omsk
Novo-sibirsk
Khar'kov
Volgograd
Caspian Sea
Aral Sea
Baku
Tashkent
Urumchi
SINKIANG
TIBET
Oslo
Stockholm
Baltic Sea
NORWAY
SWEDEN
FINLAND
DEN.
North Sea
UNITED KINGDOM
IRELAND
Glasgow
London
Berlin
POL.
Warsaw
GER.
Paris
FRANCE
Madrid
PORTUGAL
Lisbon
SPAIN
ITALY
Rome
YUGO.
HUN.
RUM.
BUL.
GRE.
Black Sea
Istanbul
TURKEY
SYRIA
ISR.
IRAQ
IRAN
Tehran
AFGH.
Kabul
Islamabad
PAK.
Karachi
New Delhi
INDIA
Calcutta
Bombay
Madras
SRI LANKA (CEYLON)
B. of Bengal
ARABIAN SEA
MALDIVES
Mediterranean Sea
Algiers
Rabat
MOROCCO
TUNISIA
Tripoli
ALGERIA
LIBYA
Cairo
EGYPT
SAUDI ARABIA
Mecca
Persian G.
OMAN
Red Sea
SAHARA
AFRICA
MAURITANIA
SEN.
Dakar
MALI
NIGER
CHAD
SUDAN
UPP. VOLTA
NIGERIA
GUINEA
SIERRA LEONE
LIBERIA
IVORY COAST
GHANA
TOGO
Lagos
CAM.
CENT. AFR. EMP.
ETHIOPIA
Addis Ababa
G. of Aden
SOMALIA
Mogadishu
UGANDA
KENYA
Nairobi
GABON
CONGO
Brazzaville
Kinshasa
ZAIRE
TANZANIA
Dar es Salaam
Luanda
ANGOLA
ZAMBIA
MOZAMBIQUE
Mozambique Chan.
MADAGASCAR
RHO.
S.W. AFRICA
Windhoek
BOTS.
Johannesburg
Maputo
SOUTH AFRICA
LESOTHO
Cape Town
C. of Good Hope
SEYCHELLES
MAURITIUS
Reunion
(Fr.)
INDIAN OCEAN
Amsterdam I.
(Fr.)
St. Paul I.
(Fr.)
Crozet Is.
(Fr.)
Kerguélen
(Fr.)
McDonald Is.
(Australia)
Heard I.
(Australia)
Pr. Edward Is.
(S. Afr.)
Bouvet I.
(Nor.)
HUDSON BAY
CANADA
Newfoundland
St. John's
C. Race
Montréal
Ottawa
Halifax
Boston
New York
Philadelphia
Washington
UNITED STATES
Atlanta
Gulf of Mexico
NORTH ATLANTIC OCEAN
Azores
(Port.)
Canary Is.
(Sp.)
Tropic of Cancer
BAHAMAS
WEST INDIES
CUBA
Hispaniola
PUERTO RICO
(U.S.)
Caribbean Sea
CAPE VERDE
Caracas
Panama
VEN.
GUYANA
SUR.
FR. G.
Bogota
COL.
Quito
ECU.
Equator
Manaus
Belém
Amazon
SOUTH AMERICA
BRAZIL
PERU
Lima
Natal
Salvador
BOLIVIA
La Paz
Brasília
PAR.
Rio de Janeiro
São Paulo
Antofagasta
Asunción
Tropic of Capricorn
Córdoba
URU.
Santiago
CHILE
ARGENTINA
Montevideo
Buenos Aires
Ascension
(Br.)
St. Helena (Br.)
SOUTH ATLANTIC OCEAN
Tristan da Cunha
(Br.)
Gough I.
(Br.)
FALKLAND IS.
(Br.)
Punta Arenas
Cape Horn
S. Georgia
(Br.)
SCOTIA SEA
SOUTH SANDWICH IS.
(Br.)
SOUTH ORKNEY IS. (Br.)
Longitude West of Greenwich
Longitude East of Greenwich
Copyright by C.S. HAMMOND & Co., N.Y.
G
H
J
K
L
M
60°
30°
0°
80°
40°
20°
1
2
3
4
5
6
7
8

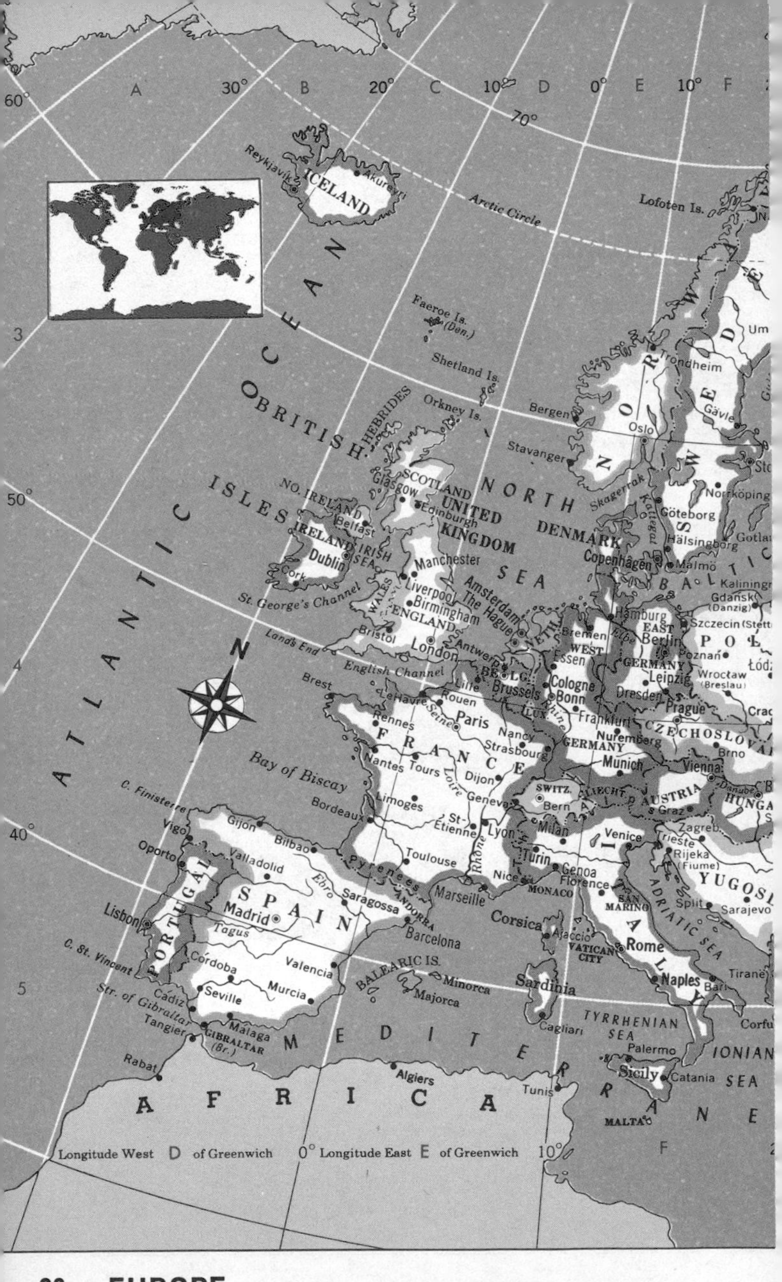
ATLANTIC OCEAN
ICELAND
Reykjavik
Akureyri
Arctic Circle
Lofoten Is.
Faeroe Is. (Den.)
Shetland Is.
Orkney Is.
HEBRIDES
BRITISH ISLES
SCOTLAND
UNITED KINGDOM
NO. IRELAND
IRELAND
Glasgow
Edinburgh
Belfast
Dublin
Cork
IRISH SEA
St. George's Channel
Manchester
Liverpool
Birmingham
WALES
ENGLAND
Bristol
London
Land's End
English Channel
NORTH SEA
NORWAY
Bergen
Oslo
Stavanger
Trondheim
SWEDEN
Gävle
Norrköping
Göteborg
Hälsingborg
Malmö
Skagerrak
Kattegat
DENMARK
Copenhagen
BALTIC
Gdansk (Danzig)
Kaliningr
Gotlar
Amsterdam
The Hague
NETH.
Antwerp
BELG.
Brussels
Lille
LUX.
Hamburg
Bremen
Elbe
Essen
Cologne
Bonn
Rhine
Frankfurt
Nuremberg
Munich
WEST GERMANY
EAST GERMANY
Berlin
Leipzig
Dresden
Szczecin (Stett
Poznań
Łódź
Wrocław (Breslau)
POL
Prague
CZECHOSLOVA
Brno
Vienna
AUSTRIA
Graz
Danube
HUNGA
LIECHT.
SWITZ.
Bern
Geneva
Brest
Le Havre
Rouen
Seine
Paris
Rennes
FRANCE
Nancy
Strasbourg
Nantes
Tours
Loire
Dijon
Limoges
Bordeaux
St-Étienne
Lyon
Rhone
Toulouse
Nice
MONACO
Marseille
Bay of Biscay
C. Finisterre
Vigo
Gijon
Bilbao
Oporto
Valladolid
Ebro
Pyrenees
Saragossa
ANDORRA
Barcelona
PORTUGAL
SPAIN
Madrid
Tagus
Lisbon
C. St. Vincent
Córdoba
Valencia
Murcia
Seville
Cádiz
Málaga
Str. of Gibraltar
GIBRALTAR (Br.)
Tangier
Rabat
BALEARIC IS.
Minorca
Majorca
Corsica
Ajaccio
Sardinia
Cagliari
Milan
Turin
Genoa
Florence
Venice
SAN MARINO
VATICAN CITY
Rome
Naples
Bari
ITALY
ADRIATIC SEA
Trieste
Zagreb
Rijeka (Fiume)
Split
Sarajevo
YUGOSL
Tirane
Corfu
TYRRHENIAN SEA
Palermo
Sicily
Catania
IONIAN SEA
MALTA
MEDITERRANEAN
Algiers
Tunis
AFRICA
Longitude West D of Greenwich
0° Longitude East E of Greenwich
10°
60°
50°
40°
70°
30°
20°
A
B
C
D
E
F
3
4
5

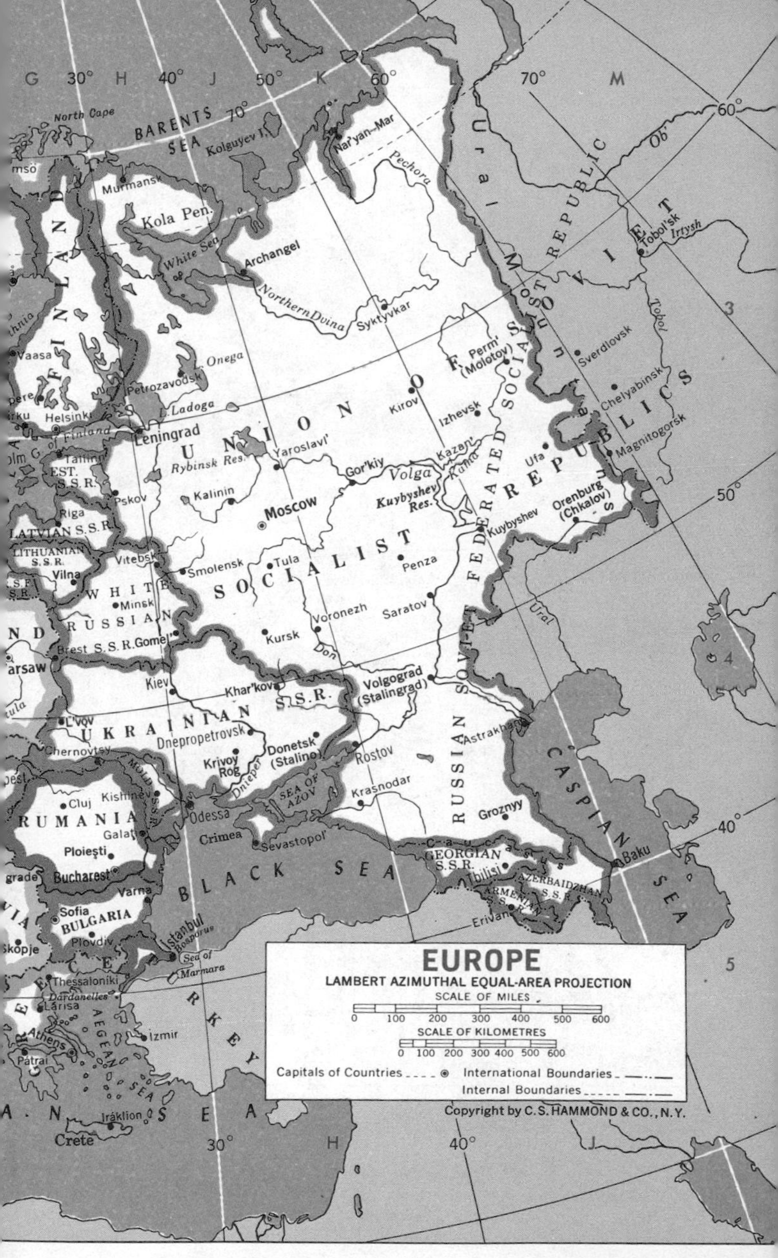
EUROPE
LAMBERT AZIMUTHAL EQUAL-AREA PROJECTION
SCALE OF MILES
SCALE OF KILOMETRES
Capitals of Countries
International Boundaries
Internal Boundaries
Copyright by C. S. HAMMOND & CO., N.Y.
UNION OF SOVIET SOCIALIST REPUBLICS
RUSSIAN SOVIET FEDERATED SOCIALIST REPUBLIC
BARENTS SEA
BLACK SEA
CASPIAN SEA
SEA OF AZOV
AEGEAN SEA
Moscow
Leningrad
Archangel
Murmansk
Kola Pen.
White Sea
Kiev
Khar'kov
Volgograd (Stalingrad)
Odessa
Crimea
Sevastopol'
Rostov
Krasnodar
Groznyy
Baku
Tbilisi
Erivan
Astrakhan
Saratov
Kuybyshev
Penza
Tula
Kursk
Voronezh
Smolensk
Minsk
Vilna
Riga
Tallinn
Helsinki
Vaasa
FINLAND
RUMANIA
BULGARIA
TURKEY
UKRAINIAN S.S.R.
WHITE RUSSIAN S.S.R.
Gor'kiy
Kazan'
Kirov
Perm' (Molotov)
Izhevsk
Ufa
Sverdlovsk
Chelyabinsk
Magnitogorsk
Orenburg (Chkalov)
Tobol'sk
Syktyvkar
Nar'yan-Mar
Yaroslavl'
Kalinin
Pskov
Vitebsk
Gomel'
Brest
Kishinev
Cluj
Galați
Ploiești
Bucharest
Varna
Sofia
Plovdiv
Istanbul
Izmir
Athens
Crete
Dnepropetrovsk
Krivoy Rog
Donetsk (Stalino)
Chernovtsy
L'vov
Petrozavodsk
Volga
Kuybyshev Res.
Rybinsk Res.
Northern Dvina
Pechora
Ural
Don
Dnieper

UNITED KINGDOM
and
IRELAND
BONNE PROJECTION
Copyright by C.S. HAMMOND & Co., N.Y.
SCALE OF MILES
SCALE OF KILOMETERS
Capitals of Countries
Canals
SHETLAND ISLANDS
Same scale as main map
ORKNEY ISLANDS
OUTER HEBRIDES
ATLANTIC
OCEAN
IRISH SEA
Lerwick
Kirkwall
Wick
Thurso
Stornoway
Inverness
Aberdeen
Dundee
Perth
Fort William
Oban
Glasgow
Edinburgh
Paisley
Berwick-upon-Tweed
Newcastle
South Shields
Londonderry
Blackpool
Preston
Manchester
Liverpool
Bradford
Huddersfield
Oldham
Stockport

IRELAND
NORTH SEA
IRISH SEA
ENGLISH CHANNEL
ENGLAND
WALES
FRANCE
Dublin
Limerick
Cork
Galway
Waterford
Wexford
Dundalk
Drogheda
Sligo
Lough Allen
Lough Conn
Lough Mask
Lough Corrib
Lough Ree
Lough Derg
R. Shannon
Erris Head
Achill I.
Clare I.
Aran Is.
Galway Bay
Dingle Bay
Bantry Bay
Cape Clear
Wicklow Head
St. George's Channel
Isle of Man
Anglesey
Cardigan Bay
Caernarvon Bay
Bristol Channel
Lundy I.
LONDON
Birmingham
Manchester
Liverpool
Leeds
Bradford
Sheffield
Bristol
Cardiff
Newport
Rhondda
Swansea
York
Hull
Nottingham
Leicester
Coventry
Northampton
Norwich
Ipswich
Cambridge
Oxford
Luton
Southend-on-Sea
Reading
Croydon
Southampton
Portsmouth
Bournemouth
Brighton
Plymouth
Exeter
Stoke-on-Trent
Derby
Wolverhampton
Walsall
Blackpool
Preston
Blackburn
Birkenhead
Scarborough
Flamborough Hd.
The Wash
Thames
R. Severn
Strait of Dover
Isle of Wight
Land's End
Lizard Head
Scilly Is.
Eddystone Rocks
Lyme Bay
Portland Bill
CHANNEL ISLANDS
Alderney
Guernsey
Jersey
Cherbourg
Bay of the Seine
Le Havre
Rouen
Dieppe
Boulogne
Calais
GREATER LONDON
Watford
Enfield
Cheshunt
Brentwood
Barnet
Haringey
Redbridge
Havering
Harrow
Westminster
Newham
Barking
Ealing
Thurrock
R. Thames
Greenwich
Bexley
Hounslow
Kingston upon T.
Bromley
Gravesend
Egham
Croydon
Esher
Woking
Epsom & Ewell
R. Colne
MILES
54°
52°
50°
10°
8°
6° Longitude
West of
4° Greenwich
2°
0°

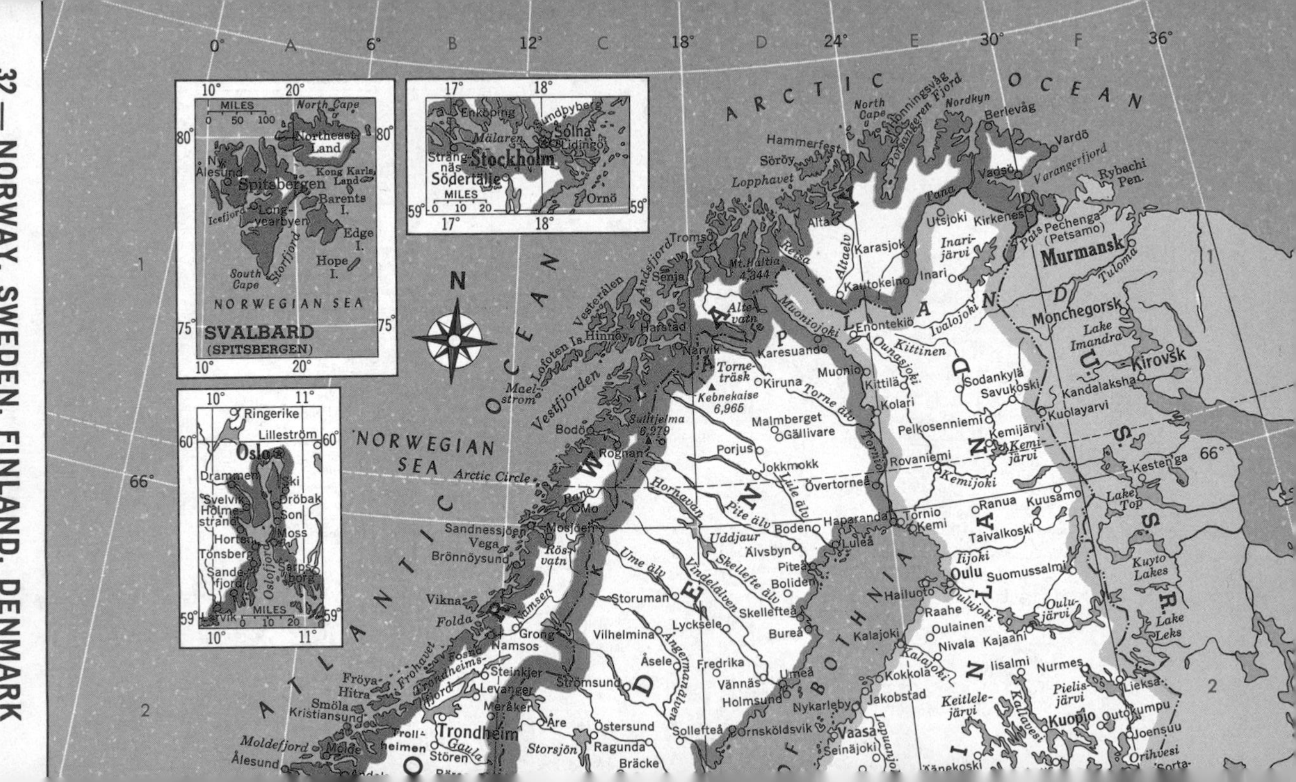
ARCTIC OCEAN
NORWEGIAN SEA
ATLANTIC OCEAN
NORWAY
SWEDEN
FINLAND
LAPLAND
U.S.S.R.
GULF OF BOTHNIA
Arctic Circle
SVALBARD (SPITSBERGEN)
Spitsbergen
Northeast Land
North Cape
Kong Karls Land
Barents I.
Edge I.
Hope I.
Storfjord
South Cape
Longyearbyen
Ny Ålesund
Icefjord
Stockholm
Södertälje
Solna
Sundbyberg
Lidingö
Mälaren
Enköping
Strängnäs
Ornö
Oslo
Ringerike
Lilleström
Drammen
Ski
Dröbak
Son
Moss
Svelvik
Holmestrand
Horten
Tönsberg
Sandefjord
Sarpsborg
Oslofjord
Larvik
MILES
Hammerfest
Söröy
Lopphavet
North Cape
Honningsvåg
Porsangeren Fjord
Nordkyn
Berlevåg
Vardö
Vadsö
Varangerfjord
Rybachi Pen.
Kirkenes
Pechenga (Petsamo)
Murmansk
Tuloma
Monchegorsk
Lake Imandra
Kirovsk
Kandalaksha
Kuolayarvi
Kestenga
Lake Top
Kuyto Lakes
Lake Leks
Tromsö
Senja
Andsfjord
Vesterålen
Hinnöy
Harstad
Lofoten Is.
Maelstrom
Vestfjorden
Narvik
Mt. Haltia 4,344
Alta
Altaelv
Reisa
Karasjok
Kautokeino
Utsjoki
Tana
Inari
Inarijärvi
Ivalojoki
Alte vatn
Torneträsk
Kiruna
Kebnekaise 6,965
Karesuando
Muonio
Muoniojoki
Enontekiö
Ounasjoki
Kittinen
Kittilä
Kolari
Sodankylä
Savukoski
Torne älv
Malmberget
Gällivare
Sulitjelma 6,279
Bodö
Rognan
Porjus
Jokkmokk
Lule älv
Övertorneå
Tornio
Rovaniemi
Pelkosenniemi
Kemijärvi
Kemijoki
Ranua
Kuusamo
Taivalkoski
Kemi
Haparanda
Boden
Luleå
Rana
Mo
Mosjöen
Sandnessjöen
Vega
Brönnöysund
Rösvatn
Hornavan
Pite älv
Uddjaur
Älvsbyn
Piteå
Boliden
Skellefte älv
Vindelälven
Ume älv
Storuman
Skellefteå
Lycksele
Vilhelmina
Åsele
Fredrika
Ångermanälven
Bureå
Umeå
Vännäs
Holmsund
Vikna
Folda
Namsen
Grong
Namsos
Steinkjer
Strömsund
Frohavet
Fosna
Trondheimsfjord
Fröya
Hitra
Smöla
Kristiansund
Levanger
Meråker
Åre
Östersund
Storsjön
Ragunda
Bräcke
Trondheim
Gaula
Stören
Trollheimen
Moldefjord
Molde
Ålesund
Sollefteå
Örnsköldsvik
Nykarleby
Jakobstad
Vaasa
Seinäjoki
Lapuanjoki
Kokkola
Kalajoki
Hailuoto
Raahe
Oulainen
Nivala
Oulu
Oulujoki
Iijoki
Suomussalmi
Oulujärvi
Kajaani
Iisalmi
Nurmes
Pielisjärvi
Lieksa
Keitelejärvi
Kallavesi
Kuopio
Outokumpu
Joensuu
Orihvesi
Äänekoski

NORWAY, SWEDEN, FINLAND and DENMARK
CONIC PROJECTION
Copyright by C. S. HAMMOND & Co., N. Y.
SCALE OF MILES
0 50 100 150 200
KILOMETRES
0 50 100 150 200
Capitals of Countries
International Boundaries
Canals
OSLO
STOCKHOLM
HELSINKI
COPENHAGEN
LENINGRAD
Bergen
Stavanger
Kristiansand
Göteborg
Malmö
Uppsala
Västerås
Örebro
Norrköping
Linköping
Tampere
Turku
Tallinn
Riga
Kaliningrad
Gdańsk
Gdynia
Hamburg
Kiel
Lübeck
Rostock
Aalborg
Aarhus
Odense
Esbjerg
Flensburg
Gotland
Öland
Bornholm
SKAGERRAK
KATTEGAT
NORTH SEA
BALTIC SEA
GULF
Gulf of Finland
Gulf of Riga
ESTONIA
DENMARK
POLAND
EAST GERMANY
WEST GERMANY
NETHERLANDS
Longitude 18° East of Greenwich

GERMANY
CONIC PROJECTION
SCALE OF MILES
0 10 20 40 60
SCALE OF KILOMETRES
0 10 20 40 60
Capitals of Countries
State and District Capitals
Canals
Copyright by C. S. HAMMOND & Co., N.Y.
DENMARK
BALTIC SEA
NORTH SEA
POLAND
NETHERLANDS
SCHLESWIG-HOLSTEIN
MECKLENBURG
BRANDENBURG
LOWER SAXONY
NORTH RHINE-WESTPHALIA
EAST GERMANY
HAMBURG
BREMEN
BERLIN
Kiel Bay
Mecklenburg Bay
Pomeranian Bay
Helgoland Bay
NORTH FRISIAN ISLANDS
WEST FRISIAN IS.

WESTPHALIA
HESSE
THURINGIA
SAXONY
RHINELAND-PALATINATE
BADEN-WÜRTTEMBERG
BAVARIA
GERMANY
CZECHOSLOVAKIA
FRANCE
SWITZERLAND
AUSTRIA
BELGIUM
LUXEMBOURG
Bohemian Forest
Bavarian Forest
Bavarian Alps
Fichtel Gebirge
Erzgebirge
Westerwald
Odenwald
Prague
Dresden
Chemnitz (Karl Marx Stadt)
Plauen
Erfurt
Weimar
Jena
Gotha
Frankfurt
Wiesbaden
Mainz
Koblenz
Cologne
Bonn
Aachen
Trier
Luxembourg
Saarbrücken
Kaiserslautern
Ludwigshafen
Mannheim
Heidelberg
Darmstadt
Würzburg
Bamberg
Bayreuth
Nuremberg (Nürnberg)
Fürth
Regensburg
Ingolstadt
Augsburg
MUNICH (München)
Ulm
Stuttgart
Heilbronn
Karlsruhe
Pforzheim
Freiburg
Basel
Zürich
Strasbourg
Metz
Mulhouse
Pilsen (Plzen)
Linz
Salzburg
Passau
Danube
Rhine
Main
Neckar
Lake Constance
50°
48°
8°
10°
12°
14°
Longitude
East of Greenwich

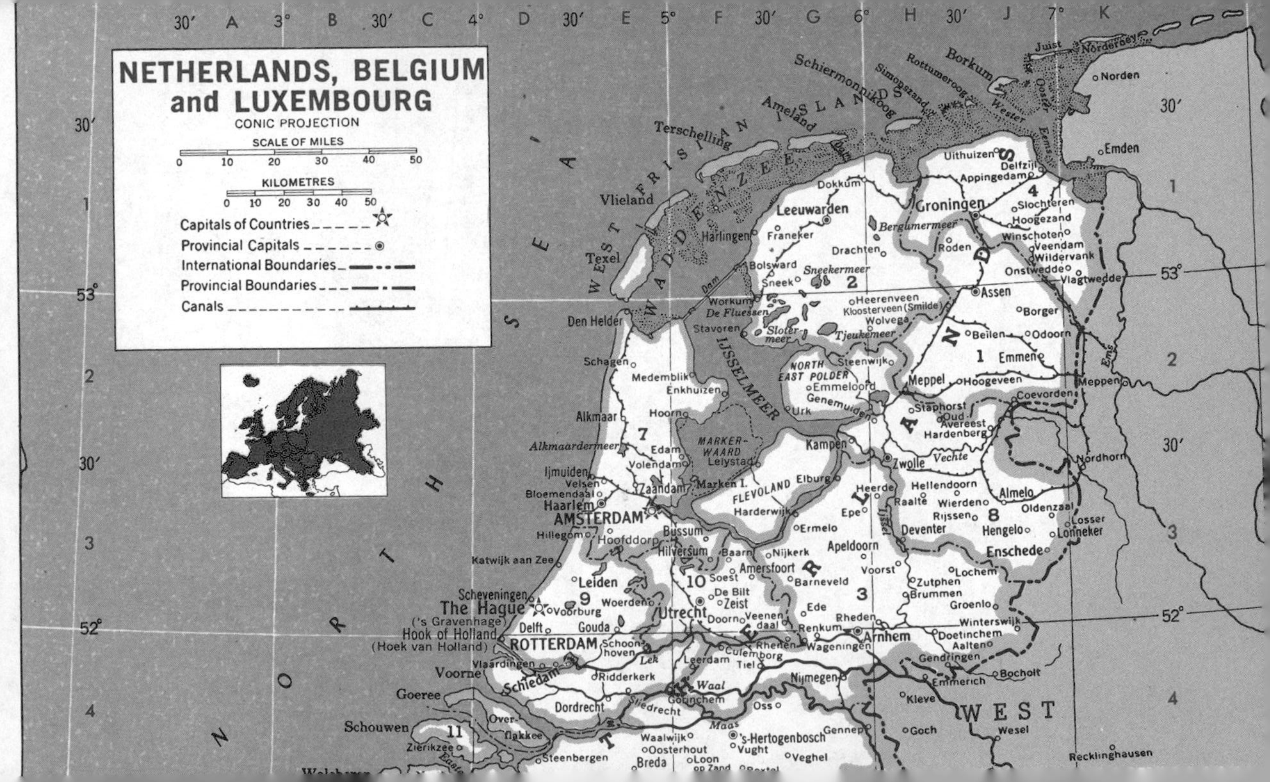
NETHERLANDS, BELGIUM and LUXEMBOURG
CONIC PROJECTION
SCALE OF MILES
0 10 20 30 40 50
KILOMETRES
0 10 20 30 40 50
Capitals of Countries
Provincial Capitals
International Boundaries
Provincial Boundaries
Canals
NORTH SEA
WEST FRISIAN ISLANDS
WADDEN ZEE
IJSSELMEER
NORTH EAST POLDER
FLEVOLAND
MARKER-WAARD
NETHERLANDS
WEST
AMSTERDAM
ROTTERDAM
The Hague
('s Gravenhage)
Hook of Holland
(Hoek van Holland)
Utrecht
Haarlem
Leeuwarden
Groningen
Assen
Zwolle
Arnhem
Emden
Norden
Meppen
Nordhorn
Kleve
Goch
Wesel
Recklinghausen

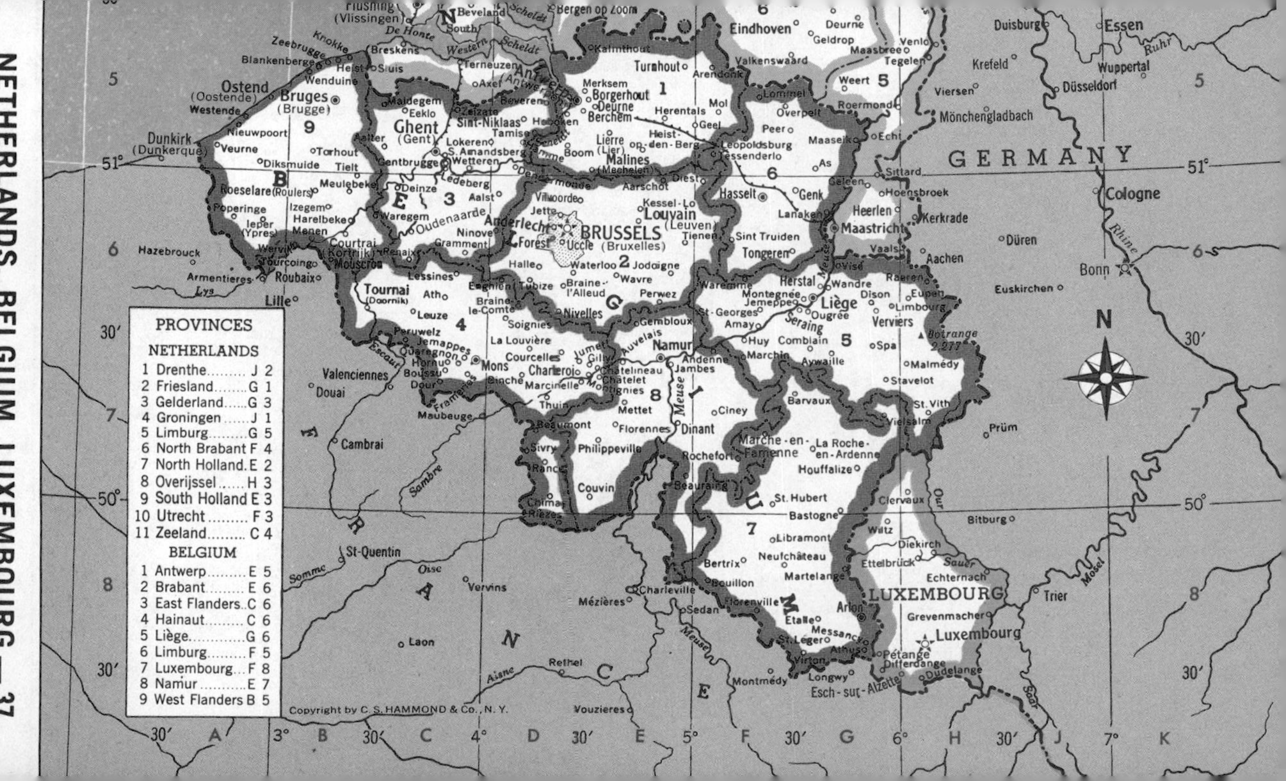
PROVINCES
NETHERLANDS
1 Drenthe........J 2
2 Friesland........G 1
3 Gelderland......G 3
4 Groningen......J 1
5 Limburg........G 5
6 North Brabant F 4
7 North Holland. E 2
8 Overijssel......H 3
9 South Holland E 3
10 Utrecht........F 3
11 Zeeland........C 4
BELGIUM
1 Antwerp........E 5
2 Brabant........E 6
3 East Flanders..C 6
4 Hainaut........C 6
5 Liège............G 6
6 Limburg........F 5
7 Luxembourg...F 8
8 Namur..........E 7
9 West Flanders B 5
GERMANY
FRANCE
BELGIUM
LUXEMBOURG
BRUSSELS (Bruxelles)
Copyright by C. S. HAMMOND & Co., N. Y.

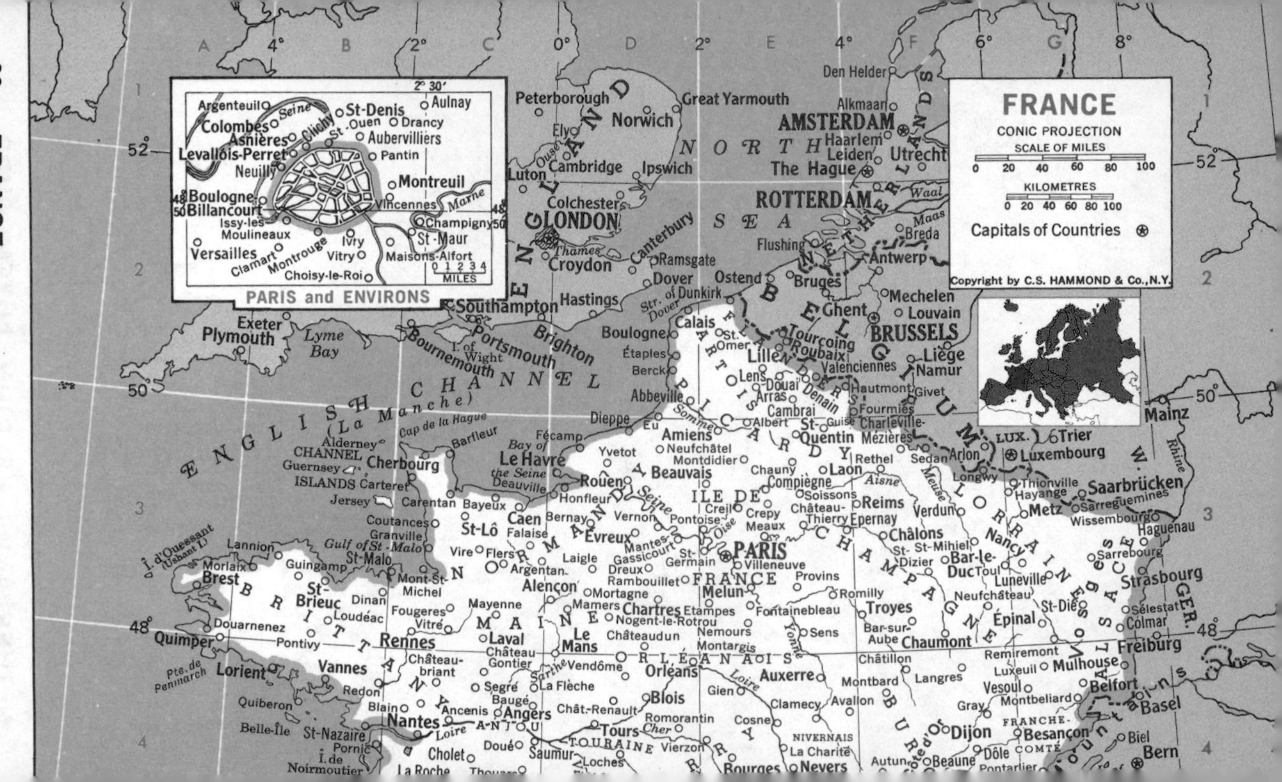
FRANCE
CONIC PROJECTION
SCALE OF MILES
0 20 40 60 80 100
KILOMETRES
0 20 40 60 80 100
Capitals of Countries
Copyright by C.S. HAMMOND & Co., N.Y.
PARIS and ENVIRONS
MILES
0 1 2 3 4
ENGLAND
NORTH SEA
ENGLISH CHANNEL
(La Manche)
NETHERLANDS
BELGIUM
LUX.
GER.
CHANNEL ISLANDS
BRITTANY
NORMANDY
MAINE
ANJOU
TOURAINE
ORLÉANAIS
ILE DE FRANCE
PICARDY
ARTOIS
FLANDERS
CHAMPAGNE
LORRAINE
ALSACE
BURGUNDY
FRANCHE-COMTÉ
NIVERNAIS
LONDON
PARIS
BRUSSELS
AMSTERDAM
The Hague
ROTTERDAM
Luxembourg
Bern
Seine
Loire
Rhine
Meuse
Somme
Marne
Aisne
Oise
Yonne
Cher
Bay of the Seine
Gulf of St-Malo
Str. of Dover
Lyme Bay

BAY OF BISCAY
MEDITERRANEAN SEA
SPAIN
Gulf of Lions
Riviera
Pyrenees
Î. de Re
Î. d'Oléron
Pte. de Grave
Étang de Carcans
Étang de Cazau
AUNIS
SAINTONGE
ANGOUMOIS
MARCHE
LIMOUSIN
Mts. du Limousin
AUVERGNE
LYONNAIS
SAVOY
DAUPHINÉ
PROVENCE
GUIENNE
GASCONY
BÉARN
LANGUEDOC
ROUSSILLON
Auvergne
Cévennes
Gironde
Garonne
Dordogne
Lot
Tarn
Adour
Rhône
Saône
Loire
Isère
Durance
La Rochelle
Niort
Rochefort
Saintes
Royan
Angoulême
Cognac
Pauillac
Bordeaux
Libourne
Coutras
Pessac
Arcachon
Marmande
Bergerac
Périgueux
Sarlat
Thiviers
St-Yrieix
Limoges
Ruffec
Civray
Bellac
Montmorillon
Guéret
Boussac
Argenton
St-Junien
Aubusson
Montluçon
Clermont-Ferrand
Vichy
Roanne
Mâcon
Paray
Thiers
Montbrison
Villefranche
Lyons
Villeurbanne
Oullins
Vienne
St-Étienne
Firminy
Annonay
Romans
Valence
Crest
Montélimar
Privas
Le Puy
Brioude
St-Flour
Langogne
Mende
Le Teil
Aurillac
St-Céré
Figeac
Decazeville
Rodez
Millau
Alès
La Grand'Combe
Uzès
Nyons
Orange
Carpentras
Avignon
Nîmes
Arles
Salon
Aix
Marseille
Martigues
Port-St-Louis
Toulon
La Seyne
Hyères
Draguignan
St-Tropez
St-Raphaël
Cannes
Antibes
Nice
Monaco
Menton
Manosque
Sisteron
Digne
Barcelonnette
Gap
Grenoble
Les Écrins 13,461
Voiron
Chambéry
Aix-les-Bains
Annecy
Chamonix
M. Blanc 15,821
Geneva
Thonon-les-Bains
Oyonnax
Bourg
Ambérieu
Briançon
St-Jean
Turin
Aosta
Ivrea
Cuneo
Bra
Pennine Alps
Graian Alps
Maritime Alps
Tulle
Brive
Bort
Ussel
Monts Dore
Puy de Dôme 4,872
Plomb du Cantal 6,096
Cahors
Villeneuve
Agen
Nérac
Castel Sarrasin
Montauban
Carmaux
Albi
Castres
Mazamet
Toulouse
Muret
Condom
Auch
Mirande
Mont-de-Marsan
Dax
Aire
Orthez
Bayonne
Biarritz
Hendaye
Pau
Oloron
Tarbes
Lourdes
Bagnères de Luchon
St-Gaudens
Pamiers
Foix
Ax
Carcassonne
Limoux
Sigean
Narbonne
Béziers
Sète
Montpellier
Perpignan
Prades
Port-Vendres
Vignemale 10,820
Pico de Aneto 11,168
ANDORRA
Seo de Urgel
San Sebastián
Santander
Bilbao
Reinosa
Tolosa
Jaca
Huesca
Cinca
Segre
Tamarite
Berga
Figueras
Gerona
Cabo de Creus
Cabo San Sebastián
Manresa
Sabadell
Badalona
BARCELONA
Lérida
Saragossa
Ebro
Beichite
Alcañiz
Reus
Tarragona
Tortosa
Cabo de Tortosa
CORSICA
Same scale as main map
C. Corse
G. of St-Florent
Bastia
Calvi
Golo
M. Cinto 8,891
Corte
Bastélica
Ajaccio
Sartène
Porto-Vecchio
Bonifacio
Str. of Bonifacio
Longitude 2° West of Greenwich 0° 2° Longitude 4° East of Greenwich 6°

40 — SPAIN, PORTUGAL

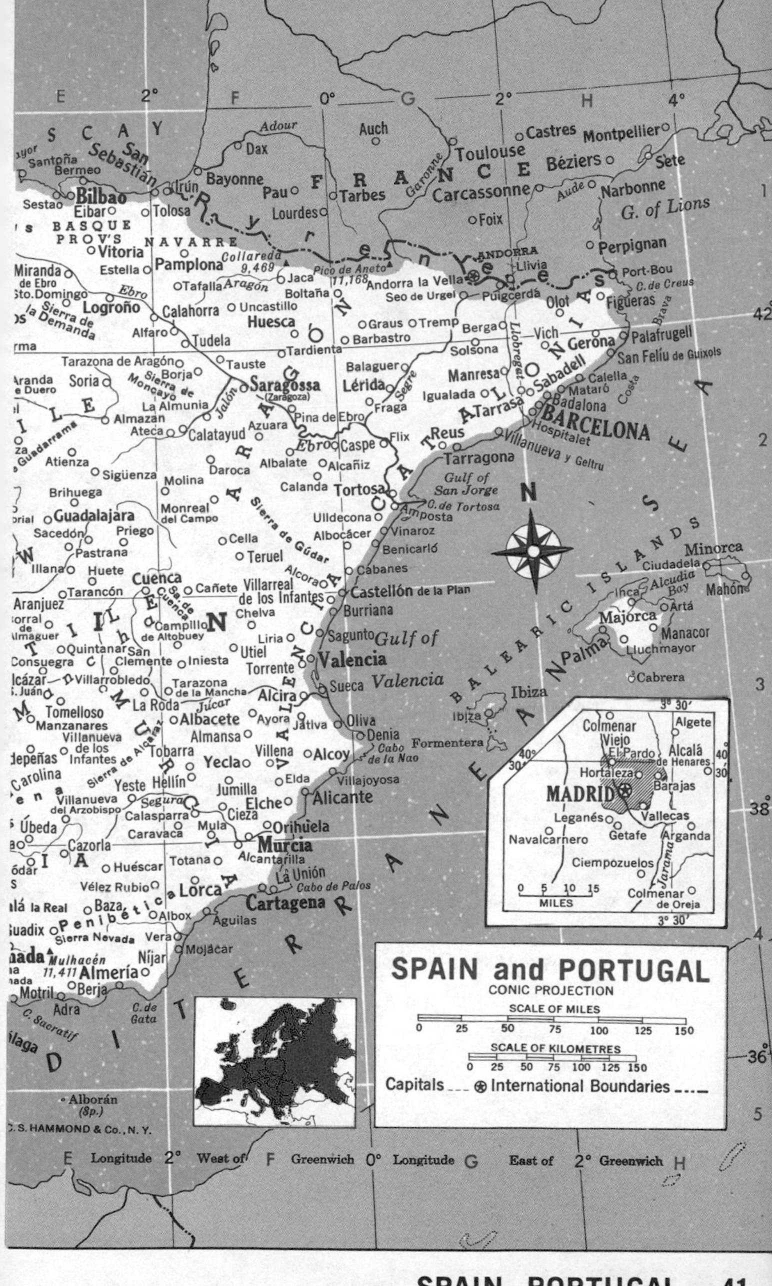
SPAIN and PORTUGAL
CONIC PROJECTION
SCALE OF MILES
0 25 50 75 100 125 150
SCALE OF KILOMETRES
0 25 50 75 100 125 150
Capitals ⊛ International Boundaries
F R A N C E
Auch
Castres
Montpellier
Toulouse
Béziers
Sète
Dax
Adour
Bayonne
Pau
Tarbes
Garonne
Carcassonne
Aude
Narbonne
G. of Lions
Lourdes
Foix
Perpignan
ANDORRA
Andorra la Vella
Llivia
Puigcerdá
Seo de Urgel
Port-Bou
C. de Creus
Figueras
Olot
Costa Brava
Pico de Aneto 11,168
Collarada 9,469
P y r e n e e s
San Sebastián
Irún
Santoña
Bermeo
Bilbao
Sestao
Eibar
Tolosa
BASQUE PROV'S
NAVARRE
Vitoria
Pamplona
Estella
Miranda de Ebro
Sto. Domingo
Logroño
Sierra de la Demanda
Tafalla
Aragón
Ebro
Calahorra
Alfaro
Tudela
Jaca
Boltaña
Uncastillo
Huesca
Graus
Tremp
Berga
Barbastro
Tardienta
Solsona
Vich
Llobregat
Gerona
Palafrugell
San Feliu de Guixols
Tarazona de Aragón
Tauste
Borja
Sierra de Moncayo
Soria
Saragossa (Zaragoza)
Lérida
Balaguer
Segre
Manresa
Sabadell
Calella
Mataró
Badalona
Costa
Igualada
Terrassa
BARCELONA
Hospitalet
Villanueva y Geltru
La Almunia
Jalón
Almazan
Ateca
Calatayud
Azuara
Pina de Ebro
Fraga
Ebro
Caspe
Flix
Reus
Tarragona
Gulf of San Jorge
Atienza
Sigüenza
Daroca
Albalate
Alcañiz
Molina
Calanda
Tortosa
C. de Tortosa
Amposta
Brihuega
Guadalajara
Monreal del Campo
Sierra de Gúdar
Ulldecona
Vinaroz
Albocácer
Benicarló
Sacedón
Priego
Pastrana
Cella
Teruel
Illana
Huete
Cuenca
Alcora
Cabanes
Tarancón
Cañete
Villarreal de los Infantes
Castellón de la Plan
Burriana
Aranjuez
Chelva
Campillo de Altobuey
Liria
Sagunto
Gulf of Valencia
Quintanar
San Clemente
Iniesta
Utiel
Torrente
Valencia
Consuegra
Villarrobledo
Tarazona de la Mancha
Sueca
Alcira
Tomelloso
Manzanares
La Roda
Júcar
Albacete
Ayora
Játiva
Oliva
Denia
Cabo de la Nao
Formentera
Villanueva de los Infantes
Almansa
Tobarra
Yecla
Villena
Alcoy
Villajoyosa
Carolina
Sierra de Alcaraz
Yeste
Hellín
Jumilla
Elda
Alicante
Villanueva del Arzobispo
Segura
Calasparra
Cieza
Elche
Úbeda
Caravaca
Mula
Orihuela
Cazorla
Murcia
Alcantarilla
Huéscar
Totana
La Unión
Cabo de Palos
Vélez Rubio
Lorca
Cartagena
Baza
Albox
Aguilas
Penibética
Sierra Nevada
Vera
Mojácar
Guadix
Mulhacén 11,411
Níjar
Almería
Motril
Berja
Adra
C. de Gata
C. Sacratif
Málaga
Alborán (Sp.)
BALEARIC ISLANDS
Minorca
Ciudadela
Alcudia Bay
Mahón
Inca
Artá
Majorca
Manacor
Palma
Lluchmayor
Cabrera
Ibiza
MEDITERRANEAN SEA
N
Colmenar Viejo
Algete
El Pardo
Alcalá de Henares
Hortaleza
MADRID
Barajas
Leganés
Vallecas
Navalcarnero
Getafe
Arganda
Ciempozuelos
Jarama
Colmenar de Oreja
0 5 10 15
MILES
C. S. HAMMOND & Co., N. Y.
Longitude 2° West of Greenwich 0° Longitude East of 2° Greenwich

ITALY and SWITZERLAND
CONIC PROJECTION
SCALE OF MILES
0 20 40 60 80 100 120
SCALE OF KILOMETRES
0 20 40 60 80 100 120
Capitals of Countries
Regional Capitals
SWITZERLAND
LIECHTENSTEIN
AUSTRIA
W. GERMANY
YUGOSLAVIA
FRANCE
MONACO
SAN MARINO
VATICAN CITY
CORSICA (Fr.)
ADRIATIC SEA
LIGURIAN SEA
LOMBARDY
PIEDMONT
LIGURIA
EMILIA-ROMAGNA
TUSCANY
UMBRIA
MARCHE
LATIUM
ABRUZZI
MOLISE
TRENTINO-ALTO ADIGE
VENETIA
FRIULI-VENEZIA GIULIA
AOSTA
Bern
Zürich
Basel
Lucerne
Geneva
Lausanne
Innsbruck
Graz
Klagenfurt
Ljubljana
Rijeka (Fiume)
Zadar (Zara)
Split
Sarajevo
Mostar
Dubrovnik
Cetinje
Trieste
Venice
Milan
Turin
Genoa
Bologna
Florence
Rome
Ancona
Perugia
Pescara
Nice
Ajaccio
Bastia
Leghorn (Livorno)
Elba
Julian Alps
Carnic Alps
Dolomite Alps
Lepontine Alps
Cottian Alps
Gulf of Venice
Gulf of Genoa
M. Blanc 15,781
M. Corno 9,560
M. Cinto 8,891

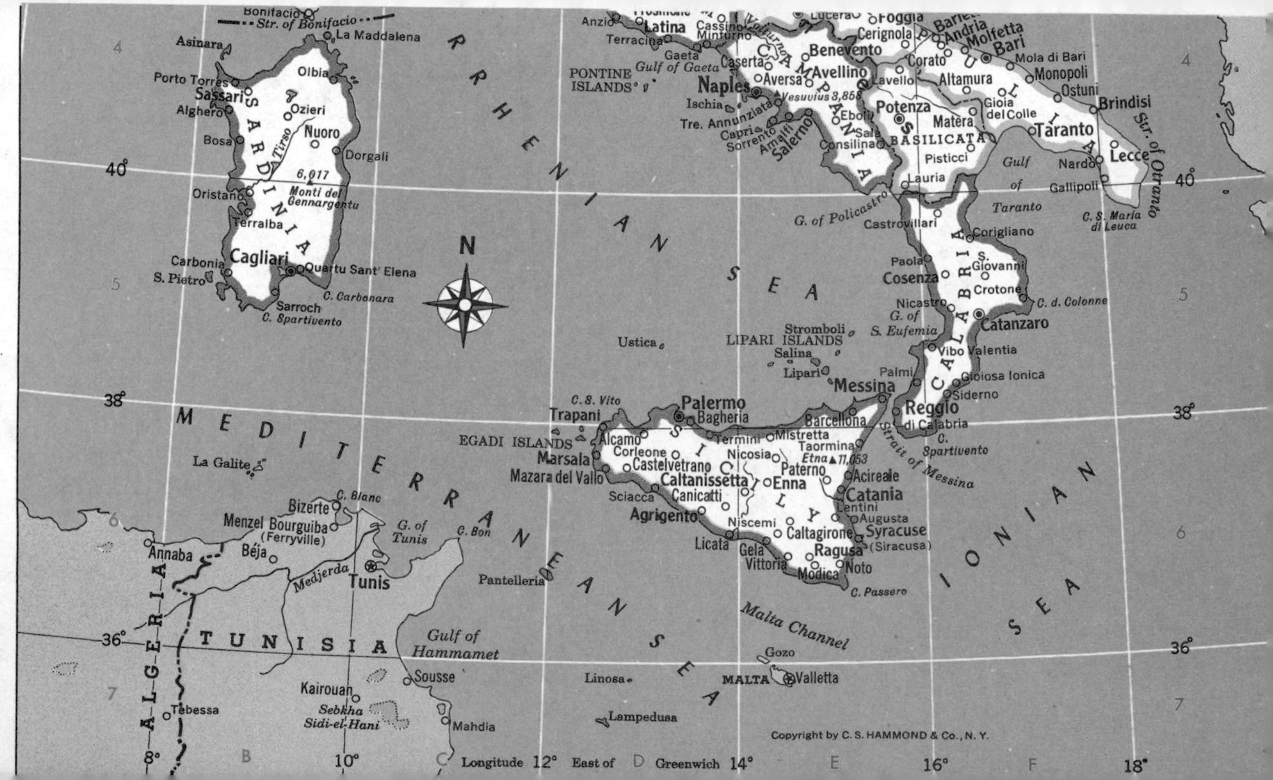

Bonifacio
Str. of Bonifacio
La Maddalena
Asinara
Porto Torres
Sassari
Alghero
Olbia
Ozieri
Nuoro
Tirso
Bosa
Dorgali
6,017
Monti del Gennargentu
Oristano
Terralba
SARDINIA
Carbonia
S. Pietro
Cagliari
Quartu Sant' Elena
C. Carbonara
Sarroch
C. Spartivento
TYRRHENIAN SEA
N
Anzio
Latina
Terracina
Cassino
Minturno
Volturno
Gaeta
Gulf of Gaeta
PONTINE ISLANDS
Lucera
Foggia
Cerignola
Barletta
Andria
Molfetta
Bari
Mola di Bari
Monopoli
Ostuni
Brindisi
Str. of Otranto
Benevento
Corato
Caserta
Aversa
Avellino
CAMPANIA
Naples
Vesuvius 3,858
Ischia
Tre. Annunziata
Capri
Sorrento
Amalfi
Salerno
Eboli
Sala Consilina
Lavello
Altamura
Potenza
BASILICATA
Matera
Gioia del Colle
Taranto
APULIA
Lecce
Nardò
Gallipoli
Pisticci
Lauria
Gulf of Taranto
C. S. Maria di Leuca
G. of Policastro
Castrovillari
Corigliano
Paola
Cosenza
S. Giovanni
Crotone
CALABRIA
C. d. Colonne
Nicastro
G. of S. Eufemia
Catanzaro
Stromboli
LIPARI ISLANDS
Salina
Lipari
Ustica
Vibo Valentia
Palmi
Gioiosa Ionica
Siderno
Messina
Reggio di Calabria
C. Spartivento
Strait of Messina
C. S. Vito
Trapani
Palermo
Bagheria
Barcellona
Termini
Mistretta
Taormina
EGADI ISLANDS
Alcamo
Corleone
Nicosia
Etna 11,053
Marsala
Castelvetrano
Paterno
Acireale
Mazara del Vallo
Caltanissetta
Enna
Catania
SICILY
Sciacca
Canicatti
Lentini
Agrigento
Niscemi
Augusta
Caltagirone
Syracuse (Siracusa)
Licata
Gela
Vittoria
Ragusa
Modica
Noto
C. Passero
MEDITERRANEAN SEA
IONIAN SEA
La Galite
C. Blanc
Bizerte
Menzel Bourguiba (Ferryville)
G. of Tunis
C. Bon
Annaba
Béja
Medjerda
Tunis
Pantelleria
Malta Channel
ALGERIA
TUNISIA
Gulf of Hammamet
Gozo
Sousse
Linosa
MALTA
Valletta
Kairouan
Sebkha Sidi-el-Hani
Tebessa
Lampedusa
Mahdia
Copyright by C. S. HAMMOND & Co., N. Y.
40°
38°
36°
8°
10°
Longitude 12° East of Greenwich 14°
16°
18°

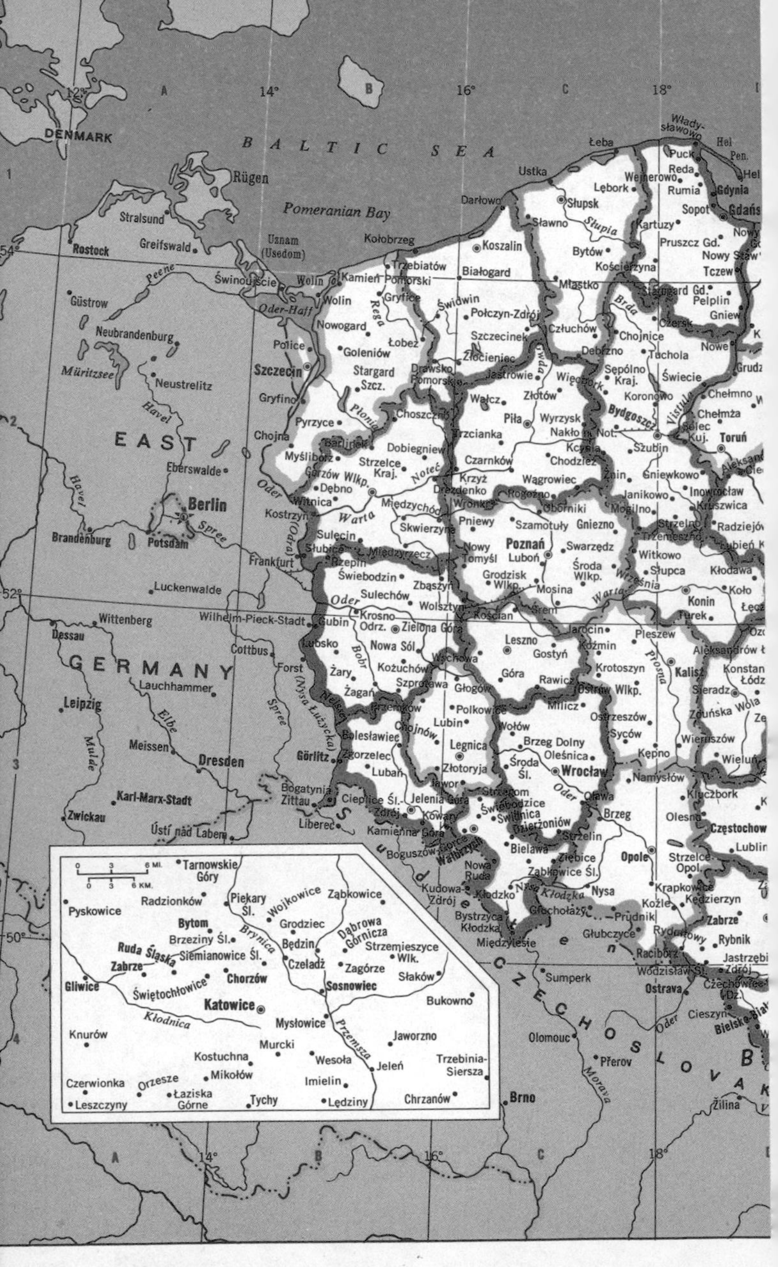

BALTIC SEA
Pomeranian Bay
DENMARK
EAST
GERMANY
CZECHOSLOVAKIA
Rügen
Stralsund
Greifswald
Rostock
Güstrow
Neubrandenburg
Müritzsee
Neustrelitz
Peene
Havel
Eberswalde
Berlin
Potsdam
Brandenburg
Spree
Luckenwalde
Wittenberg
Dessau
Leipzig
Lauchhammer
Cottbus
Forst
Elbe
Mulde
Meissen
Dresden
Karl-Marx-Stadt
Zwickau
Ústí nad Labem
Liberec
Zittau
Görlitz
Uznam (Usedom)
Świnoujście
Wolin
Oder-Haff
Police
Szczecin
Gryfino
Chojna
Kostrzyn
Frankfurt
Wilhelm-Pieck-Stadt
Kołobrzeg
Trzebiatów
Kamień Pomorski
Gryfice
Rega
Nowogard
Goleniów
Łobez
Stargard Szcz.
Płonia
Pyrzyce
Myślibórz
Barlinek
Gorzów Wlkp.
Dębno
Witnica
Słubice
Sulęcin
Rzepin
Świebodzin
Sulechów
Krosno Odrz.
Gubin
Lubsko
Żary
Żagań
Nowa Sól
Kożuchów
Szprotawa
Zielona Góra
Bobr
Nysa Łużycka
Bolesławiec
Zgorzelec
Lubań
Bogatynia
Cieplice Śl.-Zdrój
Jelenia Góra
Kamienna Góra
Kowary
Wałbrzych
Boguszów-Gorce
Sudeten
Kudowa-Zdrój
Kłodzko
Nowa Ruda
Bystrzyca Kłodzka
Międzylesie
Nysa Kłodzka
Koszalin
Białogard
Świdwin
Połczyn-Zdrój
Szczecinek
Złocieniec
Drawsko Pomorskie
Choszczno
Dobiegniew
Strzelce Kraj.
Noteć
Międzychód
Skwierzyna
Warta
Międzyrzecz
Zbąszyń
Wolsztyn
Darłowo
Ustka
Sławno
Słupsk
Słupia
Łeba
Lębork
Bytów
Miastko
Człuchów
Debrzno
Jastrowie
Wałcz
Piła
Trzcianka
Czarnków
Krzyż
Drezdenko
Wronki
Złotów
Wyrzysk
Nakło n. Not.
Więcbork
Sępólno Kraj.
Chojnice
Tuchola
Brda
Władysławowo
Hel
Hel Pen.
Puck
Reda
Rumia
Wejherowo
Gdynia
Sopot
Gdańsk
Kartuzy
Pruszcz Gd.
Kościerzyna
Starogard Gd.
Pelplin
Tczew
Czersk
Gniew
Nowe
Świecie
Vistula
Koronowo
Bydgoszcz
Chełmno
Chełmża
Solec Kuj.
Toruń
Szubin
Kcynia
Chodzież
Wągrowiec
Rogoźno
Oborniki
Szamotuły
Pniewy
Nowy Tomyśl
Poznań
Luboń
Grodzisk Wlkp.
Mosina
Swarzędz
Gniezno
Środa Wlkp.
Września
Żnin
Gniewkowo
Janikowo
Inowrocław
Kruszwica
Mogilno
Strzelno
Trzemeszno
Radziejów
Witkowo
Słupca
Konin
Koło
Kłodawa
Turek
Kościan
Śrem
Jarocin
Pleszew
Leszno
Gostyń
Koźmin
Krotoszyn
Prosna
Kalisz
Góra
Rawicz
Ostrów Wlkp.
Milicz
Ostrzeszów
Syców
Kępno
Wieruszów
Wieluń
Sieradz
Zduńska Wola
Łódź
Głogów
Polkowice
Lubin
Chojnów
Legnica
Złotoryja
Jawor
Wołów
Brzeg Dolny
Oleśnica
Środa Śl.
Wrocław
Oława
Oder
Strzegom
Świebodzice
Świdnica
Dzierżoniów
Bielawa
Strzelin
Ząbkowice Śl.
Ziębice
Brzeg
Namysłów
Kluczbork
Olesno
Częstochowa
Opole
Strzelce Opol.
Krapkowice
Kędzierzyn
Koźle
Nysa
Głuchołazy
Prudnik
Głubczyce
Racibórz
Rydułtowy
Rybnik
Zabrze
Wodzisław Śl.
Jastrzębie-Zdrój
Ostrava
Cieszyn
Bielsko-Biała
Šumperk
Olomouc
Přerov
Morava
Brno
Žilina
Tarnowskie Góry
Pyskowice
Radzionków
Piekary Śl.
Wojkowice
Ząbkowice
Bytom
Grodziec
Dąbrowa Górnicza
Brzeziny Śl.
Będzin
Strzemieszyce Wlk.
Ruda Śląska
Siemianowice Śl.
Czeladź
Zagórze
Zabrze
Chorzów
Sosnowiec
Sławków
Gliwice
Świętochłowice
Katowice
Bukowno
Kłodnica
Brynica
Mysłowice
Przemsza
Knurów
Jaworzno
Murcki
Kostuchna
Wesoła
Jeleń
Trzebinia-Siersza
Mikołów
Orzesze
Imielin
Czerwionka
Łaziska Górne
Tychy
Lędziny
Chrzanów
Leszczyny

44 — POLAND

POLAND
CONIC PROJECTION
SCALE OF MILES
0 10 20 40 60 80
SCALE OF KILOMETERS
0 10 20 40 60 80
Capitals of Countries
Other Capitals
International Boundaries
Internal Boundaries
Canals
Poland is divided into 49 provinces (bearing the same name as their capitals) and the autonomous cities of Warsaw, Łódź and Cracow.
U. S. S. R.
Kaliningrad
Baltiysk
Chernyakhovsk
Kapsukas
Warsaw
Łódź
Cracow
Gdańsk
Olsztyn
Białystok
Lublin
Kielce
Radom
Rzeszów
Tarnów
Brest
Kobrin
Kovel'
Lutsk
L'vov
Vistula
Bug
Narew
Pripyat'
Dniester
High Tatra
Gerlachovka 8,707
Rysy 8,199
Longitude 20° East of Greenwich 22°
© Copyright HAMMOND INCORPORATED, Maplewood, N. J.

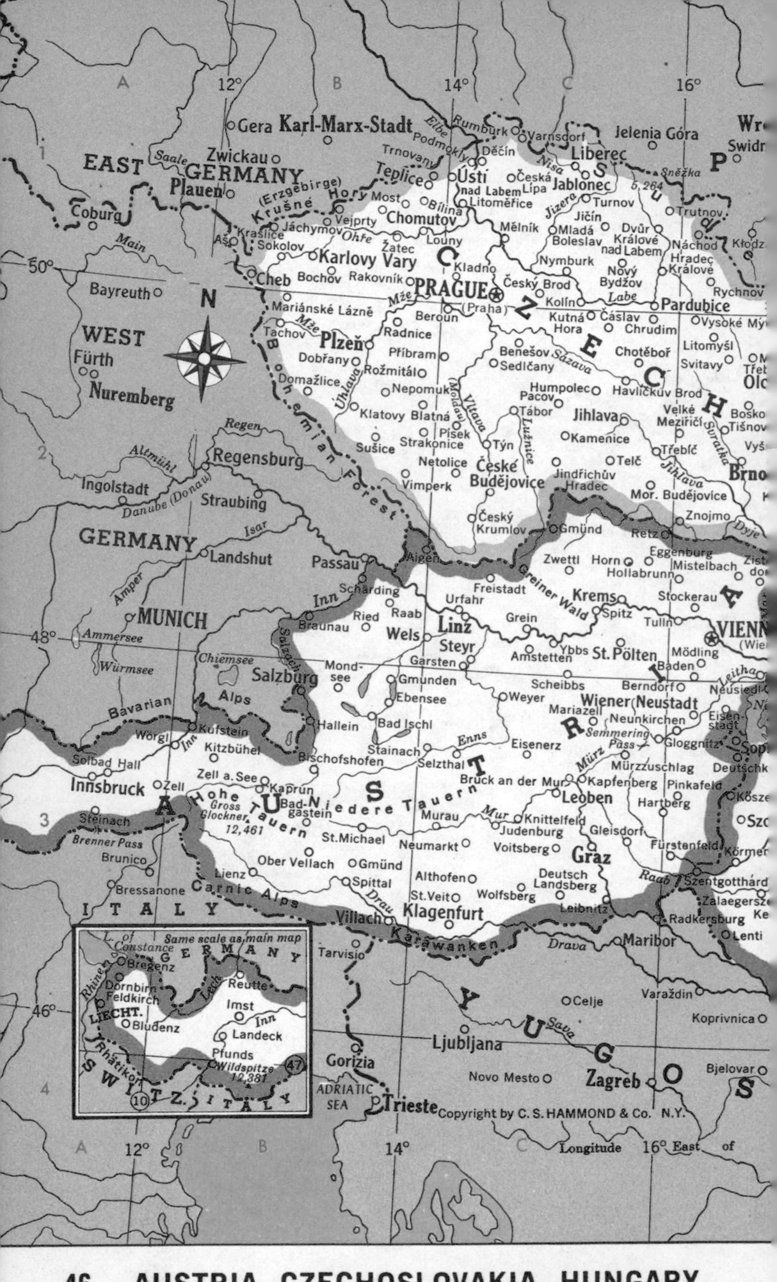

46 — AUSTRIA, CZECHOSLOVAKIA, HUNGARY

AUSTRIA, CZECHOSLOVAKIA and HUNGARY
Conic Projection
SCALE OF MILES
0 20 40 60 80 100
SCALE OF KILOMETRES
0 20 40 60 80 100
Capitals of Countries
International Boundaries
Canals
BUDAPEST
POLAND
SLOVAKIA
HUNGARY
RUMANIA
YUGOSLAVIA
U.S.S.R.

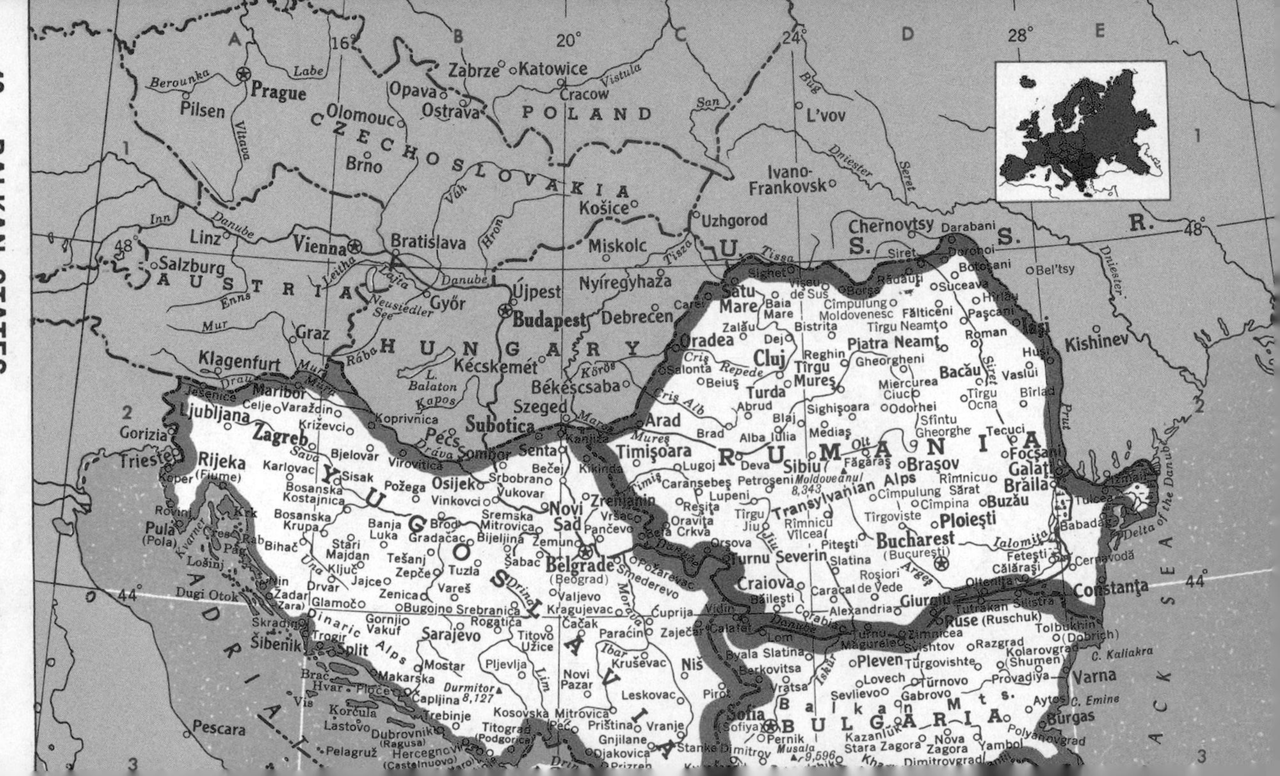

48 — BALKAN STATES

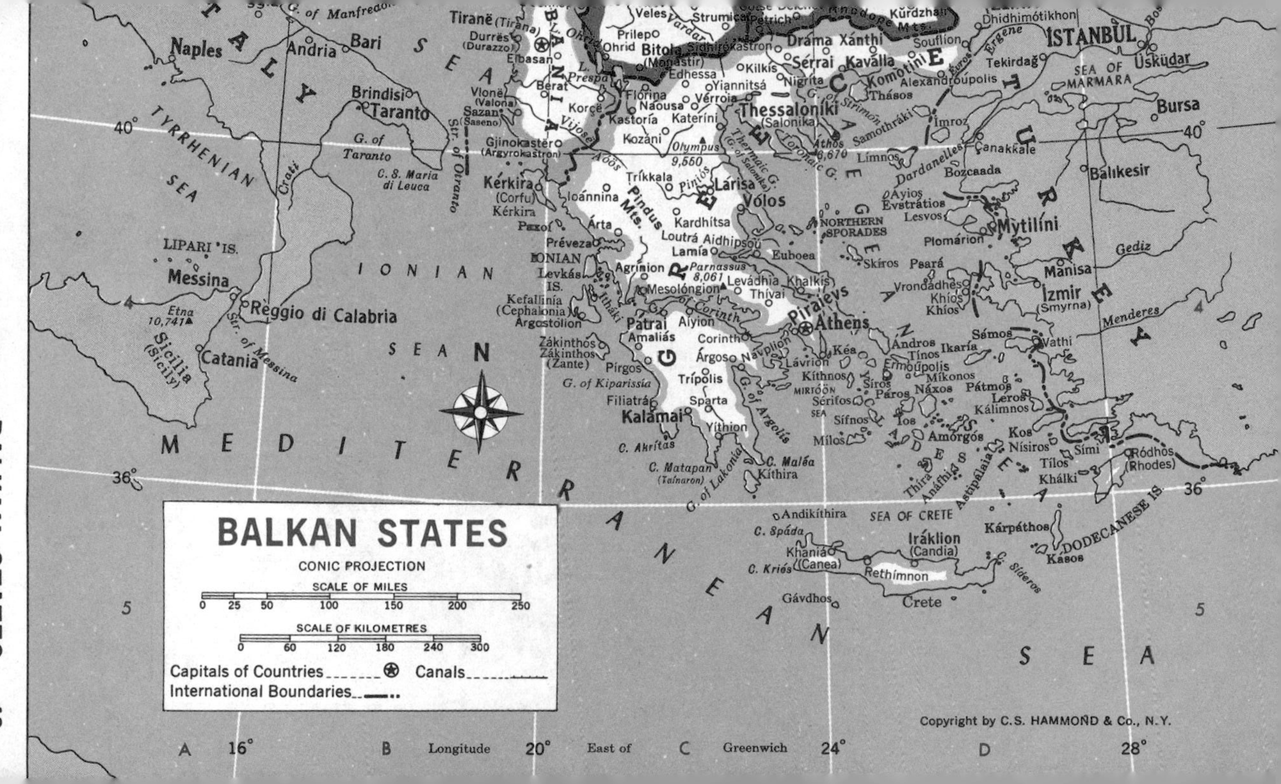
BALKAN STATES
CONIC PROJECTION
SCALE OF MILES
0 25 50 100 150 200 250
SCALE OF KILOMETRES
0 60 120 180 240 300
Capitals of Countries
Canals
International Boundaries
Copyright by C.S. HAMMOND & Co., N.Y.
A 16° B Longitude 20° East of C Greenwich 24° D 28°
ITALY
Naples
Andria
Bari
Brindisi
Taranto
G. of Taranto
C. S. Maria di Leuca
Str. of Otranto
TYRRHENIAN SEA
LIPARI IS.
Messina
Reggio di Calabria
Str. of Messina
Etna 10,741
Sicilia (Sicily)
Catania
IONIAN SEA
MEDITERRANEAN SEA
ALBANIA
Tiranë (Tirana)
Durrës (Durazzo)
Elbasan
Berat
Vlonë (Valona)
Sazan (Saseno)
Korçë
Gjinokastër (Argyrokastron)
L. Ohrid
L. Prespa
Ohrid
Prilep
Veles
Strumica
Bitola (Monastir)
Petrich
Rhodope Mts.
Kŭrdzhali
GREECE
Drama
Xánthi
Sérrai
Kavalla
Komotini
Alexandroúpolis
Souflíon
Thásos
Samothráki
Édhessa
Kilkís
Yiannitsá
Nigríta
Flórina
Náousa
Véroia
Kastoría
Kateríni
Kozáni
Thessaloníki (Salonika)
Thermaic G. (G. of Salonika)
Olympus 9,550
Áthos 6,670
Límnos
Tríkkala
Lárisa
Vólos
Pindus Mts.
Ioánnina
Kérkira (Corfu)
Árta
Paxoí
Préveza
Kardhítsa
Lamía
Euboea
NORTHERN SPORADES
Skíros
Levkás
Agrínion
Parnassus 8,061
Mesolóngion
Levádhia
Thívai
Khalkís
Kefallinía (Cephalonia)
Argostólion
Itháki
Pátrai
Aíyion
Corinth
G. of Corinth
Piraiévs
Athens
Zákinthos (Zante)
Amaliás
Pírgos
Árgos
Návplion
Trípolis
Sparta
G. of Kiparissía
Filiatrá
Kalámai
G. of Argolís
Yíthion
C. Akrítas
C. Matapan (Taínaron)
G. of Lakonía
C. Maléa
Kíthira
Andikíthira
C. Spáda
Khaniá (Canea)
C. Kriós
Gávdhos
Rethímnon
Iráklion (Candia)
Crete
SEA OF CRETE
Kárpathos
Kásos
C. Síderos
DODECANESE IS.
AEGEAN SEA
Kéa
Lávrion
Kíthnos
MIRTÓON SEA
Sérifos
Sífnos
Mílos
Ándros
Tínos
Ermoúpolis
Síros
Míkonos
Páros
Náxos
Íos
Amorgós
Thíra
Anáfi
Astipálaia
Ikaría
Sámos
Vathí
Pátmos
Léros
Kálimnos
Kos
Nísiros
Tílos
Sími
Khálki
Ródhos (Rhodes)
Áyios Evstrátios
Lésvos
Mytilíni
Plomárion
Psará
Vrondádhes
Khíos
TURKEY
Dhidhimótikhon
ISTANBUL
Üsküdar
Tekirdağ
SEA OF MARMARA
Bursa
İmroz
Dardanelles
Çanakkale
Bozcaada
Balıkesir
Gediz
Manisa
İzmir (Smyrna)
Menderes

UNION OF SOVIET SOCIALIST REPUBLICS
European Part
CONIC PROJECTION
MILES
0 50 100 200 300
KILOMETRES
0 100 200 300
National Capitals
Capitals of Union Republics
International boundaries
Union Republic boundaries
Canals
ARCTIC OCEAN
KARA SEA
BALTIC SEA
FINLAND
NORWAY
SWEDEN
Novaya Zemlya
Yamal Pen.
Gulf of Ob'
Kolguyev I.
Kanin Pen.
Kola Pen.
White Sea
Murmansk
Archangel (Arkhangel'sk)
LENINGRAD
Helsinki
Stockholm
Tallinn
Riga
Petrozavodsk
Vologda
Yaroslavl
Kostroma
Kirov
Perm (Molotov)
Izhevsk
Sverdlovsk
Chelyabinsk
Nizhniy Tagil
Syktyvkar
Vorkuta
Salekhard
Pechora
Northern Dvina
Onega
Mezen'
Kalinin
Pskov
Novgorod
Tartu
ESTONIAN S.S.R.
LATVIAN S.S.R.
LITHUANIAN S.S.R.
G. of Finland
Gulf of Riga
Gulf of Bothnia
Arctic Circle
Timan Ridge
North Ural Mts.
Central Ural Mts.
SOCIALIST REPUBLIC

The government of the United States does not recognize the incorporation of Estonia, Latvia and Lithuania into the Soviet Union.
Post-war territorial changes shown on this map do not necessarily represent the final status of such boundaries. Only after the signing of peace treaties can changes be considered official and definite.
Copyright by C.S. HAMMOND & Co., N.Y.
Longitude 40° East of Greenwich 45°
BLACK SEA
CASPIAN SEA
Sea of Azov
Sea of Marmara
TURKEY
IRAN
RUMANIA
BULGARIA
POLAND
UKRAINIAN S.S.R.
WHITE RUSSIAN S.S.R.
MOLDAVIAN S.S.R.
GEORGIAN S.S.R.
ARMENIAN S.S.R.
AZERBAIDZHAN S.S.R.
Caucasus Mts.
Kiev
Minsk
Odessa
Volgograd (Stalingrad)
Astrakhan
Rostov
Kuybyshev
Saratov
Tbilisi
Baku
Istanbul
Ankara
Bucharest

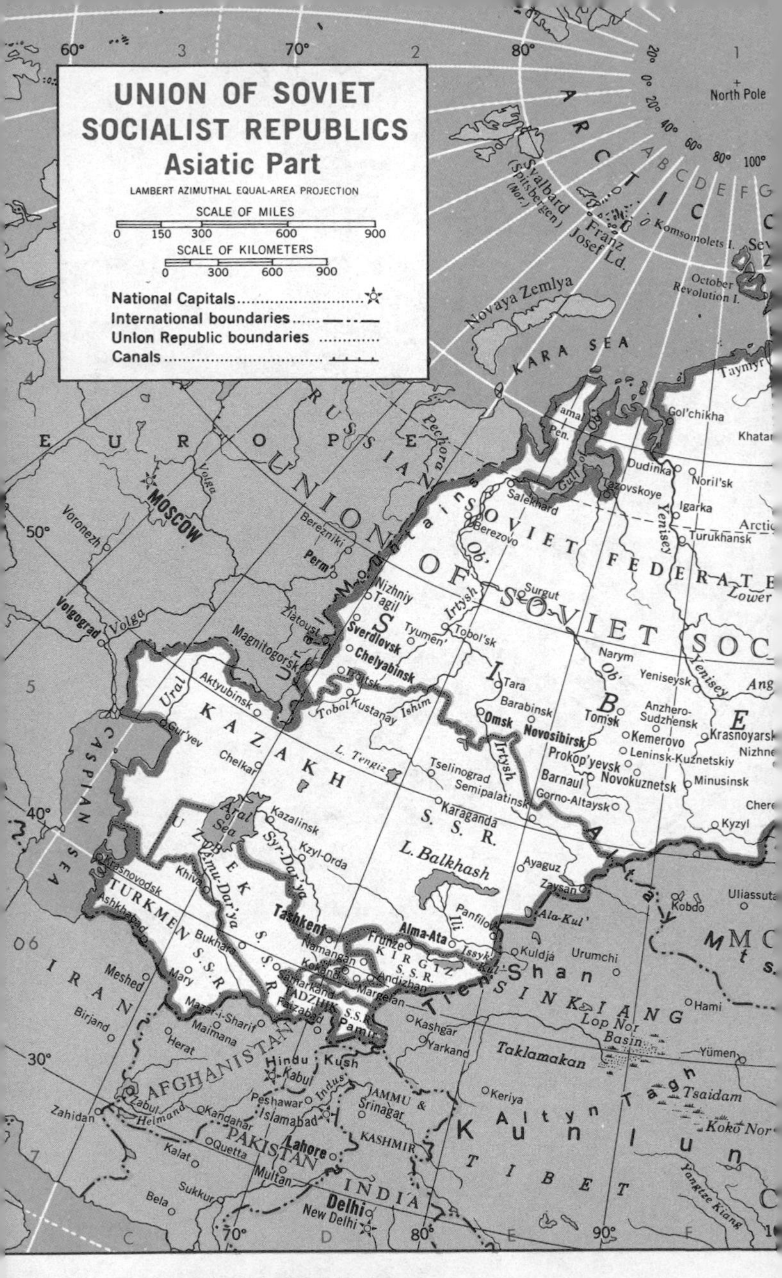

52 — U.S.S.R. - Asiatic Part

UNITED STATES
ALASKA
Bering Str.
C. Dezhnev
Wrangel I.
Nunivak I.
St. Lawrence I.
Chukchi Pen.
G. of Anadyr'
BERING SEA
Aleutian Is.
New Siberian Is.
Bol'shoy Lyakhov I.
Chelyuskin
LAPTEV SEA
Cherskiy
Anadyr'
Kolyma Range
Omolon
Kolyma
Srednekolymsk
Olyutorskiy
Komandorskiye Is.
Ust'-Kamchatsk
Bulun
Yana
Indigirka
Zyryanka
Dzhelinde
Verkhoyansk
Shelekhov Gulf
Magadan
Kamchatka Pen.
Tigil'
Petropavlovsk-Kamchatskiy
Lena
REPUBLIC
SOCIALIST REPUBLICS
Vilyuysk
Yakutsk
Okhotsk
SEA OF OKHOTSK
Ust'-Maya
C. Lopatka
Paramushir
Vilyuy
Olëkminsk
Tommot
Aldan
Nel'kan
Ayan
Shantar Is.
C. Yelizaveta
Okha
Peleduy
Bodaybo
Kirensk
Stanovoy Ra.
Chumikan
Nikolayevsk
Aleksandrovsk-Sakhalinskiy
Sakhalin
Poronaysk
Olekma
Skovorodino
Zeya
Amur
Tatar Str.
Kuril Is.
Urup
L. Baykal
Shilka
Svobodnyy
Belogorsk
Komsomol'sk
Sikhote-Alin
Yuzhno-Sakhalinsk
Korsakov
Iturup
Zima
Sretensk
Chita
Nerchinsk
Blagoveshchensk
Obluch'ye
Khabarovsk
La Pérouse Str.
Ulan-Ude
Aigun
Kyakhta
Gt. Khingan
Nonni
Ussuri
Sapporo
Ulan Bator
Kerulen
Tsitsihar
MANCHURIA
Sungari
L. Khanka
Harbin
Hakodate
KARAKORUM
MONGOLIA
Changchun
Kirin
Vladivostok
Najin
Ch'ŏngjin
SEA OF JAPAN
Sendai
Gobi (Shamo)
INNER MONGOLIA
Mukden
Yalu
N. KOREA
P'yŏngyang
Seoul
TOKYO
Yokohama
PEKING (Peiping)
Nagoya
Kyoto
Kobe
Osaka
Yinchwan
Tientsin
Lüta
S. KOREA
Taegu
GREAT WALL
Hwang-Ho
Taiyüan
Tsinan
Tsingtao
YELLOW SEA
Pusan
Hiroshima
Kitakyushu
Sinsiang
Lanchow
Sian
Kaifeng
Cheju I.
Nagasaki
CHINA
Nanking
SHANGHAI
PACIFIC OCEAN
© Copyright HAMMOND INCORPORATED, Maplewood, N. J.

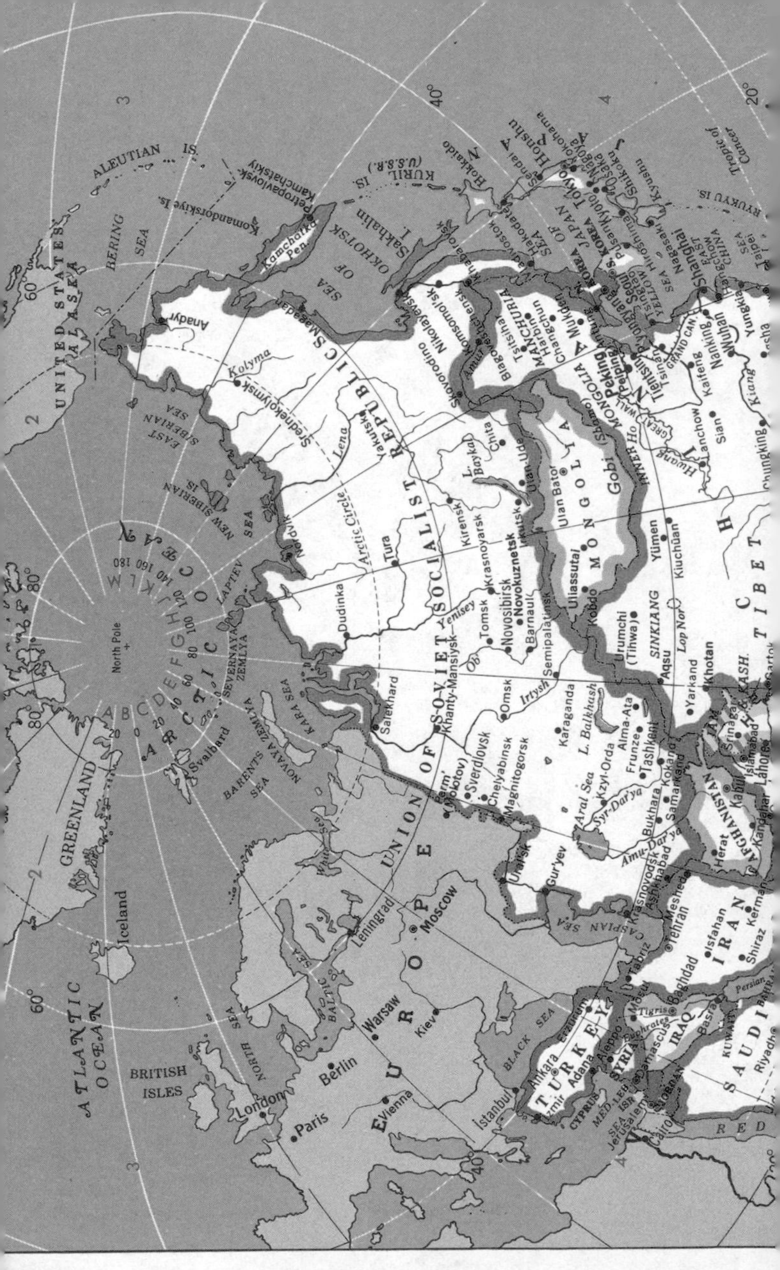

ATLANTIC OCEAN
BRITISH ISLES
London
Paris
Berlin
Vienna
Warsaw
Kiev
Leningrad
Moscow
E U R O P E
Istanbul
BLACK SEA
NORTH SEA
BALTIC SEA
Iceland
GREENLAND
Svalbard
BARENTS SEA
NOVAYA ZEMLYA
KARA SEA
SEVERNAYA ZEMLYA
North Pole
ARCTIC OCEAN
LAPTEV SEA
NEW SIBERIAN IS.
EAST SIBERIAN SEA
UNITED STATES
ALASKA
BERING SEA
ALEUTIAN IS.
Komandorskiye Is.
Petropavlovsk-Kamchatskiy
Kamchatka Pen.
SEA OF OKHOTSK
Sakhalin I.
KURIL IS. (U.S.S.R.)
Hokkaido
Honshu
Sendai
Tokyo
Yokohama
Nagoya
Kyoto
Osaka
Shikoku
Kyushu
Hiroshima
Nagasaki
RYUKYU IS.
Tropic of Cancer
SEA OF JAPAN
Vladivostok
Khabarovsk
Komsomolsk
Nikolayevsk
Magadan
Anadyr
Kolyma
Srednekolymsk
Yakutsk
Lena
Arctic Circle
Nordvik
Tura
Dudinka
Salekhard
Khanty-Mansiysk
Yenisey
Ob
Irtysh
UNION OF SOVIET SOCIALIST REPUBLICS
Perm' (Molotov)
Sverdlovsk
Chelyabinsk
Magnitogorsk
Ural'sk
Gur'yev
Omsk
Tomsk
Novosibirsk
Novokuznetsk
Barnaul
Semipalatinsk
Krasnoyarsk
Kirensk
Irkutsk
L. Baykal
Ulan-Ude
Chita
Skovorodino
Blagoveshchensk
Amur
Karaganda
L. Balkhash
Alma-Ata
Frunze
Tashkent
Kokand
Samarkand
Bukhara
Kzyl-Orda
Aral Sea
Syr-Dar'ya
Amu-Dar'ya
Krasnovodsk
Ashkhabad
CASPIAN SEA
MONGOLIA
Ulan Bator
Uliassutai
Kobdo
Gobi
INNER MONGOLIA
SINKIANG
Urumchi (Tihwa)
Aqsu
Lop Nor
Yarkand
Khotan
Yümen
Kiuchüan
MANCHURIA
Tsitsihar
Harbin
Changchun
Mukden
KOREA
Seoul
Pyongyang
Pusan
Peking (Peiping)
Tientsin
Tsinan
GREAT WALL
Lanchow
Sian
Kaifeng
Nanking
Wuhan
Shanghai
Hangchow
Chungking
Kiang
Huang Ho
GRAND CAN.
YELLOW SEA
EAST CHINA SEA
Taipei
TIBET
Srinagar
Islamabad
Lahore
AFGHANISTAN
Kabul
Kandahar
Herat
IRAN
Tehran
Meshed
Isfahan
Shiraz
Kerman
Tabriz
Baghdad
IRAQ
Mosul
Tigris
Euphrates
Basra
KUWAIT
SAUDI ARABIA
Riyadh
SYRIA
Aleppo
Damascus
JORDAN
LEB.
ISR.
Jerusalem
Cairo
TURKEY
Ankara
Adana
Erzurum
Izmir
CYPRUS
MED. SEA
RED SEA
Persian

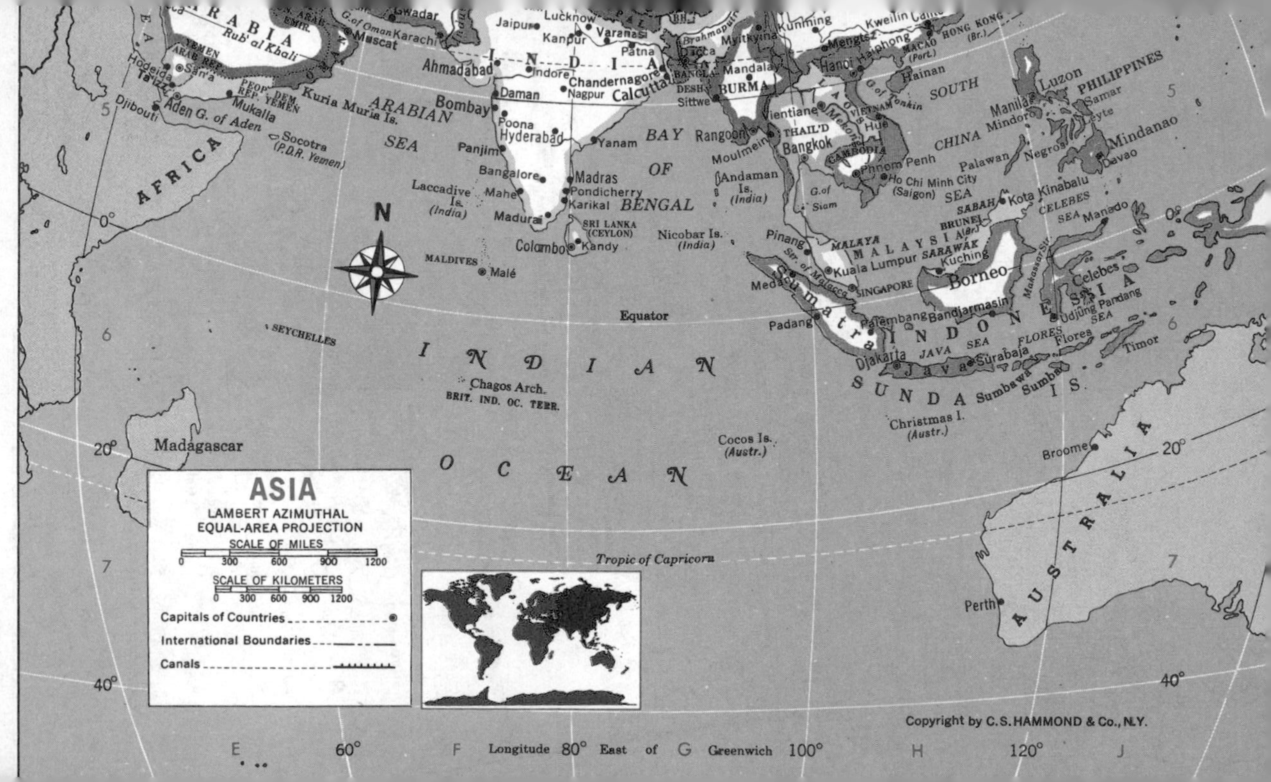

ASIA
LAMBERT AZIMUTHAL
EQUAL-AREA PROJECTION
SCALE OF MILES
0 300 600 900 1200
SCALE OF KILOMETERS
0 300 600 900 1200
Capitals of Countries
International Boundaries
Canals
INDIAN OCEAN
ARABIAN SEA
BAY OF BENGAL
SOUTH CHINA SEA
INDIA
BURMA
ARABIA
AFRICA
AUSTRALIA
INDONESIA
MALAYSIA
PHILIPPINES
SUNDA IS.
Borneo
Celebes
Sumatra
Java
Madagascar
Equator
Tropic of Capricorn
Bombay
Calcutta
Madras
Karachi
Colombo
Rangoon
Bangkok
Singapore
Djakarta
Manila
Perth
Longitude 80° East of Greenwich
Copyright by C.S. HAMMOND & Co., N.Y.

56 — NEAR AND MIDDLE EAST

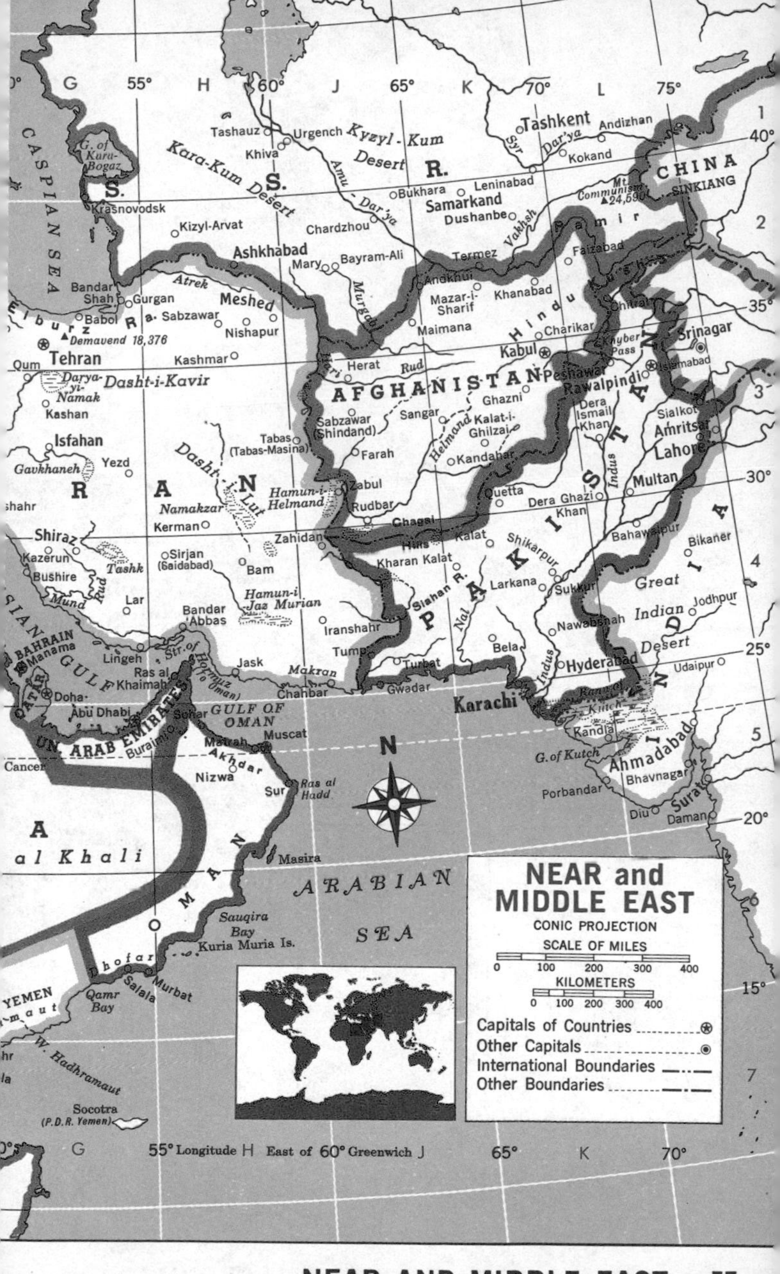
NEAR and MIDDLE EAST
CONIC PROJECTION
SCALE OF MILES
0 100 200 300 400
KILOMETERS
0 100 200 300 400
Capitals of Countries
Other Capitals
International Boundaries
Other Boundaries
CASPIAN SEA
Kara-Kum Desert
Kyzyl-Kum Desert
Tashkent
Samarkand
Dushanbe
Ashkhabad
Meshed
Tehran
Isfahan
Shiraz
Kerman
Herat
Kabul
AFGHANISTAN
PAKISTAN
IRAN
CHINA
INDIA
Karachi
Hyderabad
Rawalpindi
Peshawar
Lahore
Multan
Quetta
Kandahar
Ahmadabad
GULF OF OMAN
Muscat
OMAN
ARABIAN SEA
UN. ARAB EMIRATES
BAHRAIN
QATAR
YEMEN
Socotra
55° Longitude H East of 60° Greenwich J
65° K 70°

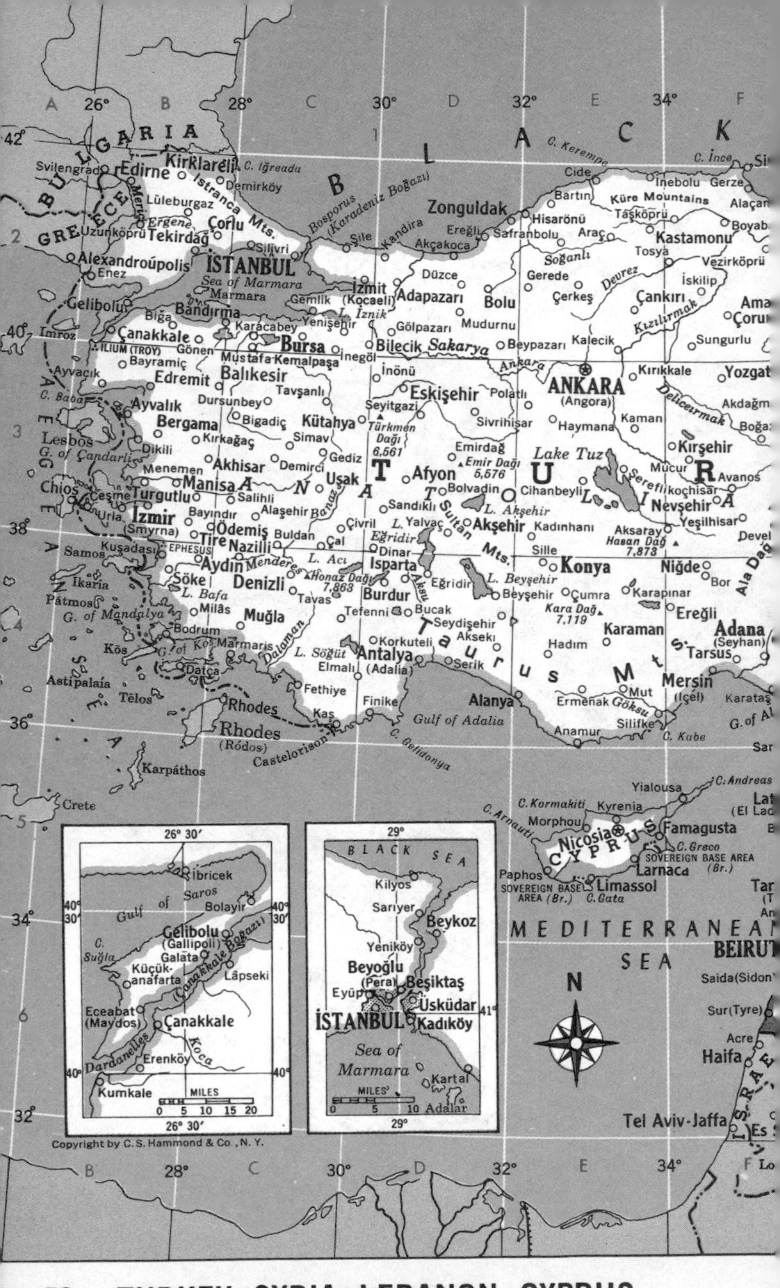

58 — TURKEY, SYRIA, LEBANON, CYPRUS

TURKEY, SYRIA, LEBANON and CYPRUS
CONIC PROJECTION
MILES
0 25 50 75 100 125 150
KILOMETRES
0 25 50 75 100 125 150
Capitals of Countries
International Boundaries
Ruins
SYRIA
IRAQ
JORDAN
LEBANON
U.S.S.R.
IRAN
TURKEY
DAMASCUS
Aleppo
Homs
Hama
Antâkya
Urfa
Diyarbakır
Erzurum
Trabzon
Samsun
Kayseri
Sivas
Malatya
Elâzığ
Van
Lake Van
Tbilisi
Erivan
Batumi
Mosul
Deir ez Zor
Tadmor (Palmyra)
Amman
Pontic Mountains
Eastern Taurus Mts.
Anti-Taurus Mts.
Euphrates
Tigris
Longitude 36° East of Greenwich 38° 40° 42° 44°

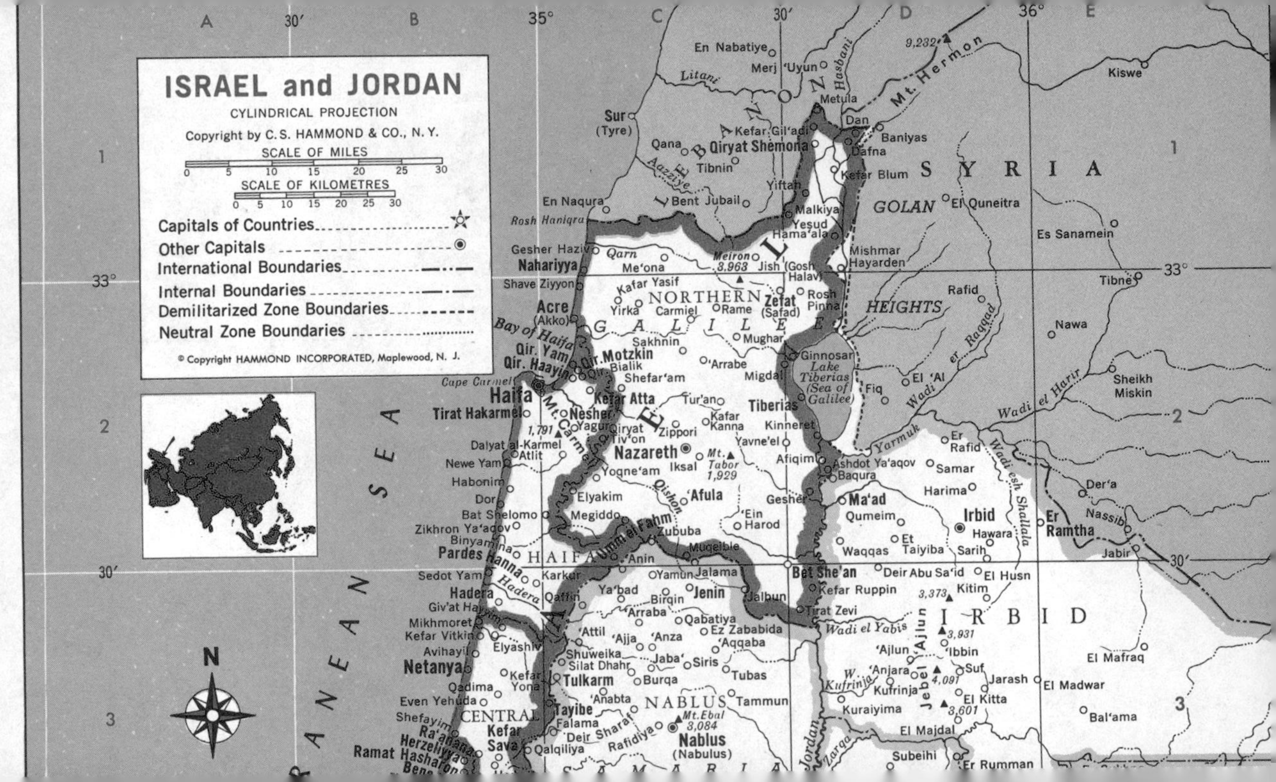
ISRAEL and JORDAN
CYLINDRICAL PROJECTION
Copyright by C.S. HAMMOND & CO., N.Y.
SCALE OF MILES
0 5 10 15 20 25 30
SCALE OF KILOMETRES
0 5 10 15 20 25 30
Capitals of Countries
Other Capitals
International Boundaries
Internal Boundaries
Demilitarized Zone Boundaries
Neutral Zone Boundaries
© Copyright HAMMOND INCORPORATED, Maplewood, N. J.
N
SYRIA
LEBANON
GOLAN HEIGHTS
NORTHERN
GALILEE
HAIFA
CENTRAL
NABLUS
IRBID
Mt. Hermon
9,232
Lake Tiberias (Sea of Galilee)
Bay of Haifa
Cape Carmel
Mt. Carmel
Mt. Tabor
1,929
Mt. Ebal
3,084
Meiron
3,963
Haifa
Acre (Akko)
Nahariyya
Nazareth
Tiberias
Zefat (Safad)
'Afula
Qiryat Shemona
Sur (Tyre)
Netanya
Hadera
Tulkarm
Jenin
Nablus (Nabulus)
Bet She'an
Irbid
Er Ramtha
El Quneitra
Es Sanamein
Kiswe
Der'a
El Mafraq
Jarash
Wadi el Yabis
Wadi esh Shallala
Wadi el Harir
Yarmuk
Jordan
Zarqa
Litani
Hasbani
Qishon

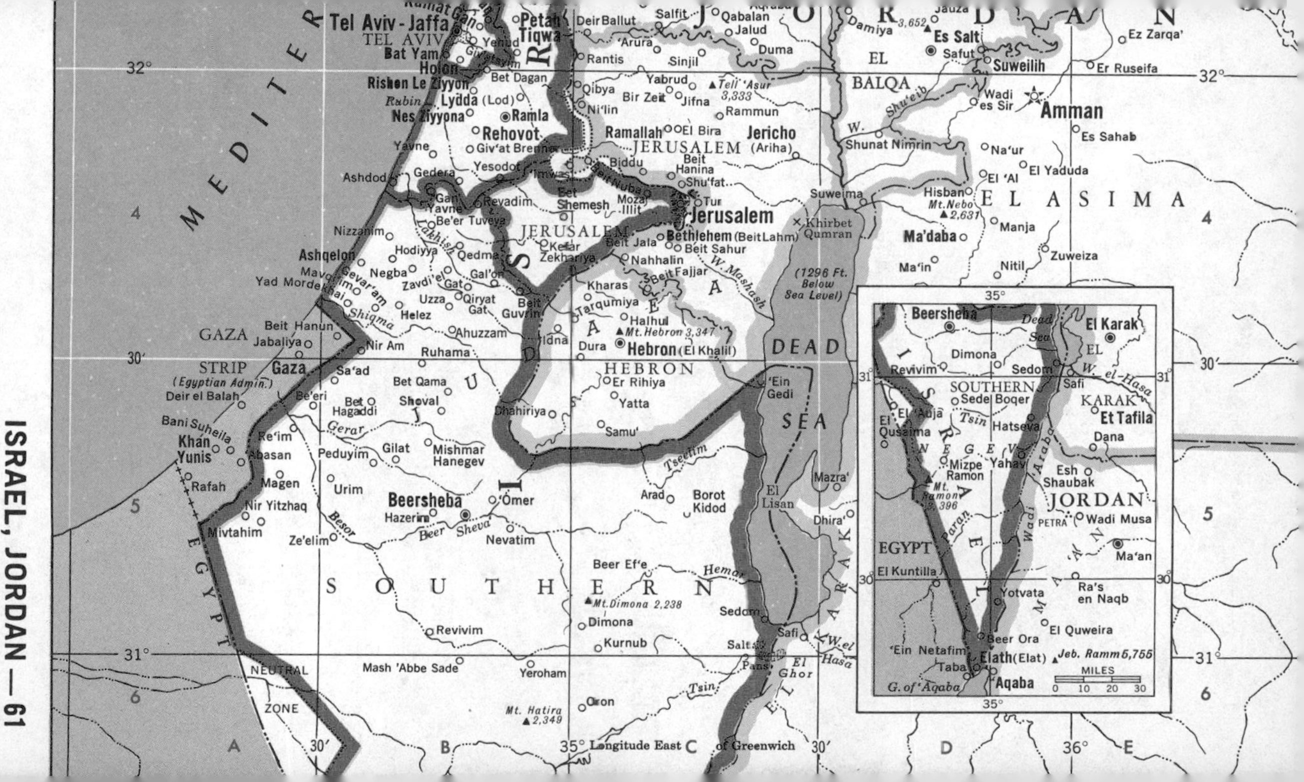

Tel Aviv - Jaffa
TEL AVIV
Bat Yam
Holon
Rishon Le Ziyyon
Lydda (Lod)
Ramla
Nes Ziyyona
Rehovot
Petah Tiqwa
Ashdod
Ashqelon
Yad Mordekhai
GAZA STRIP
(Egyptian Admin.)
Gaza
Jabaliya
Beit Hanun
Deir el Balah
Bani Suheila
Khan Yunis
Rafah
Beersheba
Hebron (El Khalil)
Mt. Hebron 3,347
HEBRON
Bethlehem (Beit Lahm)
Jerusalem
JERUSALEM
Jericho (Ariha)
Ramallah
El Bira
Amman
Es Salt
Suweilih
Ez Zarqa'
Er Ruseifa
Es Sahab
BALQA
Ma'daba
Mt. Nebo 2,631
Hisban
Khirbet Qumran
(1296 Ft. Below Sea Level)
DEAD SEA
'Ein Gedi
El Lisan
Mazra'
Dhira'
Sedom
Safi
Dimona
Mt. Dimona 2,238
Mt. Hatira 2,349
Yeroham
Arad
Kurnub
Revivim
Nevatim
'Omer
Mash 'Abbe Sade
MEDITER
JORDAN
EL ASIMA
SOUTHERN
NEUTRAL ZONE
EGYPT
35° Longitude East C of Greenwich
32°
31°
30'
36°
El Karak
Et Tafila
KARAK
Dana
Ma'an
Wadi Musa
PETRA
Esh Shaubak
Ra's en Naqb
El Quweira
Jeb. Ramm 5,755
Aqaba
Elath (Elat)
Taba
G. of 'Aqaba
El Kuntilla
'Ein Netafim
Yotvata
Beer Ora
Mt. Ramon 3,396
Mizpe Ramon
Sede Boqer
Hatseva
Yahav
El Qusaima
MILES
0 10 20 30

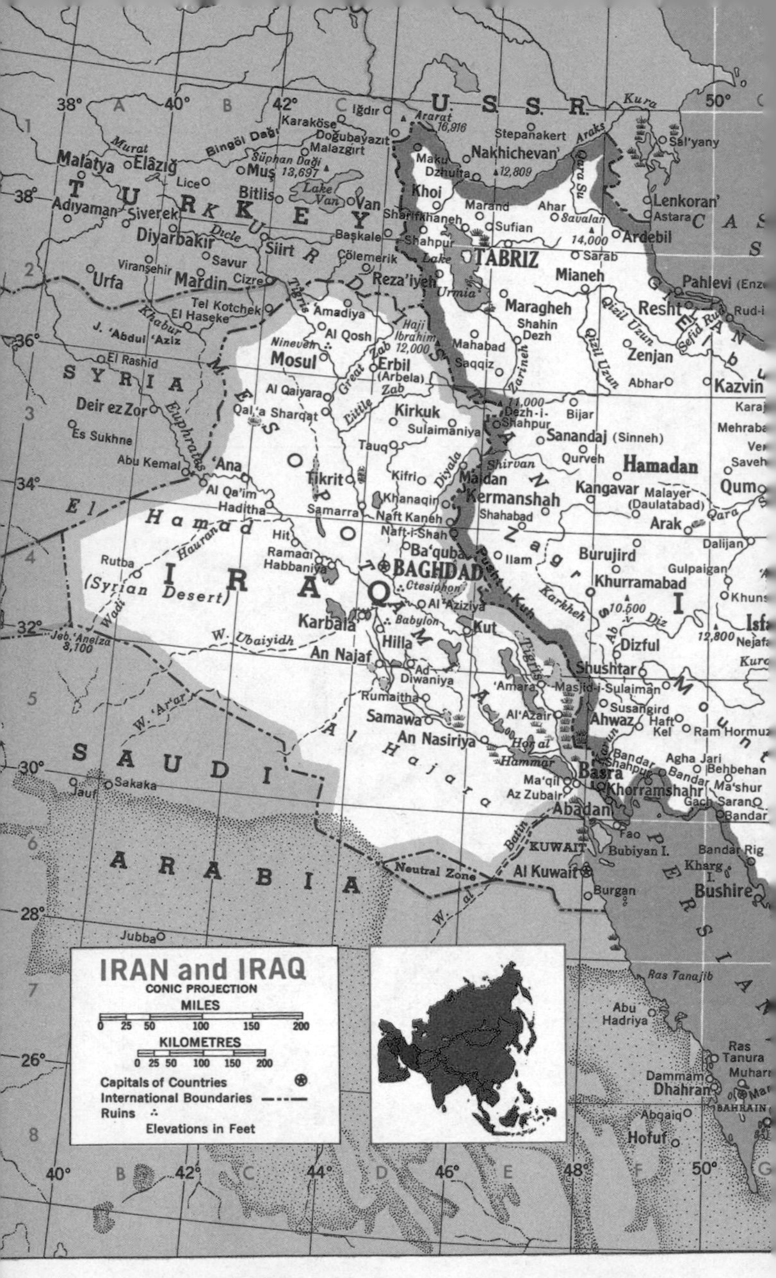
IRAN and IRAQ
CONIC PROJECTION
MILES
0 25 50 100 150 200
KILOMETRES
0 25 50 100 150 200
Capitals of Countries
International Boundaries
Ruins
Elevations in Feet
U. S. S. R.
T U R K E Y
K U R D I S T A N
S Y R I A
I R A Q
I
S A U D I
A R A B I A
M E S O P O T A M I A
Z a g r o s
KUWAIT
Neutral Zone
El Hamad
(Syrian Desert)
Al Hajara
P E R S I A N
Kura
Ararat 16,916
Iğdır
Karaköse
Doğubayazıt
Malazgirt
Bingöl Dağı
Süphan Dağı 13,697
Murat
Malatya
Elâzığ
Muş
Lice
Bitlis
Lake Van
Van
Adıyaman
Siverek
Diyarbakır
Dicle
Siirt
Başkale
Çölemerik
Savur
Viranşehir
Urfa
Mardin
Cizre
Tel Kotchek
El Haseke
Khabur
J. 'Abdul 'Aziz
El Rashid
Deir ez Zor
Es Sukhne
Euphrates
Abu Kemal
'Ana
Al Qa'im
Haditha
Hit
Ramadi
Habbaniya
Rutba
Hauran
Wadi
Jeb. 'Anelza 3,100
W. Ubaiyidh
W. 'Ar'ar
Sakaka
Jauf
Jubba
Tigris
'Amadiya
Al Qosh
Nineveh
Mosul
Great Zab
Erbil (Arbela)
Little Zab
Al Qaiyara
Qal'a Sharqat
Kirkuk
Sulaimaniya
Tauq
Tikrit
Kifri
Samarra
Khanaqin
Naft Kaneh
Naft-i-Shah
Ba'quba
BAGHDAD
Ctesiphon
Al 'Aziziya
Karbala
Babylon
Hilla
Kut
An Najaf
Ad Diwaniya
Rumaitha
Samawa
An Nasiriya
'Amara
Al 'Azair
Hor al Hammar
Ma'qil
Az Zubair
Basra
Abadan
Khorramshahr
Fao
Bubiyan I.
Al Kuwait
Burgan
Batin
W. al Batin
Stepanakert
Nakhichevan'
Maku
Dzhulfa
12,809
Araks
Qara Su
Sal'yany
Lenkoran'
Astara
Khoi
Marand
Ahar
Savalan 14,000
Sharifkhaneh
Sufian
Ardebil
Shahpur
Lake Urmia
TABRIZ
Sarab
Reza'iyeh
Mianeh
Pahlevi
Resht
Sefid Rud
Rud-i
Maragheh
Shahin Dezh
Qizil Uzun
Zenjan
Haji Ibrahim 12,000
Mahabad
Saqqiz
Zarineh
Abhar
Kazvin
11,000
Dezh-i-Shahpur
Bijar
Karaj
Mehraba
Sanandaj (Sinneh)
Shirvan
Qurveh
Hamadan
Saveh
Diyala
Maidan
Kermanshah
Kangavar
Malayer (Daulatabad)
Qum
Shahabad
Qara
Arak
Dalijan
Ilam
Pusht-i-Kuh
Burujird
Gulpaigan
Khurramabad
Khuns
Karkheh
Ab-i-Diz 10,500
12,800
Nejafa
Isfa
Dizful
Shushtar
Masjid-i-Sulaiman
Susangird
Haft Kel
Ram Hormuz
Ahwaz
Karun
Bandar Shahpur
Agha Jari
Behbehan
Bandar Ma'shur
Gach Saran
Bandar
Bandar Rig
Kharg I.
Bushire
Ras Tanajib
Abu Hadriya
Ras Tanura
Dammam
Dhahran
BAHRAIN
Abqaiq
Hofuf
38°
40°
42°
44°
46°
48°
50°
36°
34°
32°
30°
28°
26°
A
B
C
D
E
F
G
1
2
3
4
5
6
7
8

52°
54°
56°
58°
60°
62°
64°
H
J
K
L
M
N
U.
S.
S.
R.
Kara-Kum Desert
Cheleken Pen.
Ogurchinskiy
Kizyl-Arvat
Kara-Kala
Kavakli
Amu Dar'ya
Farab-Pristan'
Chardzhou
(Oxus)
Repetek
Burdalyk
38°
Ashkhabad
Artyk
Kaakhka
Ravnina
Mary
Bayram-Ali
Tedzhen
Murgab
Atrek
Gurgan
Gasan-Kuli
Bandar Shah
Bujnurd
Shirvan
Darreh Gaz
Dushak
Tedzhen
Gunbad-i-Qabus
Jajarm
Quchan
Chalus
Gurgan
(Asterabad)
Amul
Sari
Babol
Mihrabi
10,360
Meshed
Sarakhs
Serakhs
36°
Range
Shahrud
Sabzawar
Demavend
18,934
Nezwar
13,500
Damghan
Qal'eh-i-Mura
Nishapur
Kashaf Rud
Takhta-Bazar
Bala Murghab
Firuzkuh
Turun
Turbat-i-Haidari
Turbat-i-Shaikh Jam
Kushka
Kushk
TEHRAN
Samnan
Turut
Kashmar
(Sultanabad)
Hari Rud
Paropamisus
Dasht-i-Kavir
Tirpul
Herat
11,795
Obeh
HASHTADAN
Khaf
Bijistan
34°
Kashan
Firdaus
(Tun)
Kuh-i-Kalat
Qain
Namaksar
Farsi
Kuh-i-Wala
12,680
Khur
Tabas
Yazdan
Sabzawar
Daulatabad
Natanz
Ardistan
Anarak
Anardarra
IRAN
Nain
Birjand
Rud
Tabas
(Tabas-Masina)
Farah Rud
Farah
Kuhpayeh
Khusf
Naibandan
Duruh
32°
Maibud
Juwain
AFGHANISTAN
Shahriza
Khash
Gavkhaneh
Yezd
Taft
Shir
13,870
Neh
Bafq
Dasht-i-Lut
Hamun-i-Helmand
Zabul
SEISTAN
Dasht-i-Margo
Abadeh
Abarquh
Ravar
14,080
Anar
Helmand
Chahar Burjak
Deh Bid
Bahramabad
Shahdad
Namakzar
Rudbar
30°
Aj-i-Bala
11,815
Kerman
Gaud-i-Zirreh
Zarghan
Persepolis
Mahan
Robat
Shiraz
Kur Rud
L. Tashk
L. Bakhtegan
Zahidan
(Duzdab)
Chagai Hills
Lalehzar
13,000
Niriz
Sirjan
Baft
PAKISTAN
Fasa
Estahbanat
Bam
Mirjawa
Nok Kundi
Safidar
10,450
Darab
Taftan
13,262
Rigan
Talab
Mand
Rud
Jahrum
Bazman
11,446
Khash
Siahan Kuh
Ladgasht
28°
Tarum
Gusht
Ahmadi
Hamun-i-Jaz-Murian
Shahistan
Lar
LARISTAN
Bandar Abbas
Iranshahr
(Fahrej)
Bampur
Mehran
Naband
Minab
Bampur
BALUCHISTAN
Hormuz I.
Sarbaz
Nihing
Shaikh Shu'aib I.
Lingeh
Strait of Hormuz
6,800
Tump
Qishm I.
Ras Musandam
Geh
Nikshahr
26°
Qais I.
Ruus al Jibal
(To Oman)
PAKISTAN
GULF
UNITED ARAB EMIRATES
G. of Oman
Jask
Bir
Dashtiari
MAKRAN
Dasht
Gwadar
Dibah
Ras Maidani
Chahbar
Gwatar
8
52°
Longitude
54°
East
of
56°
Greenwich
58°
60°
62°
1
2
3
4
5
6

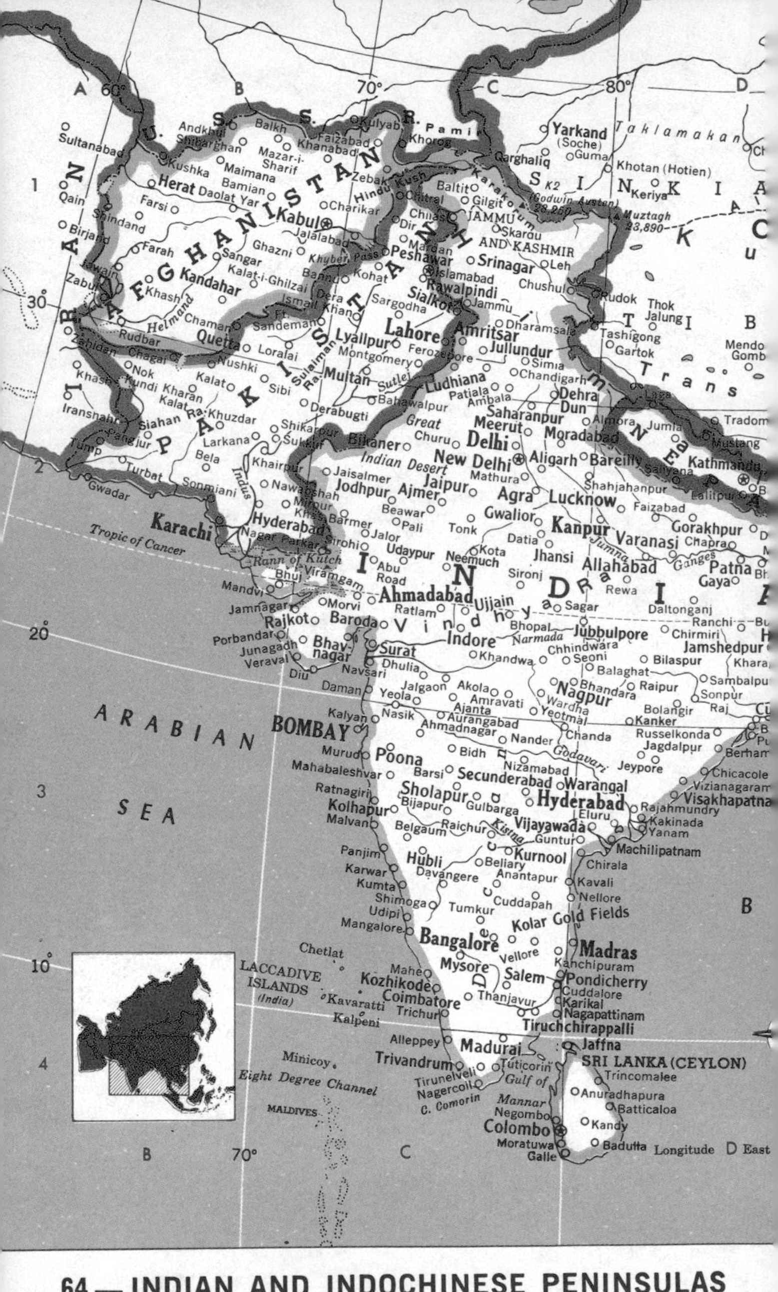

64 — INDIAN AND INDOCHINESE PENINSULAS

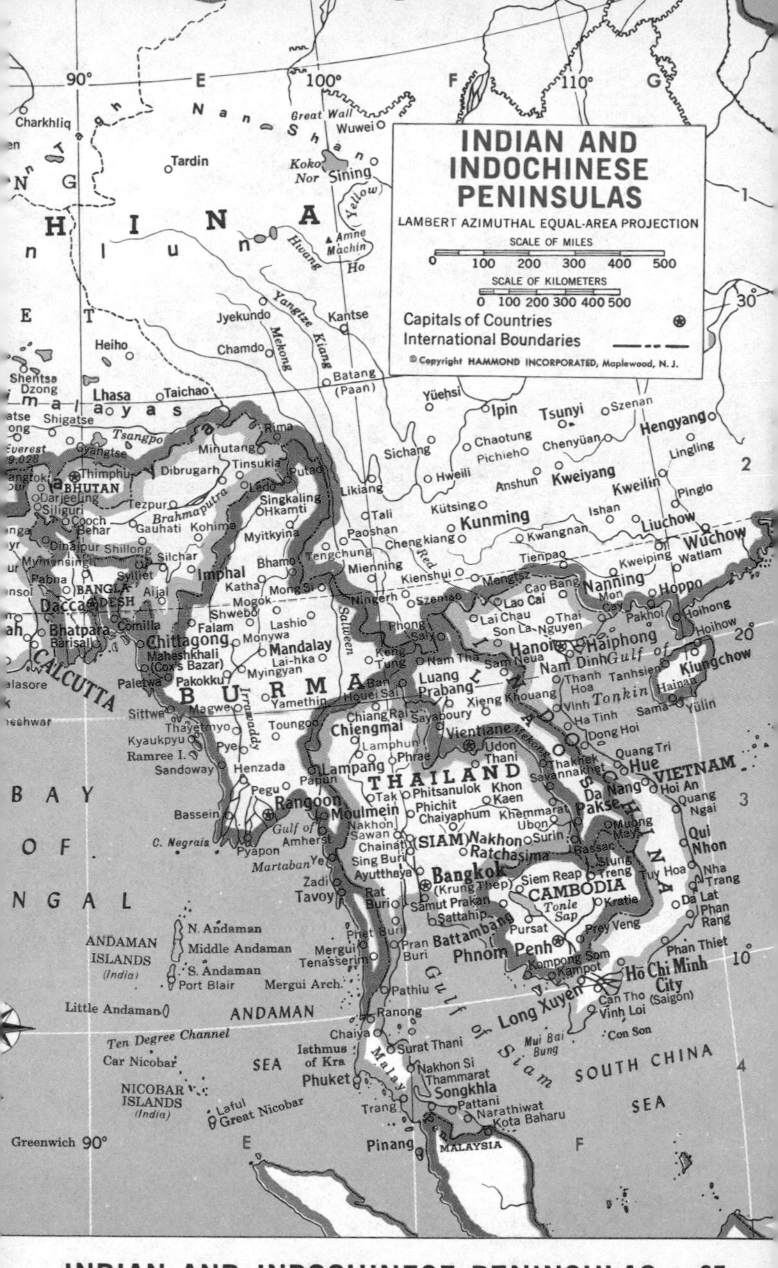
INDIAN AND INDOCHINESE PENINSULAS
LAMBERT AZIMUTHAL EQUAL-AREA PROJECTION
SCALE OF MILES
0 100 200 300 400 500
SCALE OF KILOMETERS
0 100 200 300 400 500
Capitals of Countries
International Boundaries
© Copyright HAMMOND INCORPORATED, Maplewood, N. J.
90°
100°
110°
E
F
G
Charkhliq
Great Wall
Wuwei
Nan Shan
Tardin
Koko Nor
Sining
Yellow
Amne Machin
Hwang Ho
C H I N A
Jyekundo
Yangtze Kiang
Mekong
Kantse
Heiho
Chamdo
Batang (Paan)
Shentsa Dzong
Lhasa
Taichao
Shigatse
Tsangpo
Gyangtse
Rima
Minutang
Tinsukia
Dibrugarh
Thimphu
BHUTAN
Darjeeling
Siliguri
Cooch Behar
Tezpur
Brahmaputra
Gauhati
Kohima
Ledo
Putao
Singkaling Hkamti
Dinajpur
Shillong
Silchar
Mymensingh
Sylhet
Pabna
BANGLADESH
Dacca
Aijal
Imphal
Myitkyina
Bhamo
Katha
Mogok
Shwebo
Falam
Comilla
Bhatpara
Barisal
CALCUTTA
Chittagong
Maheshkhali
(Cox's Bazar)
Paletwa
Pakokku
Monywa
Mandalay
Lai-hka
Myingyan
Lashio
Mong Si
B U R M A
Magwe
Yamethin
Sittwe
Thayetmyo
Irrawaddy
Toungoo
Kyaukpyu
Ramree I.
Pye
Sandoway
Henzada
Pegu
Rangoon
Bassein
C. Negrais
Pyapon
Gulf of Martaban
Moulmein
Amherst
Ye
Zadi
Tavoy
Salween
Yüehsi
Ipin
Tsunyi
Szenan
Hengyang
Lingling
Sichang
Chaotung
Pichieh
Chenyüan
Hweili
Anshun
Kweiyang
Kweilin
Pinglo
Likiang
Tali
Kütsing
Kunming
Ishan
Liuchow
Paoshan
Chengkiang
Kwangnan
Wuchow
Tengchung
Mienning
Red
Tienpao
Kweiping
Watlam
Kienshui
Mengtsz
Cao Bang
Nanning
Hoppo
Ningerh
Szemao
Lao Cai
Mon Cay
Pakhoi
Hoihong
Lai Chau
Thai Nguyen
Son La
Hoihow
Phong Saly
Hanoi
Haiphong
Keng Tung
Nam Tha
Sam Neua
Nam Dinh
Gulf of Tonkin
Kiungchow
Hainan
Ban Houei Sai
Luang Prabang
Thanh Hoa
Tanhsien
Xieng Khouang
Vinh
Sama
Yülin
Chiang Rai
Sayaboury
Chiengmai
Vientiane
Ha Tinh
Dong Hoi
Lamphun
Udon Thani
Mekong
Phrae
Thakhek
Quang Tri
Lampang
Savannakhet
Hue
VIETNAM
Papun
THAILAND
Tak
Phitsanulok
Khon Kaen
Da Nang
Hoi An
Phichit
Chaiyaphum
Khemmarat
Pakse
Quang Ngai
Nakhon Sawan
Chainat
(SIAM)
Nakhon Ratchasima
Ubon
Surin
Muong May
Qui Nhon
Sing Buri
Ayutthaya
Bassac
Stung Treng
Bangkok (Krung Thep)
Siem Reap
Tuy Hoa
Nha Trang
CAMBODIA
Rat Buri
Samut Prakan
Tonle Sap
Kratie
Da Lat
Sattahip
Battambang
Pursat
Phan Rang
Phet Buri
Prey Veng
Pran Buri
Phnom Penh
Phan Thiet
Mergui
Tenasserim
Kompong Som
Kampot
Ho Chi Minh City (Saigon)
Mergui Arch.
Pathiu
Long Xuyen
Can Tho
Vinh Loi
Ranong
Con Son
Chaiya
Mui Bai Bung
Surat Thani
Gulf of Siam
Isthmus of Kra
Nakhon Si Thammarat
SOUTH CHINA SEA
Phuket
Malay
Songkhla
Pattani
Narathiwat
Trang
Kota Baharu
Pinang
MALAYSIA
BAY OF BENGAL
N. Andaman
ANDAMAN ISLANDS (India)
Middle Andaman
S. Andaman
Port Blair
Little Andaman
ANDAMAN SEA
Ten Degree Channel
Car Nicobar
NICOBAR ISLANDS (India)
Laful
Great Nicobar
Greenwich 90°
1
2
3
4
30°
20°
10°

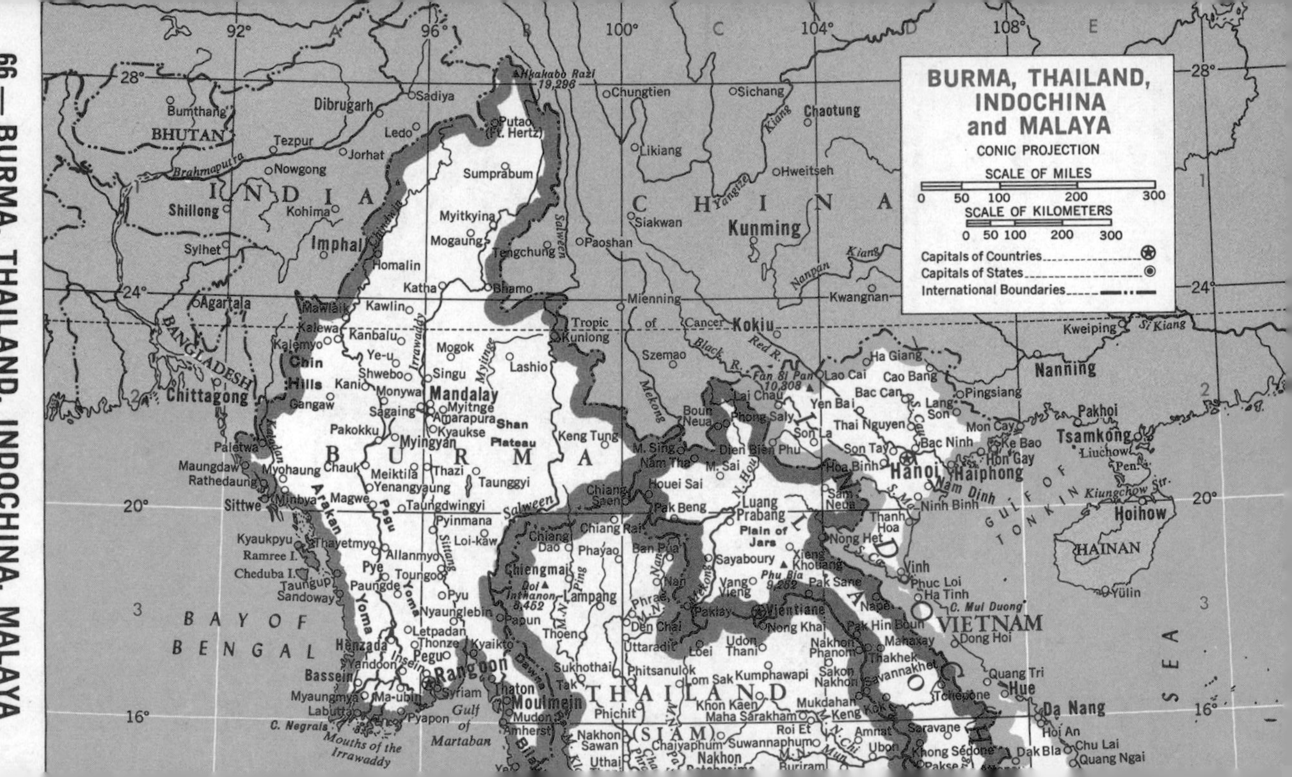

BURMA, THAILAND, INDOCHINA and MALAYA
CONIC PROJECTION
SCALE OF MILES
0 50 100 200 300
SCALE OF KILOMETERS
0 50 100 200 300
Capitals of Countries
Capitals of States
International Boundaries
INDIA
CHINA
BURMA
THAILAND
(SIAM)
INDOCHINA
VIETNAM
BHUTAN
BANGLADESH
HAINAN
BAY OF BENGAL
GULF OF TONKIN
SEA
Gulf of Martaban
Mouths of the Irrawaddy
Tropic of Cancer
Hkakabo Razi 19,296
Fan Si Pan 10,308
Doi Inthanon 8,452
Phu Bia 9,252
Kunming
Nanning
Hanoi
Haiphong
Mandalay
Rangoon
Moulmein
Imphal
Shillong
Chittagong
Vientiane
Luang Prabang
Hue
Da Nang
Hoihow
Chiengmai
Salween
Mekong
Irrawaddy
Chindwin
Brahmaputra
Sittang
Red R.
Black R.
Si Kiang
Shan Plateau
Arakan Yoma
Pegu Yoma
Chin Hills
Plain of Jars

ANDAMAN SEA
ANDAMAN IS. (India)
North Andaman
Middle Andaman
South Andaman
Port Blair
Little Andaman
Ten Degree Channel
Car Nicobar
NICOBAR IS. (India)
Great Nicobar
Great Channel
INDIAN OCEAN
MERGUI ARCHIPELAGO
BANGKOK (Krung Thep)
CAMBODIA
Phnom Penh
HO CHI MINH CITY (Saigon)
Mouths of the Mekong
GULF OF SIAM
Isthmus of Kra
MALAY PENINSULA
MALAYSIA
MALAYA
Kuala Lumpur
Pinang (George Town)
Ipoh
Strait of Malacca
SUMATRA
Medan
Banda Atjeh (Kutaradja)
INDONESIA
SINGAPORE
JOHOR
Johor Baharu
ANAMBAS IS. (Indon.)
NATUNA IS. (Indon.)
BORNEO
SOUTH CHINA
East 100° of Greenwich
Longitude

CHINA
and MONGOLIA
CONIC PROJECTION
MILES
0 100 200 300 400 500
KILOMETERS
0 100 200 300 400 500
Capitals of Countries
Provincial Capitals
International Boundaries
Provincial Boundaries
70°
80°
90°
100°
40°
30°
20°
A
B
C
D
1
2
3
4
S.
R.
Semipalatinsk
Irtysh
Yenisey
Kyzyl
Cheremkhovo
Irkutsk
Tannu-Ola Ra.
Ubsa Nor
Khubsugul
Selenga
Sukhe
Lake Balkhash
Balkhash
Taldy-Kurgan
Ili
Issyk-Kul
Alma-Ata
Bayan Ulegei
Ulangom
Kobdo
Muren
Chuguchak
Karamai
Uliassutai
Tsetserlig
KARAKORUM
Ulan Bator
Dzun Modo
Kuldja
Dzungaria
Altay
MONGOL
Arbai Khere
Mandal Gobi
Sain Sha
Altay Mts.
Tien Shan
Pobeda Pk. 24,406
Urumchi
Kitai
SINKIANG UIGHUR
Turfan
Hami
Kashgar
Faizabad
Aksu
Kucha
Kurla
Tarim
Maralbashi
Dalan Dzadagad
Gobi
Yarkand
Guma
Taklamakan
AUT.
REGION
Charkhlik
Lop Nor
Ansi
KANSU
Paiyüno
INNER
Karakoram
K2 28,250
Khotan
Cherchen
Niya
Keriya
Altyn Tagh
Yümen
Kiuchüan
NINGSIA
HUI
AUT.
REG.
Bayinhot
Yinchwan
Nan Shan
Changyeh
Mangyai
Tsaidam
Wuwei
Leh
Indus
Kun Lun
Ulugh Muztagh 25,340
Golmo
Koko Nor
Sining
Paiyin
TSINGHAI
Hwang Ho
Lanchow
Pingliang
Linsia
KANSU
Tienshui
Paoki
Chang Tang
TIBET AUT. REGION
Thok Jalung
Gartok
Bayan Kara Shan
Jyekundo
Ed Dzong
Hanchung
Wenhsien
Pingwu
Trans-Himalayas
Heiho
Chamdo
Kantse
SZECHWAN
Tsangpo
Nyenchen Tanglha Ra.
Salween
Mekong
Nanchung
Wanhsien
Shigatse
Lhasa
Brahmaputra
Batang
Chengtu
Mt. Everest 29,028
Himalaya
Pehpei
Lucknow
Gyangtse
Neikiang
Tzekung
Chungking
INDIA
NEPAL
Kathmandu
BHUTAN
INDIA
Putao
Ipin
Yangtze
Tsunyi
KWEICHOW
Kweiyang
Tuyün
Anshun
Myitkyina
Siakwan
Kunming
Tengchung
KWANGSI
Liuchow
Mienning
YUNNAN
BURMA
Kokiu
Mengtsz
AUT. R
Lao Cai
Red R.
Pakhoi
Hanoi
Haiphong
LAOS
Gulf of Tonkin
Salween
Chiengmai
Vientiane
Mekong
THAILAND
VIETNAM

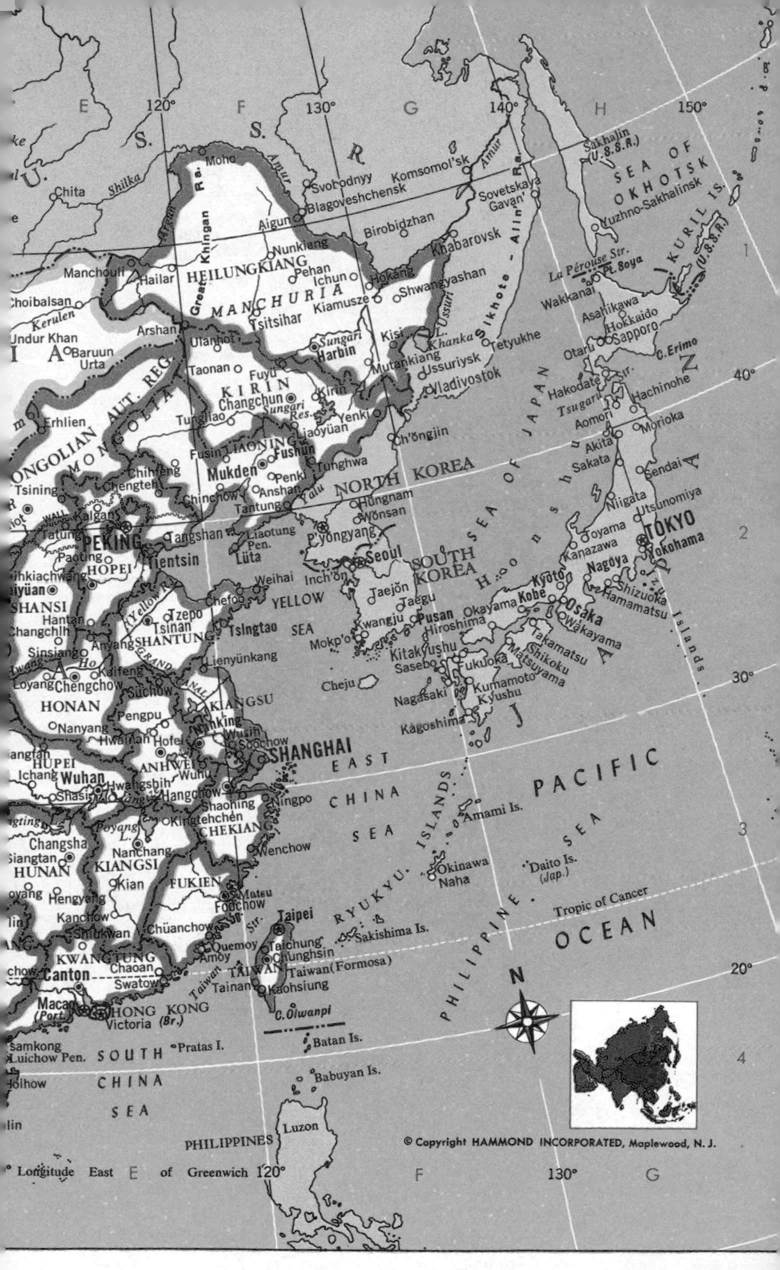
SEA OF OKHOTSK
SEA OF JAPAN
NORTH KOREA
SOUTH KOREA
YELLOW SEA
EAST CHINA SEA
PACIFIC OCEAN
SOUTH CHINA SEA
MANCHURIA
HEILUNGKIANG
KIRIN
LIAONING
HOPEI
SHANSI
SHANTUNG
HONAN
KIANGSU
ANHWEI
HUPEI
CHEKIANG
KIANGSI
HUNAN
FUKIEN
KWANGTUNG
TAIWAN
PHILIPPINES
PEKING
SHANGHAI
TOKYO
Tropic of Cancer
Longitude East of Greenwich

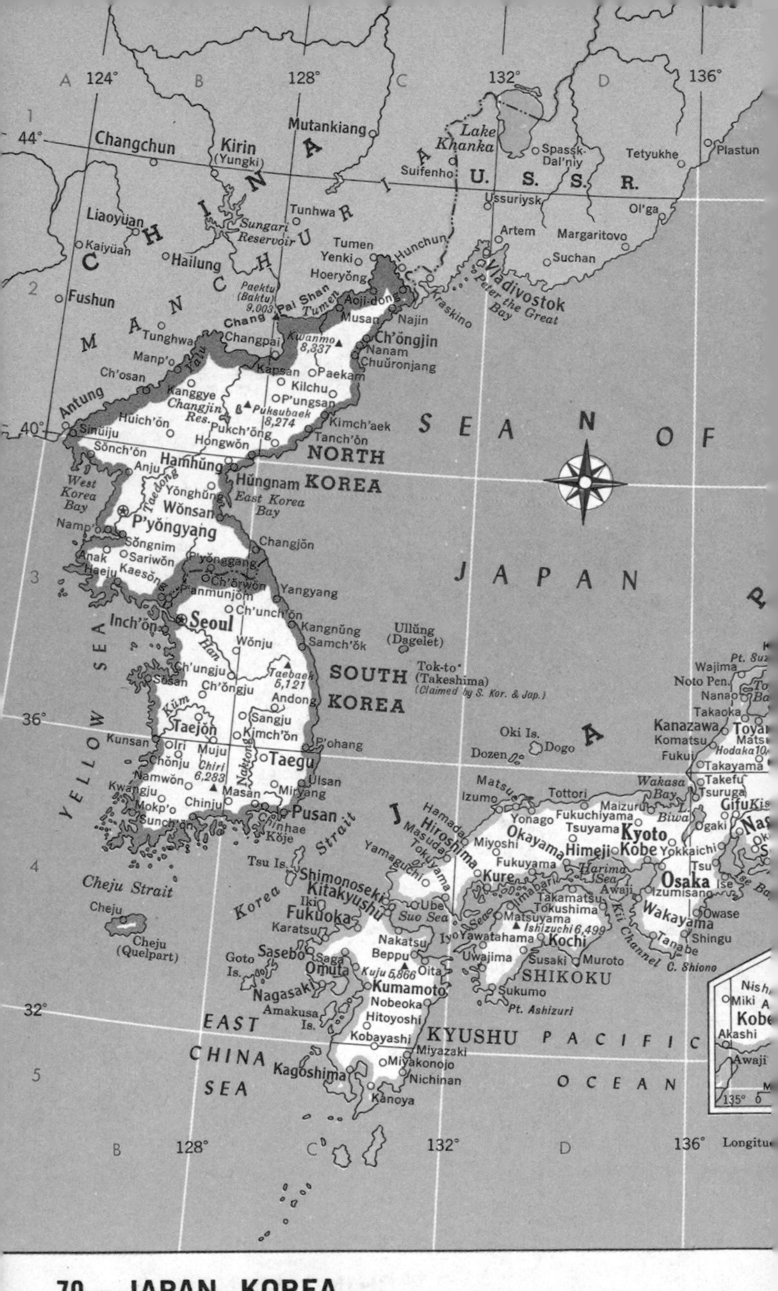

70 — JAPAN, KOREA

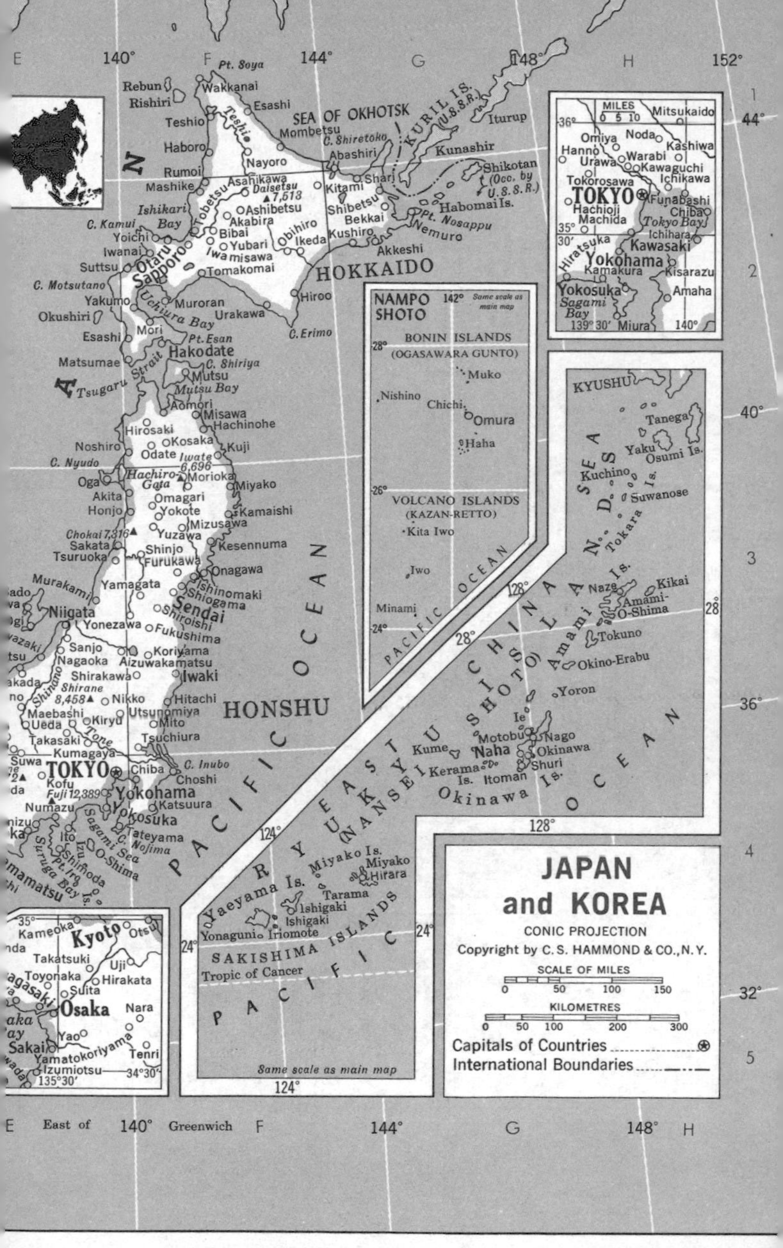

JAPAN
and KOREA
CONIC PROJECTION
Copyright by C. S. HAMMOND & CO., N. Y.
SCALE OF MILES
0 50 100 150
KILOMETRES
0 50 100 200 300
Capitals of Countries
International Boundaries
East of 140° Greenwich
140°
144°
148°
152°
44°
40°
36°
32°
28°
24°
124°
128°
SEA OF OKHOTSK
KURIL IS. (U.S.S.R.)
Iturup
Kunashir
Shikotan (Occ. by U.S.S.R.)
Habomai Is.
Pt. Nosappu
Nemuro
Pt. Soya
Wakkanai
Rebun
Rishiri
Esashi
Teshio
Teshi
Mombetsu
C. Shiretoko
Abashiri
Haboro
Nayoro
Rumoi
Mashike
Asahikawa
Daisetsu 7,513
Kitami
Shari
Ishikari Bay
Tobetsu
Ashibetsu
Shibetsu
Akabira
Bekkai
C. Kamui
Yoichi
Bibai
Obihiro
Kushiro
Iwanai
Otaru
Yubari
Ikeda
Akkeshi
Sapporo
Iwamisawa
Suttsu
Tomakomai
HOKKAIDO
C. Motsutano
Yakumo
Muroran
Hiroo
Okushiri
Uchiura Bay
Urakawa
Mori
Esashi
Pt. Esan
C. Erimo
Hakodate
Matsumae
C. Shiriya
Tsugaru Strait
Mutsu
Mutsu Bay
Aomori
Misawa
Hirosaki
Hachinohe
Noshiro
Kosaka
Kuji
Odate
Iwate 6,696
C. Nyudo
Oga
Hachiro-Gata
Morioka
Akita
Miyako
Omagari
Honjo
Yokote
Kamaishi
Mizusawa
Chokai 7,316
Yuzawa
Sakata
Kesennuma
Tsuruoka
Shinjo
Furukawa
Onagawa
Murakami
Yamagata
Ishinomaki
Shiogama
Niigata
Sendai
Shiroishi
Yonezawa
Fukushima
Sanjo
Koriyama
Nagaoka
Aizuwakamatsu
Shirakawa
Iwaki
Shinano
Shirane 8,458
Nikko
Hitachi
Maebashi
Utsunomiya
HONSHU
Kiryu
Ueda
Mito
Tone
Takasaki
Tsuchiura
Kumagaya
Suwa
TOKYO
Chiba
C. Inubo
Kofu
Choshi
Fuji 12,389
Yokohama
Numazu
Katsuura
Yokosuka
Ito
Tateyama
Sagami Sea
C. Nojima
Izu Is.
Shimoda
O-Shima
Pt. Iro
Suruga Bay
Hamamatsu
PACIFIC OCEAN
MILES
0 5 10
Mitsukaido
36°
Omiya
Noda
Hanno
Kashiwa
Urawa
Warabi
Kawaguchi
Tokorosawa
Ichikawa
TOKYO
Funabashi
Hachioji
Chiba
Machida
Tokyo Bay
35°
Ichihara
30′
Hiratsuka
Kawasaki
Yokohama
Kamakura
Kisarazu
Yokosuka
Amaha
Sagami Bay
139° 30′
Miura
140°
NAMPO SHOTO
142°
Same scale as main map
28°
BONIN ISLANDS
(OGASAWARA GUNTO)
Muko
Nishino
Chichi
Omura
Haha
26°
VOLCANO ISLANDS
(KAZAN-RETTO)
Kita Iwo
Iwo
Minami
24°
PACIFIC OCEAN
KYUSHU
Tanega
Yaku
Osumi Is.
Kuchino
Suwanose
Tokara Is.
EAST CHINA SEA
Naze
Kikai
Amami-O-Shima
Amami Is.
Tokuno
Okino-Erabu
Yoron
RYUKYU ISLANDS (NANSEI SHOTO)
Ie
Motobu
Nago
Kume
Naha
Okinawa
Kerama Is.
Shuri
Itoman
Okinawa Is.
OCEAN
Miyako Is.
Miyako
Hirara
Yaeyama Is.
Tarama
Ishigaki
Ishigaki
Yonaguni
Iriomote
SAKISHIMA ISLANDS
Tropic of Cancer
PACIFIC
Same scale as main map
35°
Kameoka
Kyoto
Otsu
Takatsuki
Uji
Toyonaka
Hirakata
Suita
Osaka
Nara
Yao
Yamatokoriyama
Tenri
Sakai
Izumiotsu
135°30′
34°30′

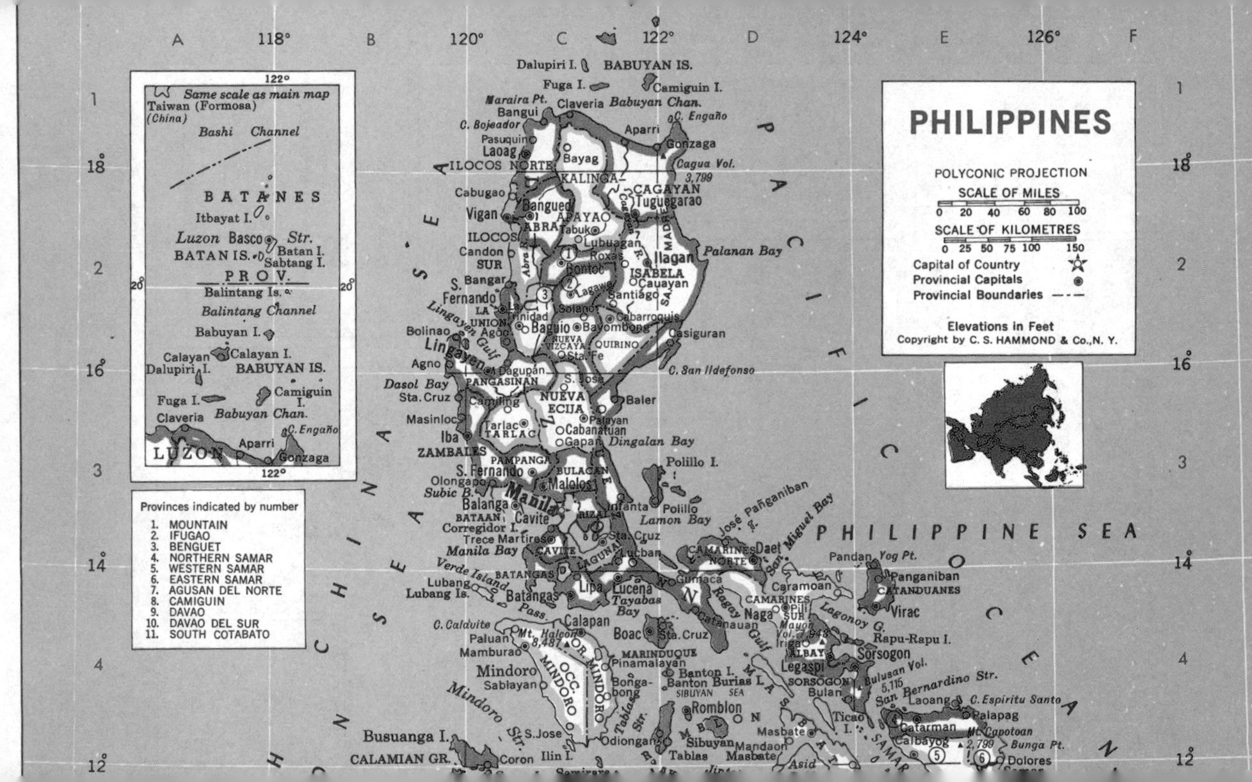
PHILIPPINES
POLYCONIC PROJECTION
SCALE OF MILES
0 20 40 60 80 100
SCALE OF KILOMETRES
0 25 50 75 100 150
Capital of Country
Provincial Capitals
Provincial Boundaries
Elevations in Feet
Copyright by C. S. HAMMOND & Co., N. Y.
Same scale as main map
Taiwan (Formosa)
(China)
Bashi Channel
BATANES
PROV.
Luzon Str.
BATAN IS.
Balintang Channel
BABUYAN IS.
Babuyan Chan.
LUZON
Provinces indicated by number
1. MOUNTAIN
2. IFUGAO
3. BENGUET
4. NORTHERN SAMAR
5. WESTERN SAMAR
6. EASTERN SAMAR
7. AGUSAN DEL NORTE
8. CAMIGUIN
9. DAVAO
10. DAVAO DEL SUR
11. SOUTH COTABATO
PACIFIC OCEAN
PHILIPPINE SEA
SOUTH CHINA SEA
SIBUYAN SEA
Manila
Lingayen Gulf
Manila Bay
Palanan Bay
Dingalan Bay
Lamon Bay
San Bernardino Str.
Mindoro Str.
OCC. MINDORO
OR. MINDORO
CALAMIAN GR.

SULU SEA
CELEBES SEA
MINDANAO SEA
BOHOL SEA
VISAYAN SEA
MINDANAO
MALAYSIA
SABAH
Palawan
Puerto Princesa
Negros
Bacolod
Iloilo
Cebu
Tacloban
Dumaguete
Surigao
Butuan
Cagayan de Oro
Davao
Cotabato
Zamboanga
Jolo
Sandakan
Moro Gulf
Davao Gulf
Sulu Archipelago
CUYO IS.
CAGAYAN IS.
Tubbataha Reefs
Balabac Str.
Basilan Str.
Longitude 122° East of Greenwich 124°

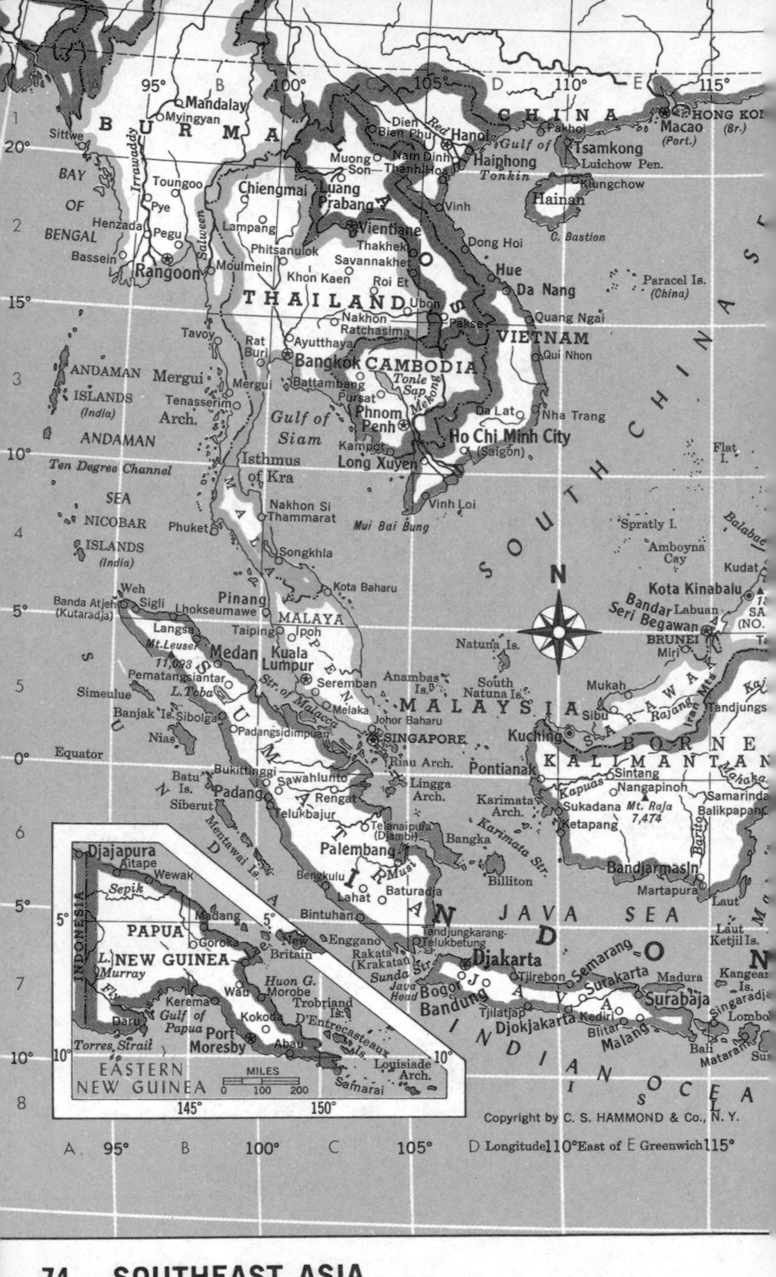

CHINA
HONG KONG (Br.)
Macao (Port.)
BURMA
Mandalay
Myingyan
Sittwe
Irrawaddy
BAY OF BENGAL
Toungoo
Pye
Henzada
Pegu
Bassein
Rangoon
Salween
Moulmein
Chiengmai
Lampang
Phitsanulok
Khon Kaen
THAILAND
LAOS
Luang Prabang
Vientiane
Thakhek
Savannakhet
Roi Et
Ubon
Pakse
Dien Bien Phu
Red
Hanoi
Muong Son
Nam Dinh
Thanh Hoa
Haiphong
Gulf of Tonkin
Vinh
Pakhoi
Tsamkong
Luichow Pen.
Hainan
Kiungchow
C. Bastion
Dong Hoi
Hue
Da Nang
Quang Ngai
Paracel Is. (China)
VIETNAM
Qui Nhon
Nha Trang
Da Lat
Ho Chi Minh City (Saigon)
Nakhon Ratchasima
Tavoy
Rat Buri
Ayutthaya
Bangkok
CAMBODIA
Tonle Sap
Mekong
Battambang
Pursat
Phnom Penh
Kampot
Long Xuyen
Vinh Loi
Mui Bai Bung
Gulf of Siam
ANDAMAN ISLANDS (India)
Mergui
Tenasserim
Mergui Arch.
ANDAMAN SEA
Ten Degree Channel
NICOBAR ISLANDS (India)
Isthmus of Kra
Nakhon Si Thammarat
Phuket
Songkhla
Kota Baharu
Pinang
MALAYA
Taiping
Ipoh
Kuala Lumpur
Seremban
Melaka
Str. of Malacca
Johor Baharu
SINGAPORE
SOUTH CHINA SEA
Flat I.
Spratly I.
Amboyna Cay
Balabac
Kudat
Kota Kinabalu
Labuan
Bandar Seri Begawan
BRUNEI
Miri
Natuna Is.
South Natuna Is.
Anambas Is.
MALAYSIA
SARAWAK
Mukah
Sibu
Rajang
Iran Mts.
Kuching
BORNEO
KALIMANTAN
Pontianak
Sintang
Kapuas
Nangapinoh
Mahakam
Samarinda
Balikpapan
Sukadana
Mt. Raja 7,474
Ketapang
Barito
Bandjarmasin
Martapura
Laut
Laut Ketjil Is.
Karimata Arch.
Karimata Str.
Bangka
Billiton
Riau Arch.
Lingga Arch.
Weh
Banda Atjeh (Kutaradja)
Sigli
Lhokseumawe
Langsa
Mt. Leuser 11,093
Medan
Pematangsiantar
L. Toba
Simeulue
Banjak Is.
Sibolga
Nias
Padangsidimpuan
Equator
SUMATRA
Batu Is.
Siberut
Mentawai Is.
Bukittinggi
Sawahlunto
Padang
Rengat
Telukbajur
Telanaipura (Djambi)
Palembang
Musi
Bengkulu
Lahat
Baturadja
Bintuhan
Enggano
Tandjungkarang-Telukbetung
Rakata (Krakatau)
Sunda Str.
Java Head
INDONESIA
JAVA SEA
Djakarta
Bogor
Bandung
JAVA
Tjirebon
Semarang
Surakarta
Tjilatjap
Djokjakarta
Kediri
Blitar
Malang
Madura
Surabaja
Kangean Is.
Bali
Singaradja
Lombok
Mataram
INDIAN OCEAN
Djajapura
Aitape
Wewak
Sepik
Madang
PAPUA NEW GUINEA
Goroka
L. Murray
Fly
New Britain
Huon G.
Morobe
Wau
Kerema
Gulf of Papua
Daru
Torres Strait
Kokoda
Port Moresby
Trobriand Is.
D'Entrecasteaux Is.
Abau
Louisiade Arch.
Samarai
EASTERN NEW GUINEA
MILES 0 100 200
145° 150°
Copyright by C. S. HAMMOND & Co., N. Y.
A 95° B 100° C 105° D Longitude 110° East of E Greenwich 115°

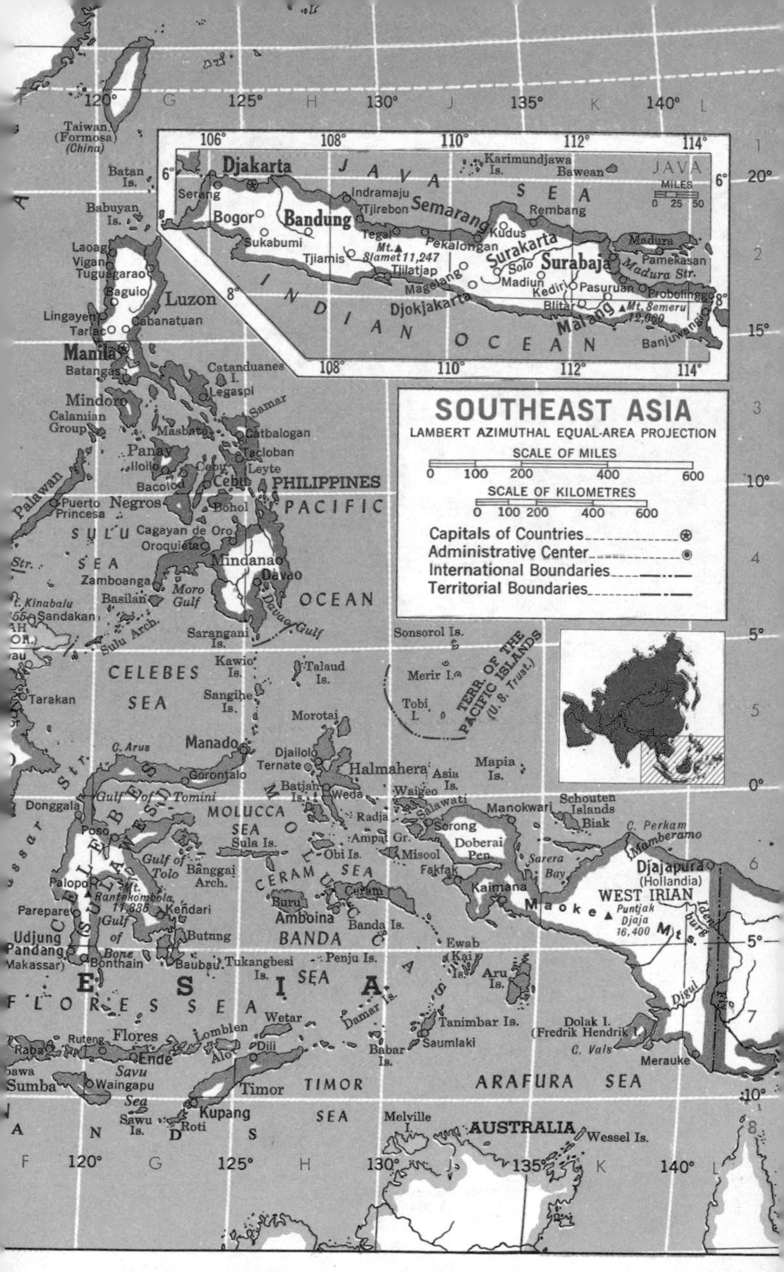

SOUTHEAST ASIA
LAMBERT AZIMUTHAL EQUAL-AREA PROJECTION
SCALE OF MILES
0 100 200 400 600
SCALE OF KILOMETRES
0 100 200 400 600
Capitals of Countries
Administrative Center
International Boundaries
Territorial Boundaries
JAVA
MILES
0 25 50
Djakarta
Serang
Bogor
Bandung
Sukabumi
Tjiamis
Indramaju
Tjirebon
Tegal
Mt. Slamet 11,247
Tjilatjap
Pekalongan
Semarang
Kudus
Magelang
Djokjakarta
Surakarta
Solo
Madiun
Kediri
Rembang
Surabaja
Madura
Pamekasan
Madura Str.
Pasuruan
Probolinggo
Blitar
Malang
Mt. Semeru 12,060
Banjuwangi
Karimundjawa Is.
Bawean
JAVA SEA
INDIAN OCEAN
Taiwan (Formosa) (China)
Batan Is.
Babuyan Is.
Laoag
Vigan
Tuguegarao
Baguio
Luzon
Lingayen
Tarlac
Cabanatuan
Manila
Batangas
Catanduanes I.
Legaspi
Mindoro
Calamian Group
Masbate
Samar
Catbalogan
Tacloban
Leyte
Panay
Iloilo
Cebu
Bacolod
Negros
Bohol
PHILIPPINES
PACIFIC OCEAN
Palawan
Puerto Princesa
SULU SEA
Cagayan de Oro
Oroquieta
Mindanao
Davao
Davao Gulf
Zamboanga
Moro Gulf
Basilan
Sulu Arch.
Sandakan
Sarangani Is.
Kawio Is.
Talaud Is.
Sonsorol Is.
Merir I.
Tobi I.
TERR. OF THE PACIFIC ISLANDS (U. S. Trust.)
CELEBES SEA
Tarakan
Sangihe Is.
Morotai
Manado
C. Arus
Gorontalo
Djailolo
Ternate
Halmahera
Batjan Is.
Weda
Asia Is.
Mapia Is.
Waigeo
Salawati
Radja
Sorong
Manokwari
Schouten Islands
Biak
C. Perkam
Mamberamo
Gulf of Tomini
Donggala
MOLUCCA SEA
Sula Is.
Poso
Ampat Gr.
Obi Is.
Misool
Doberai Pen.
Sarera Bay
Djajapura (Hollandia)
Fakfak
Kaimana
WEST IRIAN
Maoke Mts.
Puntjak Djaja 16,400
Idenburg
Gulf of Tolo
Banggai Arch.
CERAM SEA
Ceram
Buru
Amboina
Banda Is.
Palopo
Mt. Rantekombola 11,835
Kendari
Parepare
Gulf of Bone
Butung
Udjung Pandang (Makassar)
Bonthain
Baubau
Tukangbesi Is.
Penju Is.
BANDA SEA
Ewab
Kai Is.
Aru Is.
Digul
Fly
FLORES SEA
Wetar
Damar Is.
Tanimbar Is.
Saumlaki
Babar Is.
Dolak I. (Fredrik Hendrik I.)
C. Vals
Merauke
Flores
Ruteng
Ende
Lomblen
Alor
Dili
Savu Sea
Sumba
Waingapu
Timor
TIMOR SEA
ARAFURA SEA
Kupang
Sawu Is.
Roti
Melville I.
AUSTRALIA
Wessel Is.
120° 125° 130° 135° 140°

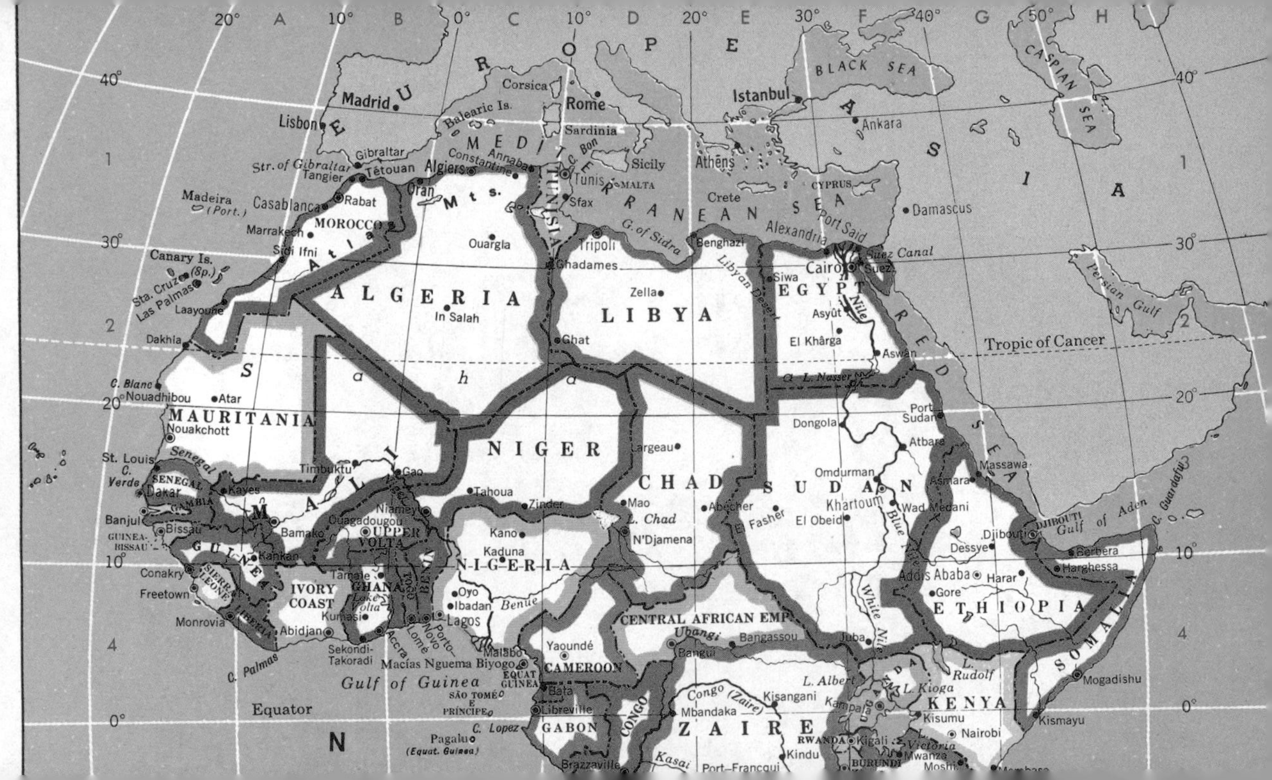

E U R O P E
A S I A
A F R I C A
CASPIAN SEA
BLACK SEA
MEDITERRANEAN SEA
RED SEA
Persian Gulf
Gulf of Aden
Gulf of Guinea
C. Guardafui
Tropic of Cancer
Equator
S a h a r a
Atlas Mts.
Libyan Desert
Suez Canal
Nile
Blue Nile
White Nile
L. Nasser
L. Chad
L. Rudolf
L. Albert
L. Kioga
L. Victoria
Congo (Zaire)
Ubangi
Kasai
Benue
Niger
Lake Volta
Senegal
Str. of Gibraltar
G. of Sidra
C. Bon
Madrid
Lisbon
Rome
Istanbul
Ankara
Athens
Damascus
Corsica
Sardinia
Sicily
MALTA
Crete
CYPRUS
Balearic Is.
Gibraltar
Madeira (Port.)
Canary Is. (Sp.)
Sta. Cruz
Las Palmas
MOROCCO
Tangier
Tétouan
Rabat
Casablanca
Marrakech
Sidi Ifni
Laayoune
Dakhla
ALGERIA
Algiers
Oran
Constantine
Annaba
Ouargla
In Salah
TUNISIA
Tunis
Sfax
LIBYA
Tripoli
Benghazi
Ghadames
Ghat
Zella
EGYPT
Cairo
Alexandria
Port Said
Suez
Siwa
Asyût
El Khârga
Aswan
SUDAN
Port Sudan
Atbara
Dongola
Omdurman
Khartoum
Wad Medani
El Obeid
El Fasher
Juba
ETHIOPIA
Massawa
Asmara
Addis Ababa
Dessye
Harar
Gore
DJIBOUTI
Djibouti
SOMALIA
Berbera
Harghessa
Mogadishu
Kismayu
KENYA
Nairobi
Kisumu
Mombasa
UGANDA
Kampala
RWANDA
Kigali
BURUNDI
Mwanza
Moshi
ZAIRE
Kisangani
Mbandaka
Kindu
Port-Francqui
CONGO
Brazzaville
GABON
Libreville
CENTRAL AFRICAN EMP.
Bangui
Bangassou
CHAD
N'Djamena
Abécher
Largeau
Mao
NIGER
Niamey
Zinder
Tahoua
NIGERIA
Kano
Kaduna
Oyo
Ibadan
Lagos
CAMEROON
Yaoundé
EQUAT. GUINEA
Bata
Malabo
Macías Nguema Biyogo
SÃO TOMÉ E PRÍNCIPE
Pagalu (Equat. Guinea)
C. Lopez
BENIN
Porto-Novo
TOGO
Lomé
GHANA
Accra
Kumasi
Tamale
Sekondi-Takoradi
UPPER VOLTA
Ouagadougou
IVORY COAST
Abidjan
LIBERIA
Monrovia
C. Palmas
SIERRA LEONE
Freetown
GUINEA
Conakry
Kankan
GUINEA-BISSAU
Bissau
GAMBIA
Banjul
SENEGAL
Dakar
St. Louis
C. Verde
MALI
Bamako
Kayes
Timbuktu
Gao
MAURITANIA
Nouakchott
Nouadhibou
Atar
C. Blanc

AFRICA
LAMBERT AZIMUTHAL
EQUAL-AREA PROJECTION
SCALE OF MILES
0 200 400 600 800 1000 1200
SCALE OF KILOMETRES
0 400 800 1200
Capitals
International Boundaries
Canals
Copyright by C.S. HAMMOND & Co., N.Y.
CAPE VERDE
SCALE OF MILES
0 50 100
Santo Antão
São Vicente
São Nicolau
Sal
Boavista
São Tiago
Fogo
Praia
ATLANTIC
OCEAN
INDIAN
OCEAN
Tropic of Capricorn
Ascension (St. Helena)
St. Helena (Br.)
ANGOLA
Luanda
Malanje
Huambo
Benguela
Lubango
Moçâmedes
Cabinda (Ang.)
Luebo
Katanga
Lubumbashi
L. Mweru
Manono
ZAMBIA
Likasi
Kabwe
Lusaka
Livingstone
RHODESIA
Salisbury
Bulawayo
L. Kariba
Zambezi
Limpopo
BOTSWANA
Maun
Serowe
Gaborone
Kalahari Desert
L. Makgadikgadi
SOUTH-WEST AFRICA
Tsumeb
Windhoek
Walvis Bay (S. Afr.)
Lüderitz
Orange
Port Nolloth
C. Fria
SOUTH AFRICA
Pretoria
Johannesburg
Kimberley
Bloemfontein
Maseru
LESOTHO
Mbabane
SWAZILAND
Pietermaritzburg
Durban
East London
Port Elizabeth
Oudtshoorn
Calvinia
Cape Town
C. of Good Hope
C. Agulhas
MOZAMBIQUE
Maputo (Lourenço Marques)
Inhambane
Beira
Quelimane
Nampula
Lilongwe
Zomba
MALAWI
L. Nyasa
L. Bangweulu
L. Tanganyika
Ruvuma
TANZANIA
Dodoma
Dar es Salaam
Zanzibar
Lindi
C. Delgado
Aldabra Is. (Seychelles)
COMORO IS.
Mozambique Channel
MADAGASCAR
Diégo-Suarez
Majunga
Tamatave
Antananarivo
Antsirabe
Tuléar
C. Ste-Marie
Réunion
MAURITIUS
Long. West of Greenwich
Long. East of Greenwich

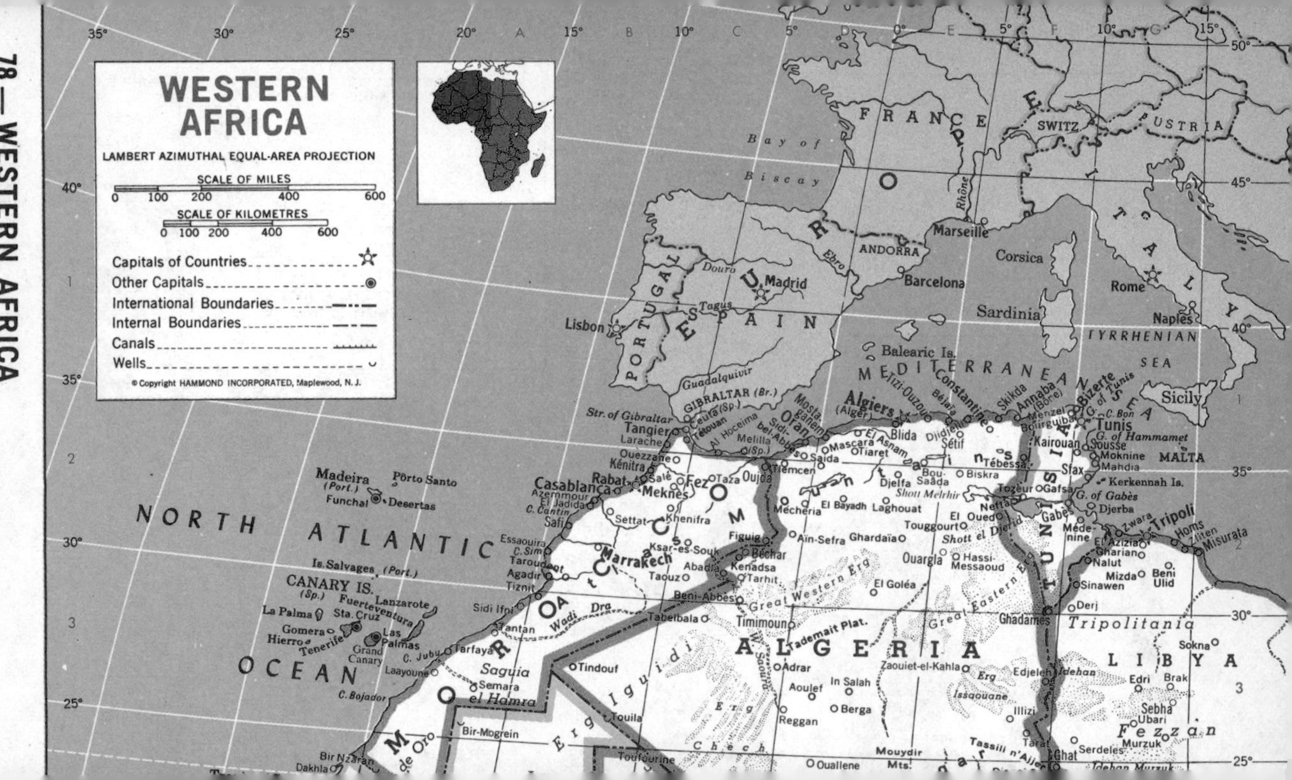
WESTERN AFRICA
LAMBERT AZIMUTHAL EQUAL-AREA PROJECTION
SCALE OF MILES
0 100 200 400 600
SCALE OF KILOMETRES
0 100 200 400 600
Capitals of Countries
Other Capitals
International Boundaries
Internal Boundaries
Canals
Wells
© Copyright HAMMOND INCORPORATED, Maplewood, N. J.
NORTH ATLANTIC OCEAN
MEDITERRANEAN SEA
TYRRHENIAN SEA
Bay of Biscay
EUROPE
FRANCE
SWITZ.
AUSTRIA
ITALY
SPAIN
PORTUGAL
ANDORRA
MOROCCO
ALGERIA
TUNISIA
LIBYA
MALTA
Madrid
Lisbon
Rome
Naples
Marseille
Barcelona
Corsica
Sardinia
Sicily
Balearic Is.
Rhône
Ebro
Douro
Tagus
Guadalquivir
GIBRALTAR (Br.)
Str. of Gibraltar
Ceuta (Sp.)
Melilla (Sp.)
Tangier
Tetouan
Larache
Al Hoceima
Ouezzane
Kénitra
Rabat
Salé
Meknès
Fez
Taza
Oujda
Casablanca
Azemmour
El Jadida
C. Cantin
Safi
Settat
Khenifra
Marrakech
Essaouira
C. Sim
Taroudant
Agadir
Tiznit
Sidi Ifni
Tantan
Wadi Dra
Ksar-es-Souk
Figuig
Béchar
Kenadsa
Abadla
Tarhit
Taouz
Beni-Abbès
Tabelbala
Tindouf
Touila
Saguia el Hamra
Semara
Tarfaya
C. Juby
Laayoune
C. Bojador
Bir-Mogrein
Bir Nzaran
Dakhla
Rio de Oro
Erg Iguidi
Erg Chech
Madeira (Port.)
Pôrto Santo
Funchal
Desertas
Is. Salvages (Port.)
CANARY IS. (Sp.)
La Palma
Gomera
Hierro
Tenerife
Sta. Cruz
Fuerteventura
Lanzarote
Las Palmas
Grand Canary
Algiers (Alger)
Oran
Mostaganem
Sidi-bel-Abbès
Tlemcen
Saïda
Mascara
Tiaret
El Asnam
Blida
Tizi-Ouzou
Bejaïa
Djidjelli
Sétif
Constantine
Skikda
Annaba (Bône)
Tébessa
Biskra
Bou-Saâda
Djelfa
Shott Melrhir
Laghouat
El Bayadh
Mecheria
Aïn-Sefra
Ghardaïa
Touggourt
El Oued
Shott el Djerid
Ouargla
Hassi-Messaoud
El Goléa
Great Western Erg
Great Eastern Erg
Timimoun
Tademait Plat.
Adrar
Aoulef
In Salah
Reggan
Berga
Wadi Saoura
Zaouiet-el-Kahla
Erg Issaouane
Illizi
Edjeleh
Mouydir Mts.
Ouallene
Tassili n'Ajjer
Toufourine
Bizerte
G. of Tunis
C. Bon
Tunis
Menzel Bourguiba
Kairouan
Sousse
G. of Hammamet
Moknine
Mahdia
Sfax
Kerkennah Is.
G. of Gabès
Gabès
Djerba
Gafsa
Tozeur
Nefta
Medenine
Zwara
Tripoli
Homs
Zliten
Misurata
El Azizia
Ghariano
Nalut
Mizda
Beni Ulid
Sinawen
Derj
Ghadames
Tripolitania
Sokna
Edri
Brak
Sebha
Ubari
Fezzan
Murzuk
Serdeles
Ghat
Idehan
Idehan Murzuk
Tarat

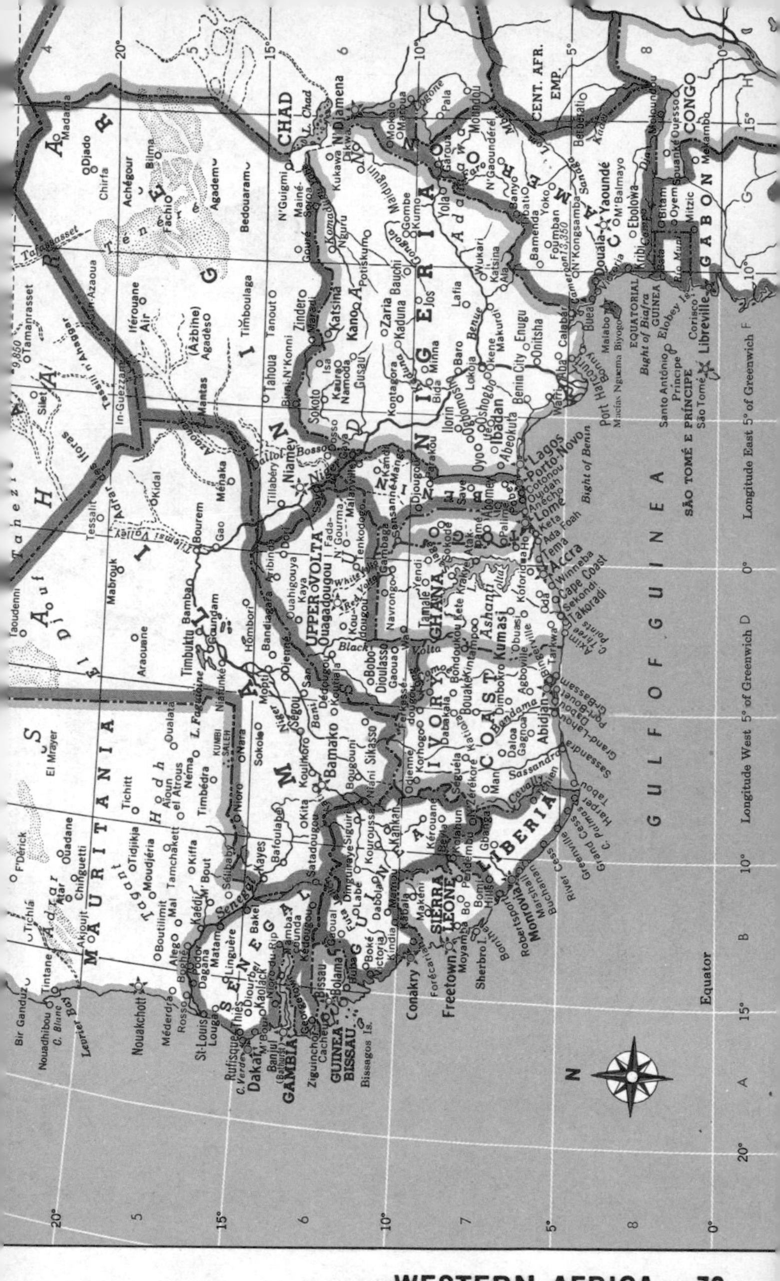

MAURITANIA
SENEGAL
GAMBIA
GUINEA BISSAU
GUINEA
SIERRA LEONE
LIBERIA
IVORY COAST
GHANA
UPPER VOLTA
MALI
NIGER
NIGERIA
CHAD
CAMEROON
CENT. AFR. EMP.
GABON
CONGO
EQUATORIAL GUINEA
SÃO TOMÉ E PRÍNCIPE
GULF OF GUINEA
Bight of Benin
Bight of Biafra
Nouakchott
Dakar
Banjul
Bissau
Conakry
Freetown
Monrovia
Abidjan
Accra
Lomé
Porto-Novo
Lagos
Ibadan
Kumasi
Ouagadougou
Bamako
Niamey
Timbuktu
Kano
N'Djamena
Yaoundé
Douala
Libreville
São Tomé
Malabo
St-Louis
Nouadhibou
Equator
Longitude West 5° of Greenwich
Longitude East 5° of Greenwich

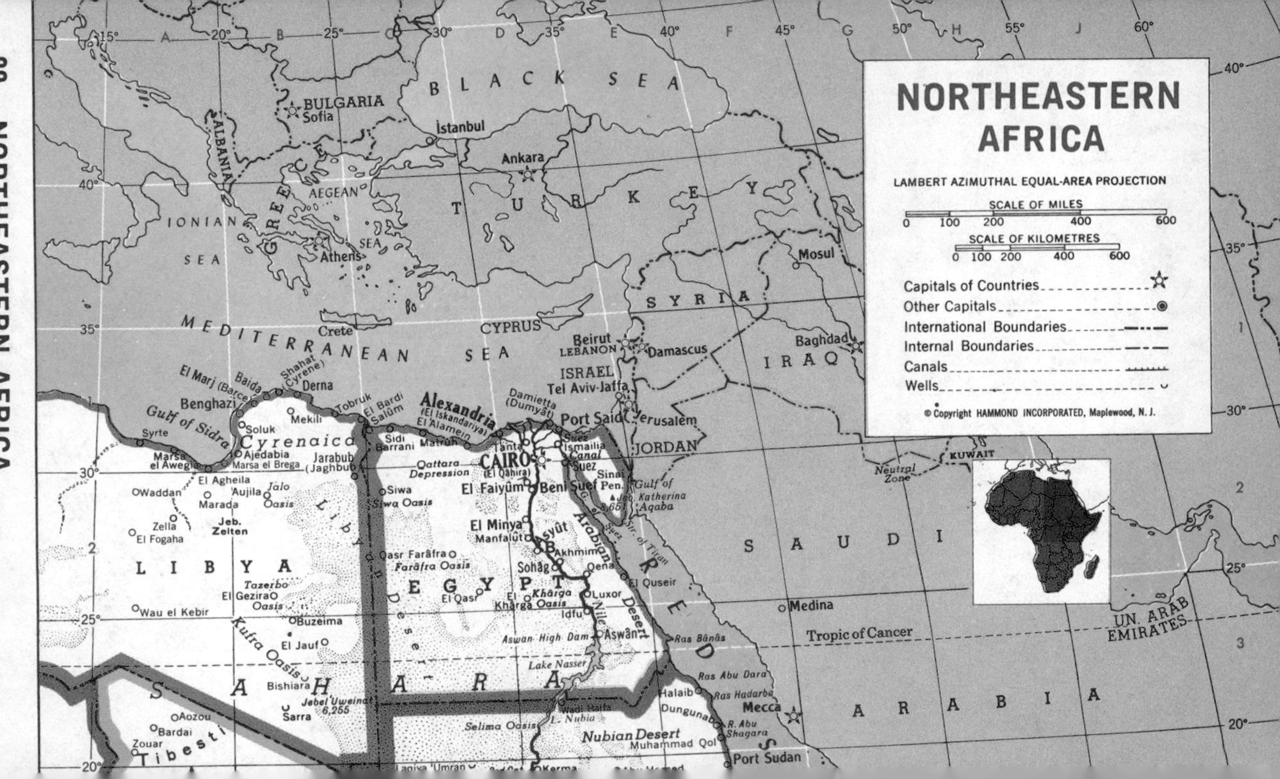

NORTHEASTERN AFRICA
LAMBERT AZIMUTHAL EQUAL-AREA PROJECTION
SCALE OF MILES
0 100 200 400 600
SCALE OF KILOMETRES
0 100 200 400 600
Capitals of Countries
Other Capitals
International Boundaries
Internal Boundaries
Canals
Wells
© Copyright HAMMOND INCORPORATED, Maplewood, N. J.
BLACK SEA
BULGARIA
Sofia
ALBANIA
GREECE
AEGEAN SEA
Athens
IONIAN SEA
Istanbul
Ankara
TURKEY
Mosul
SYRIA
Damascus
Beirut
LEBANON
ISRAEL
Tel Aviv-Jaffa
Jerusalem
JORDAN
IRAQ
Baghdad
KUWAIT
Neutral Zone
CYPRUS
Crete
MEDITERRANEAN SEA
El Marj (Barce)
Baida
Shahat (Cyrene)
Derna
Benghazi
Gulf of Sidra
Syrte
Marsa el Awegia
Soluk
Mekili
Tobruk
Bardi
Salûm
Cyrenaica
Ajedabia
Marsa el Brega
Jarabub (Jaghbub)
Sidi Barrani
Matruh
El Alamein
Alexandria (El Iskandariya)
Damietta (Dumyât)
Tanta
Port Said
Ismailia
Canal
Suez
CAIRO (El Qahira)
Qattara Depression
Sinai Pen.
Gulf of Aqaba
Jeb. Katherina 8,651
El Agheila
Aujila
Jalo Oasis
Waddan
Marada
Zella
Jeb. Zelten
El Fogaha
Siwa
Siwa Oasis
El Faiyûm
Beni Suef
El Minya
Manfalût
Asyût
Akhmim
Sohâg
Qena
El Quseir
Qasr Farâfra
Farâfra Oasis
LIBYA
Libyan Desert
EGYPT
El Qasr
El Khârga
Khârga Oasis
Luxor
Idfu
Nile
Arabian Desert
Aswan High Dam
Aswân
Lake Nasser
Tazerbo
El Gezira
Oasis
Wau el Kebir
Buzeima
Kufra Oasis
El Jauf
Bishiara
Jebel Uweinat 6,255
Sarra
SAHARA
Aozou
Bardai
Zouar
Tibesti
Wadi Halfa
L. Nubia
Selima Oasis
Nubian Desert
Halaib
Dunguna
Muhammad Qol
R. Abu Shagara
Port Sudan
Ras Bânâs
Ras Abu Dara
Ras Hadarba
RED SEA
Str. of Tiran
G. of Suez
Mecca
Medina
SAUDI ARABIA
Tropic of Cancer
UN. ARAB EMIRATES

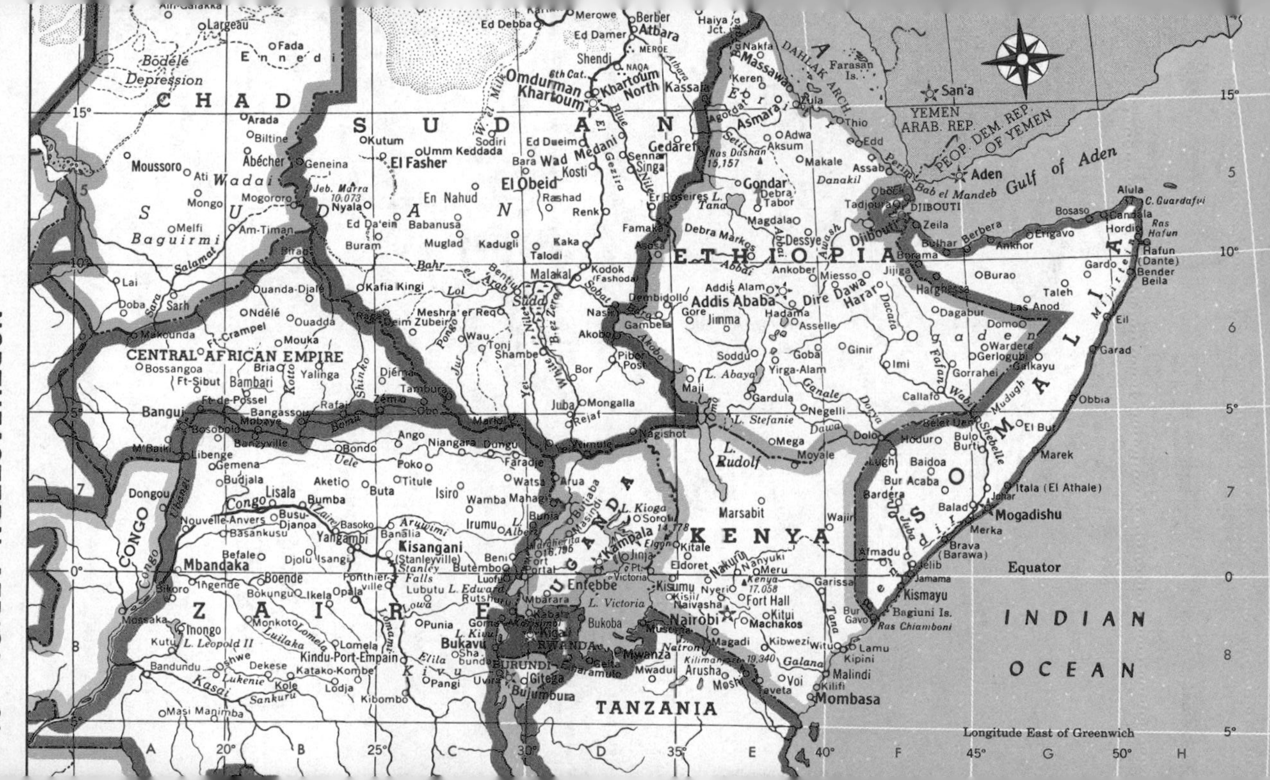
SUDAN
ETHIOPIA
SOMALIA
KENYA
UGANDA
TANZANIA
ZAIRE
CHAD
CENTRAL AFRICAN EMPIRE
CONGO
RWANDA
BURUNDI
YEMEN ARAB. REP.
PEOP. DEM. REP. OF YEMEN
DJIBOUTI
INDIAN OCEAN
Gulf of Aden
Equator
Longitude East of Greenwich
Khartoum
Omdurman
Khartoum North
Addis Ababa
Mogadishu
Nairobi
Kampala
Asmara
Aden
San'a
Djibouti
Kisangani
Mombasa
Kismayu
Juba
Bangui
Mbandaka
Bukavu
Bujumbura
Entebbe
El Obeid
El Fasher
Wad Medani
Gondar
Harar
Dire Dawa
L. Victoria
L. Rudolf
L. Tana
L. Albert
L. Edward
L. Kivu

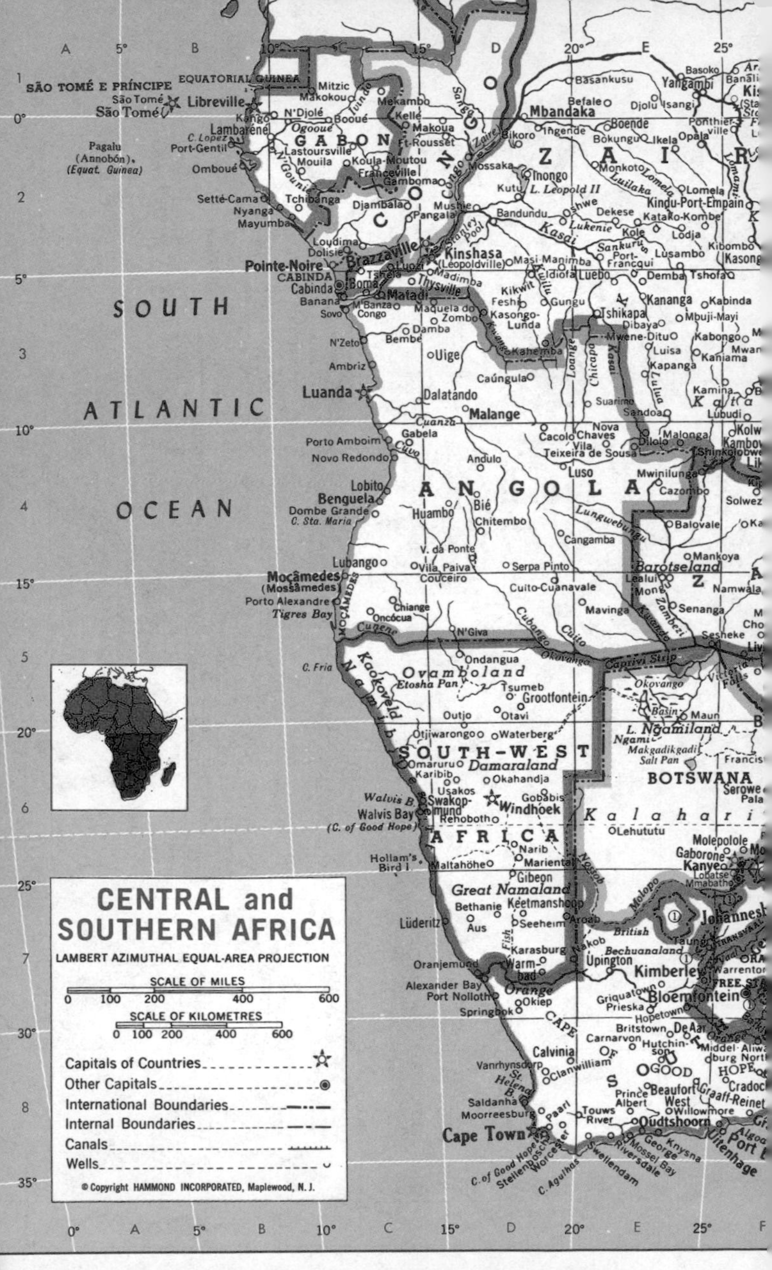
CENTRAL and SOUTHERN AFRICA
LAMBERT AZIMUTHAL EQUAL-AREA PROJECTION
SCALE OF MILES
0 100 200 400 600
SCALE OF KILOMETRES
0 100 200 400 600
Capitals of Countries
Other Capitals
International Boundaries
Internal Boundaries
Canals
Wells
© Copyright HAMMOND INCORPORATED, Maplewood, N. J.
SOUTH ATLANTIC OCEAN
SÃO TOMÉ E PRÍNCIPE
EQUATORIAL GUINEA
GABON
CONGO
ZAIRE
ANGOLA
SOUTH-WEST AFRICA
BOTSWANA
Libreville
São Tomé
Brazzaville
Kinshasa (Leopoldville)
Luanda
Windhoek
Cape Town
Johannesburg
Bloemfontein
Kimberley
Pointe-Noire
Cabinda
Benguela
Lobito
Moçâmedes (Mossâmedes)
Walvis Bay
Lüderitz
Kalahari
Ovamboland
Damaraland
Great Namaland
Barotseland
Katanga

KENYA
SOMALIA
UGANDA
RWANDA
BURUNDI
TANZANIA
MOZAMBIQUE
RHODESIA
MADAGASCAR
SWAZILAND
INDIAN OCEAN
Mozambique Channel
Equator
Tropic of Capricorn
Kampala
Nairobi
Mombasa
Dar es Salaam
Zanzibar
Dodoma
Lusaka
Salisbury
Beira
Maputo
(Lourenço Marques)
Pretoria
Durban
Pietermaritzburg
Antananarivo
Tamatave
Diégo-Suarez
Majunga
Tuléar
Fort-Dauphin
Fianarantsoa
Quelimane
Nampula
Inhambane
Lubumbashi
(Elisabethville)
Kabwe
(Broken Hill)
COMORO IS.
Moroni
Aldabra Is.
(Seychelles)
Lake Tanganyika
L. Victoria
L. Rukwa
Zambezi
Limpopo
Delagoa Bay
SOUTH AFRICAN BANTUSTANS
1 BOPHUTHATSWANA
2 TRANSKEI
30° G 35° H 40° Longitude East 45° of Greenwich 50° L 55°

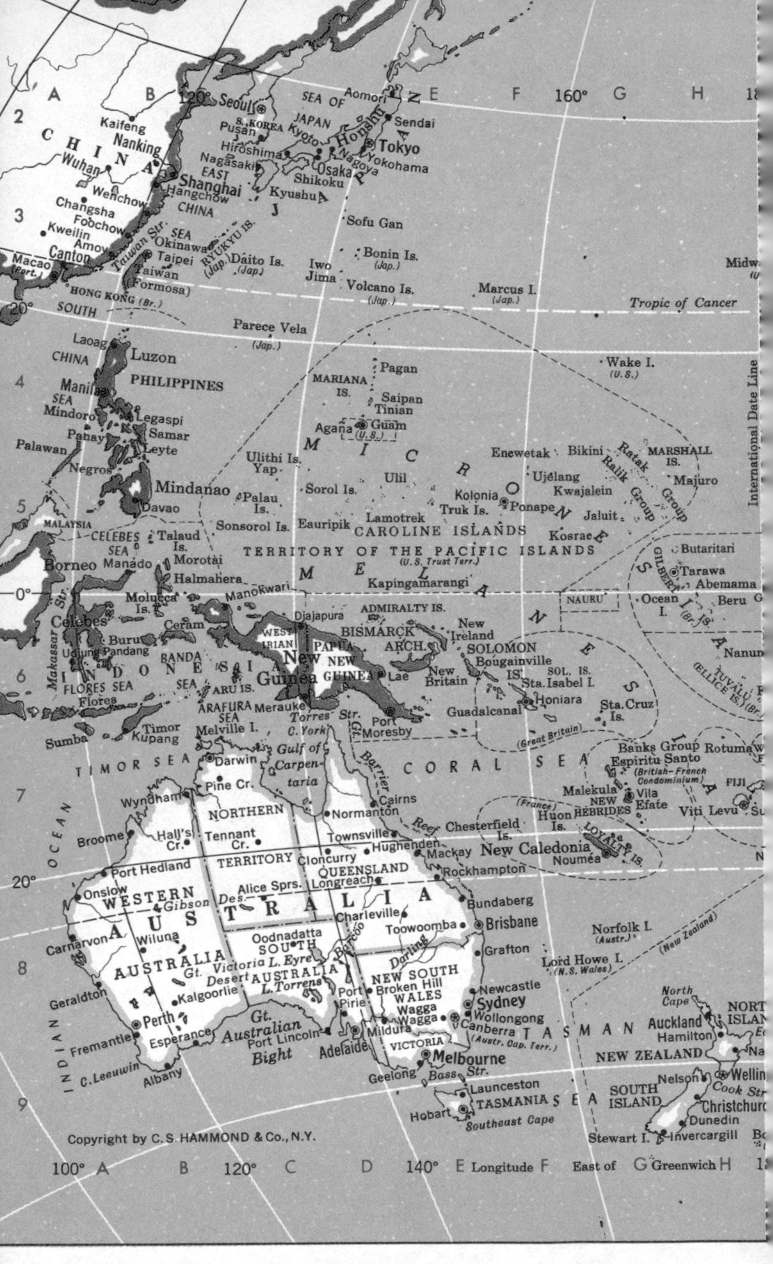

CHINA
Seoul
S. KOREA
Pusan
SEA OF JAPAN
Aomori
Sendai
Honshu
Tokyo
Yokohama
Kyoto
Nagoya
Osaka
Hiroshima
Nagasaki
Shikoku
Kyushu
JAPAN
Kaifeng
Nanking
Wuhan
Shanghai
Hangchow
EAST CHINA SEA
Wenchow
Changsha
Foochow
Kweilin
Amoy
Canton
Macao (Port.)
HONG KONG (Br.)
Taiwan Str.
Okinawa
Taipei
Taiwan (Formosa)
RYUKYU IS. (Jap.)
Daito Is. (Jap.)
Sofu Gan
Bonin Is. (Jap.)
Iwo Jima
Volcano Is. (Jap.)
Marcus I. (Jap.)
Tropic of Cancer
Midway
SOUTH CHINA SEA
Parece Vela (Jap.)
Laoag
Luzon
PHILIPPINES
Manila
Mindoro
Legaspi
Samar
Leyte
Panay
Palawan
Negros
Mindanao
Davao
Wake I. (U.S.)
International Date Line
MARIANA IS.
Pagan
Saipan
Tinian
Agana
Guam (U.S.)
MICRONESIA
Ulithi Is.
Yap
Ulul
Sorol Is.
Palau Is.
Sonsorol Is.
Eauripik
Lamotrek
Truk Is.
Kolonia
Ponape
Enewetak
Bikini
Ujelang
Kwajalein
Jaluit
Kosrae
Ratak Group
Ralik Group
MARSHALL IS.
Majuro
CAROLINE ISLANDS
TERRITORY OF THE PACIFIC ISLANDS (U.S. Trust Terr.)
MALAYSIA
CELEBES SEA
Talaud Is.
Manado
Morotai
Halmahera
Borneo
Makassar Str.
Molucca Is.
Manokwari
Celebes
Ceram
Buru
Ujung Pandang
BANDA SEA
INDONESIA
FLORES SEA
Flores
Sumba
Timor
Kupang
ARU IS.
ARAFURA SEA
Kapingamarangi
MELANESIA
Djajapura
WEST IRIAN
New Guinea
PAPUA NEW GUINEA
Lae
Merauke
Port Moresby
ADMIRALTY IS.
BISMARCK ARCH.
New Ireland
New Britain
SOLOMON IS.
Bougainville
SOL. IS.
Sta. Isabel I.
Honiara
Guadalcanal
(Great Britain)
Sta. Cruz Is.
NAURU
Butaritari
Tarawa
Abemama
Ocean I.
Beru
GILBERT IS. (Br.)
Nanumea
TUVALU (ELLICE IS.) (Br.)
Torres Str.
C. York
Gulf of Carpentaria
Great Barrier Reef
CORAL SEA
Banks Group
Espiritu Santo
(British-French Condominium)
Malekula
Vila
Efate
NEW HEBRIDES
Rotuma
FIJI
Viti Levu
(France)
Huon Is.
Chesterfield Is.
LOYALTY IS.
New Caledonia
Noumea
TIMOR SEA
Melville I.
Darwin
Pine Cr.
Wyndham
NORTHERN TERRITORY
Hall's Cr.
Tennant Cr.
Cairns
Normanton
Townsville
Hughenden
Cloncurry
Mackay
QUEENSLAND
Rockhampton
Longreach
Broome
Port Hedland
Onslow
WESTERN AUSTRALIA
Gibson Des.
Alice Sprs.
AUSTRALIA
Charleville
Bundaberg
Brisbane
Toowoomba
Carnarvon
Wiluna
Oodnadatta
SOUTH AUSTRALIA
L. Eyre
Gt. Victoria Desert
L. Torrens
Barcoo
Darling
Grafton
Norfolk I. (Austr.)
Lord Howe I. (N.S. Wales)
(New Zealand)
NEW SOUTH WALES
Broken Hill
Port Pirie
Newcastle
Sydney
Wollongong
Wagga Wagga
Canberra (Austr. Cap. Terr.)
Geraldton
Kalgoorlie
Perth
Fremantle
Esperance
Gt. Australian Bight
Port Lincoln
Adelaide
Mildura
VICTORIA
Melbourne
C. Leeuwin
Albany
Geelong
Bass Str.
Launceston
TASMANIA
Hobart
Southeast Cape
TASMAN SEA
INDIAN OCEAN
North Cape
Auckland
Hamilton
NORTH ISLAND
NEW ZEALAND
Wellington
Nelson
Cook Str.
SOUTH ISLAND
Christchurch
Dunedin
Stewart I.
Invercargill
100° A B 120° C D 140° E Longitude F East of G Greenwich H

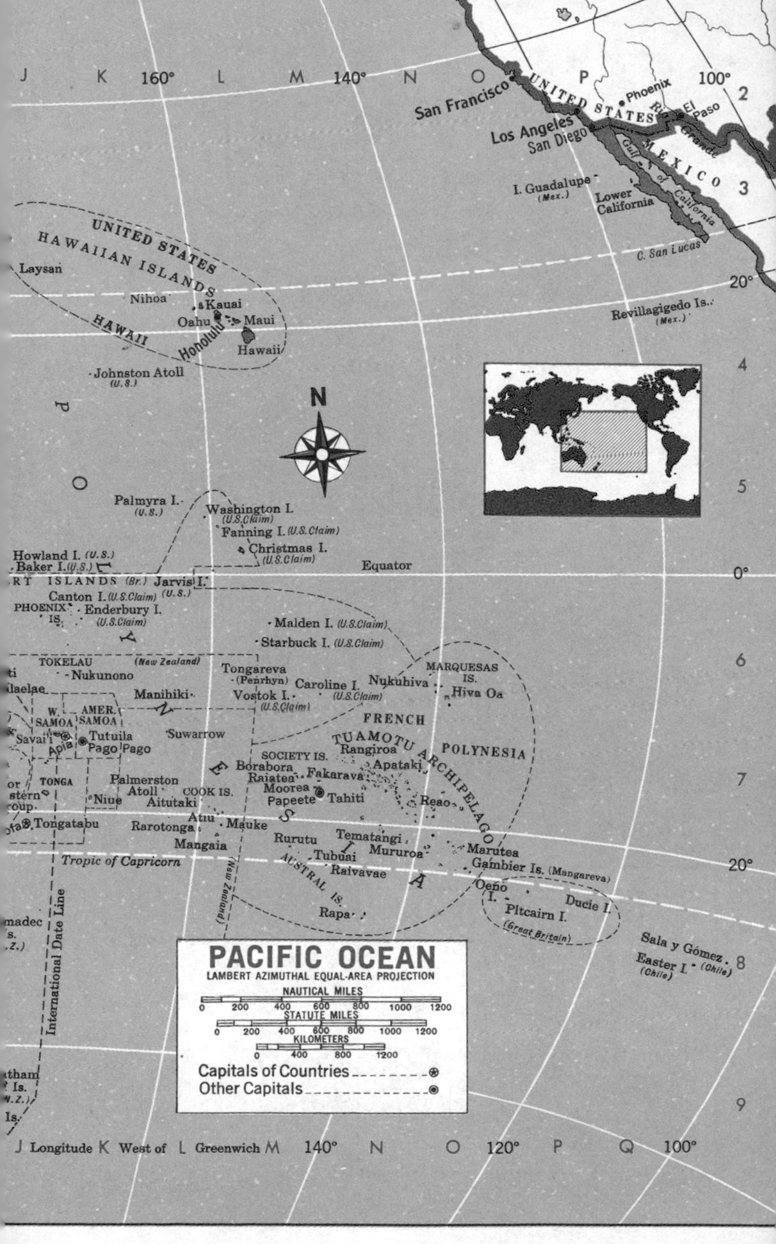
PACIFIC OCEAN
LAMBERT AZIMUTHAL EQUAL-AREA PROJECTION
NAUTICAL MILES
STATUTE MILES
KILOMETERS
Capitals of Countries
Other Capitals
San Francisco
Los Angeles
San Diego
UNITED STATES
Phoenix
El Paso
Rio Grande
MEXICO
Gulf of California
Lower California
I. Guadalupe (Mex.)
C. San Lucas
Revillagigedo Is. (Mex.)
UNITED STATES
HAWAIIAN ISLANDS
HAWAII
Laysan
Nihoa
Kauai
Oahu
Maui
Honolulu
Hawaii
Johnston Atoll (U.S.)
Palmyra I. (U.S.)
Washington I. (U.S.Claim)
Fanning I. (U.S.Claim)
Christmas I. (U.S.Claim)
Howland I. (U.S.)
Baker I. (U.S.)
Equator
Jarvis I. (U.S.)
Canton I. (U.S.Claim)
PHOENIX IS.
Enderbury I. (U.S.Claim)
Malden I. (U.S.Claim)
Starbuck I. (U.S.Claim)
TOKELAU (New Zealand)
Nukunono
Tongareva (Penrhyn)
Caroline I. (U.S.Claim)
Vostok I. (U.S.Claim)
Nukuhiva
MARQUESAS IS.
Hiva Oa
Manihiki
W. SAMOA
AMER. SAMOA
Savai'i
Apia
Tutuila
Pago Pago
Suwarrow
FRENCH POLYNESIA
TUAMOTU ARCHIPELAGO
Rangiroa
Apataki
SOCIETY IS.
Borabora
Raiatea
Fakarava
Moorea
Tahiti
Papeete
Reao
TONGA
Palmerston Atoll
COOK IS.
Niue
Aitutaki
Atiu
Mauke
Rarotonga
Mangaia
Tongatabu
Tropic of Capricorn
Rurutu
Tubuai
Raivavae
AUSTRAL IS.
Rapa
Tematangi
Mururoa
Marutea
Gambier Is. (Mangareva)
Oeno I.
Ducie I.
Pitcairn I. (Great Britain)
Sala y Gómez
Easter I. (Chile) (Chile)
International Date Line
(New Zealand)
Longitude West of Greenwich

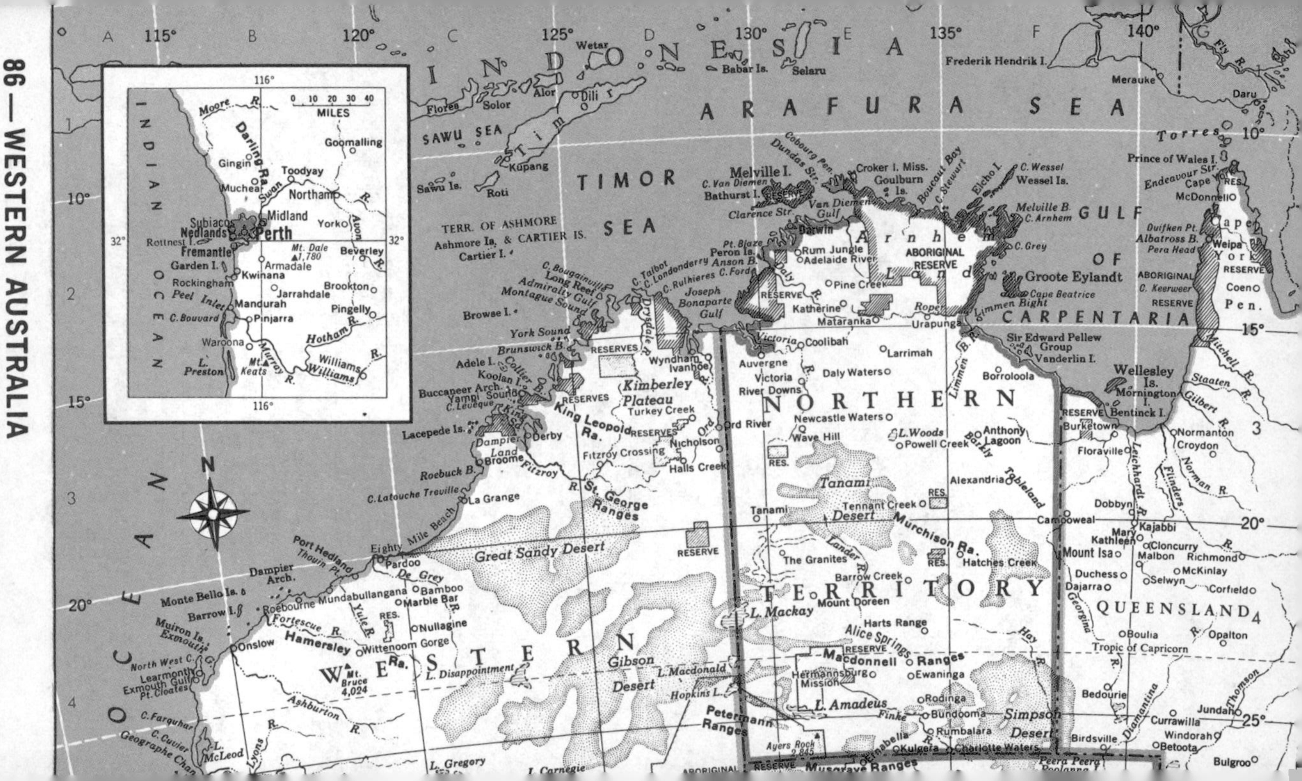

INDONESIA
ARAFURA SEA
TIMOR SEA
SAWU SEA
GULF OF CARPENTARIA
INDIAN OCEAN
OCEAN
NORTHERN TERRITORY
WESTERN
QUEENSLAND
TERR. OF ASHMORE & CARTIER IS.
Ashmore Is. & Cartier I.
Torres Str.
Cape York Pen.
Arnhem Land
ABORIGINAL RESERVE
Kimberley Plateau
King Leopold Ra.
St. George Ranges
Great Sandy Desert
Gibson Desert
Tanami Desert
Simpson Desert
Tropic of Capricorn
Darwin
Katherine
Alice Springs
Tennant Creek
Wyndham
Derby
Broome
Port Hedland
Onslow
Mount Isa
Cloncurry
Camooweal
Groote Eylandt
Melville I.
Bathurst I.
Wellesley Is.
Dampier Arch.
Hamersley Ra.
Mt. Bruce 4,024
Macdonnell Ranges
Musgrave Ranges
Petermann Ranges
Murchison Ra.
L. Amadeus
L. Mackay
L. Disappointment
Ayers Rock 2,845
Fitzroy R.
Ashburton R.
Eighty Mile Beach
Darling Ra.
Perth
Fremantle
Nedlands
Subiaco
Midland
Rottnest I.
Garden I.
Kwinana
Rockingham
Mandurah
Pinjarra
Armadale
Jarrahdale
Northam
York
Toodyay
Gingin
Muchea
Goomalling
Beverley
Brookton
Pingelly
Williams
Waroona
Mt. Dale 1,780
0 10 20 30 40 MILES

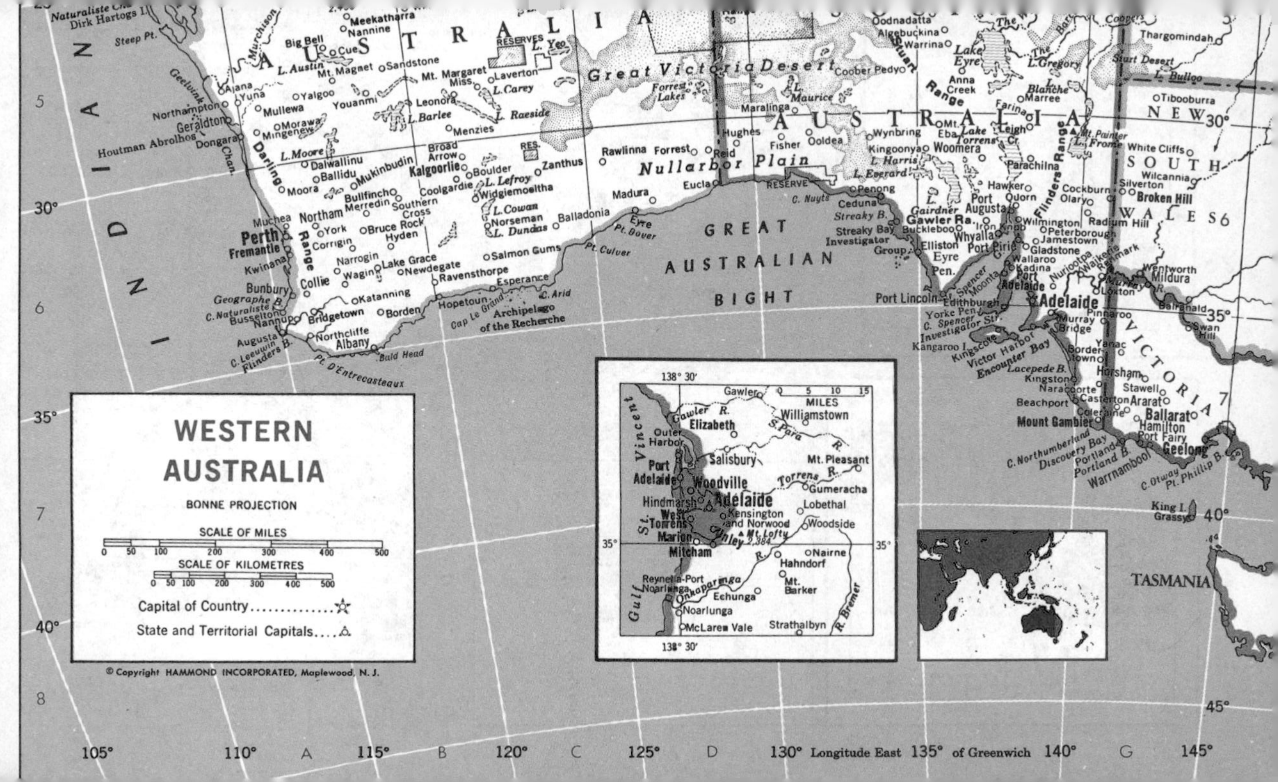

WESTERN AUSTRALIA
BONNE PROJECTION
SCALE OF MILES
0 50 100 200 300 400 500
SCALE OF KILOMETRES
0 50 100 200 300 400 500
Capital of Country
State and Territorial Capitals
© Copyright HAMMOND INCORPORATED, Maplewood, N. J.
INDIAN
AUSTRALIA
GREAT AUSTRALIAN BIGHT
SOUTH AUSTRALIA
NEW SOUTH WALES
VICTORIA
TASMANIA
Great Victoria Desert
Nullarbor Plain
Darling Range
Flinders Range
Stuart Range
Perth
Fremantle
Kwinana
Bunbury
Busselton
Albany
Geraldton
Northampton
Meekatharra
Kalgoorlie
Coolgardie
Norseman
Esperance
Archipelago of the Recherche
Eucla
Ceduna
Port Lincoln
Whyalla
Port Augusta
Port Pirie
Woomera
Lake Eyre
Adelaide
Port Adelaide
Broken Hill
Mildura
Mount Gambier
Ballarat
Geelong
Kangaroo I.
King I.
Gawler
Elizabeth
Salisbury
Williamstown
Woodville
Mt. Pleasant
Gumeracha
Lobethal
Woodside
Kensington and Norwood
Mt. Lofty 2,384
Mitcham
Marion
Hahndorf
Mt. Barker
Nairne
Echunga
Noarlunga
McLaren Vale
Strathalbyn
Gulf St. Vincent
MILES
0 5 10 15
138° 30′
35°
105° 110° 115° 120° 125° 130° Longitude East 135° of Greenwich 140° 145°
30° 35° 40° 45°
A B C D G
5 6 7 8

EASTERN AUSTRALIA and NEW ZEALAND
BONNE PROJECTION
SCALE OF MILES
SCALE OF KILOMETRES
Capital of Country
State and Territorial Capitals
ARAFURA SEA
GULF OF CARPENTARIA
Torres Str.
PAPUA NEW GUINEA
Cape York Pen.
Great Barrier Reef
CORAL SEA ISLANDS TERRITORY
NORTHERN TERRITORY
QUEENSLAND
Great Dividing Range
Gregory Range
Tropic of Capricorn
Simpson Desert
ABORIGINAL RESERVE
Groote Eylandt
Wellesley Is.
Mornington I.
Bentinck I.
Sir Edward Pellew Group
Vanderlin I.
Thursday I.
Prince of Wales I.
Wessel Is.
Cooktown
Cairns
Townsville
Mackay
Rockhampton
Gladstone
Bundaberg
Maryborough
Charters Towers
Mount Isa
Cloncurry
Longreach
Alice Springs
Willis Islets
Magdelaine Cays
Coringa Islets
Lihou Reef and Cays
Marion Reef
Frederick Reef
Kenn Reef
Swain Reefs
Saumarez Reef
Wreck Reef
Bird Islet
Cato I.
Capricorn Group
Bougainville Reef
Holmes Reef
Flinders Reef
PORT PHILLIP BAY
MELBOURNE
Geelong
Williamstown
St. Kilda
Sandringham
Frankston
Mornington
SYDNEY
Parramatta
Bankstown
Liverpool
Penrith
Sutherland
Wollongong
Port Kembla
Botany Bay
TASMAN SEA

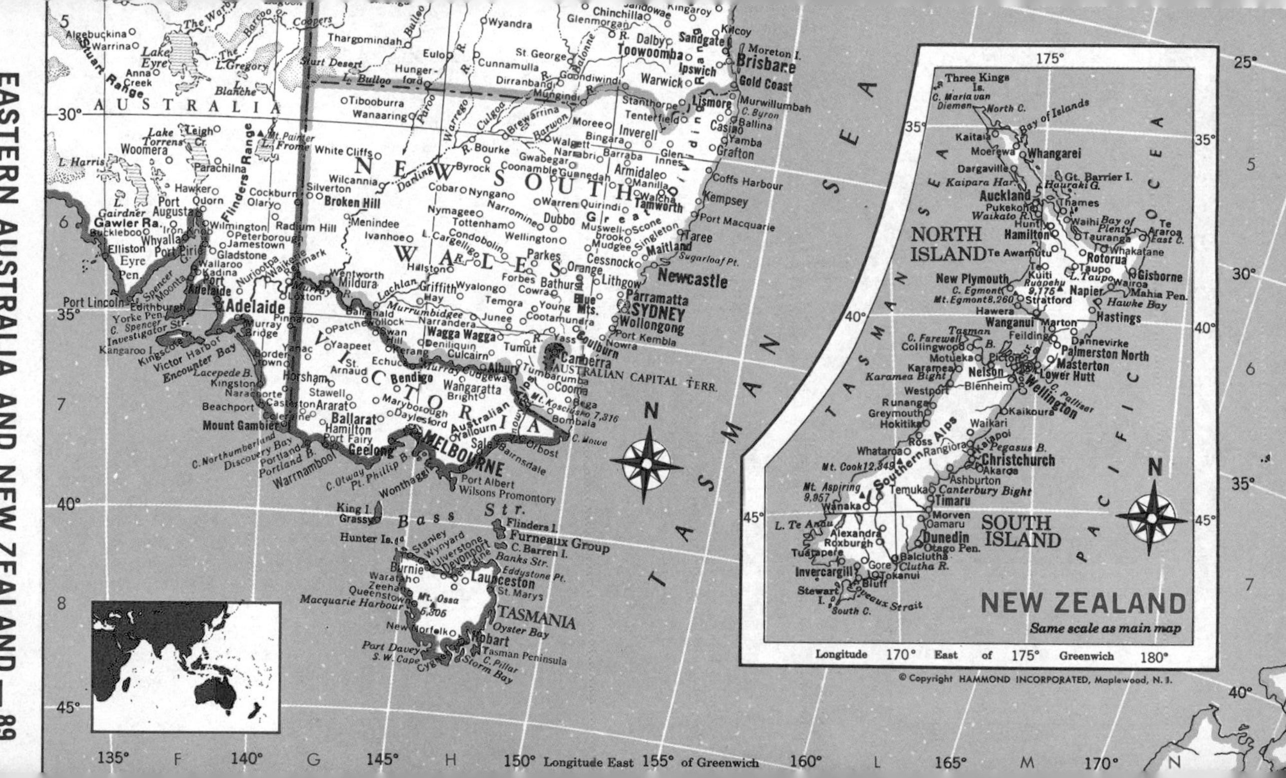

NEW ZEALAND
Same scale as main map
NORTH ISLAND
SOUTH ISLAND
TASMAN SEA
NEW SOUTH WALES
VICTORIA
AUSTRALIA
TASMANIA
Bass Str.
SYDNEY
MELBOURNE
Brisbane
Adelaide
Canberra
Hobart
Auckland
Wellington
Christchurch
Dunedin
Longitude 170° East of 175° Greenwich 180°
135° 140° 145° 150° Longitude East 155° of Greenwich 160° 165° 170°
© Copyright HAMMOND INCORPORATED, Maplewood, N.J.

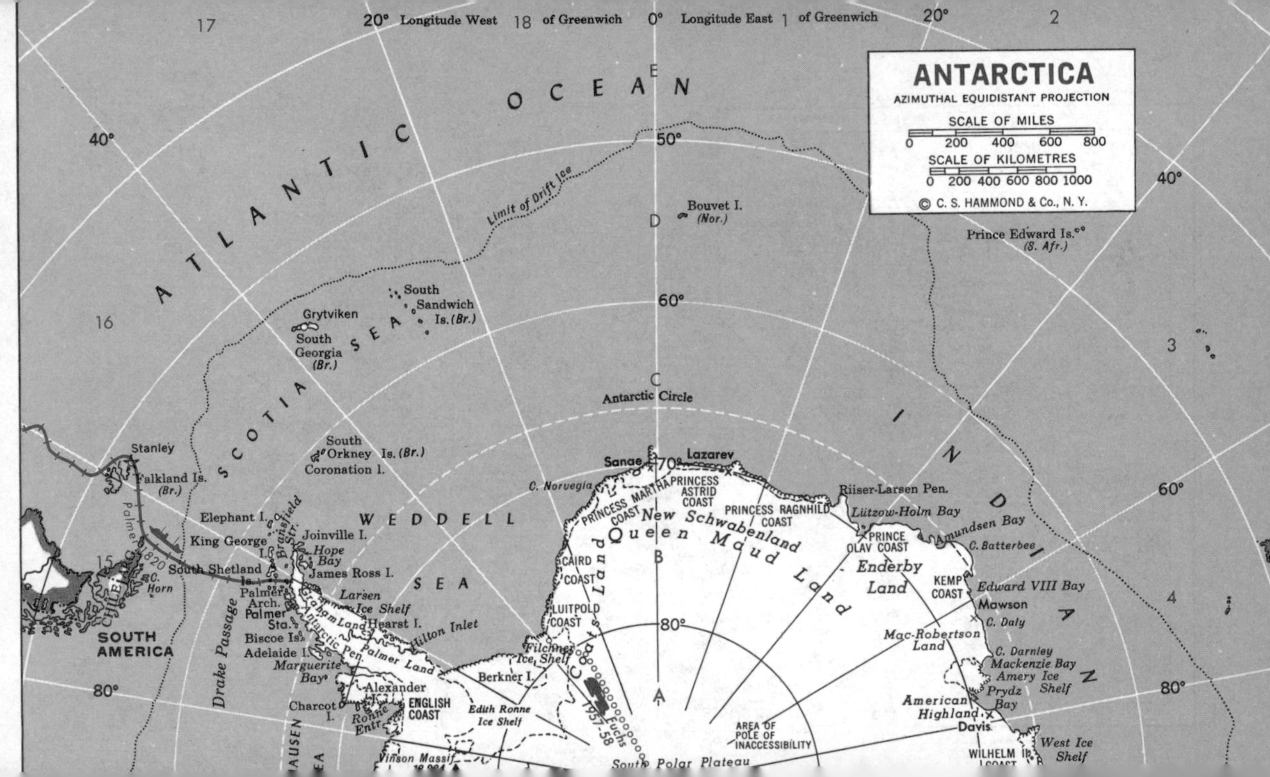

ANTARCTICA
AZIMUTHAL EQUIDISTANT PROJECTION
SCALE OF MILES
0 200 400 600 800
SCALE OF KILOMETRES
0 200 400 600 800 1000
© C. S. HAMMOND & Co., N. Y.
20° Longitude West 18 of Greenwich 0° Longitude East 1 of Greenwich
ATLANTIC OCEAN
INDIAN
SCOTIA SEA
WEDDELL SEA
Limit of Drift Ice
Antarctic Circle
Bouvet I. (Nor.)
Prince Edward Is. (S. Afr.)
South Sandwich Is. (Br.)
Grytviken
South Georgia (Br.)
South Orkney Is. (Br.)
Coronation I.
Stanley
Falkland Is. (Br.)
Palmer 1820
SOUTH AMERICA
C. Horn
Drake Passage
Elephant I.
King George I.
Bransfield Str.
South Shetland Is.
Joinville I.
Hope Bay
James Ross I.
Palmer Arch.
Palmer Sta.
Biscoe Is.
Adelaide I.
Marguerite Bay
Larsen Ice Shelf
Graham Land
Antarctic Pen.
Hearst I.
Hilton Inlet
Palmer Land
Alexander I.
Charcot I.
Ronne Entr.
ENGLISH COAST
Edith Ronne Ice Shelf
Berkner I.
Filchner Ice Shelf
Vinson Massif
LUITPOLD COAST
CAIRD COAST
Coats Land
Fuchs 1957-58
C. Norvegia
Sanae
Lazarev
PRINCESS MARTHA COAST
PRINCESS ASTRID COAST
PRINCESS RAGNHILD COAST
New Schwabenland
Queen Maud Land
Riiser-Larsen Pen.
Lützow-Holm Bay
PRINCE OLAV COAST
Amundsen Bay
C. Batterbee
Enderby Land
KEMP COAST
Edward VIII Bay
Mawson
C. Daly
Mac-Robertson Land
C. Darnley
Mackenzie Bay
Amery Ice Shelf
Prydz Bay
American Highland
Davis
West Ice Shelf
AREA OF POLE OF INACCESSIBILITY
South Polar Plateau

SOUTH POLE
Amundsen-Scott Sta.
Amundsen Dec. 14, 1911
Scott Jan. 17, 1912
Byrd Nov. 29, 1929 (airplane)
Fuchs Jan. 20, 1958
Peter I I. (Nor.)
COAST
Thurston I.
WALGREEN COAST
Hollick-Kenyon Plateau
Queen Maud Mts.
Byrd Sta.
Amundsen Sea
Marie Byrd Land
Mt. Siple 10,171
Executive Comm. Ra.
Mt. Sidley 13,717
Edward VII Pen.
Ford Ranges
C. Dart
Getz Ice Shelf
HOBBS COAST
Sulzberger Bay
C. Colbeck
Kainan Bay
Little America
Ross Ice Shelf
Roosevelt I.
Beardmore Glacier
Mt. Kirkpatrick 14,856
Mt. Markham 14,272
Mt. Lister 13,205
Scott
Ross I.
McMurdo Sd.
Victoria Land
Mt. Levick 9,101
Ross Sea
Amundsen 1911
Mt. Sabine 12,202
C. Adare
OATES COAST
GEORGE V. COAST
SOUTH MAGNETIC POLAR AREA
Ninnis Glacier Tongue
Mertz Glacier Tongue
ADELIE COAST
CLARIE COAST
C. Keltie
Dumont d' Urville
BANZARE COAST
C. Goodenough
Wilkes Land
SABRINA COAST
BUDD COAST
KNOX COAST
Vincennes Bay
Gaussberg
Davis Sea
QUEEN MARY COAST
Mirnyy
Farr Bay
Mt. Barr Smith 4,108
Shackleton Ice Shelf
Scott I.
Balleny Is.
Antarctic Circle
Byrd
1928-30
Scott 1910-13
Amundsen
Limit of Drift Ice
PACIFIC OCEAN
Macquarie I (Australia)
Campbell I. (N. Z.)
Auckland Is. (N. Z.)
Antipodes Is. (N. Z.)
Bounty Is. (N. Z.)
Stewart I.
Dunedin
NEW ZEALAND
Tasman Sea
Hobart
Tasmania
King I.
Furneaux Gr.
Bass Str.
Melbourne
AUSTRALIA
EXPLORERS' ROUTES
Palmer 1820
Amundsen 1910-12
Scott 1910-13
Byrd 1928-30
Fuchs 1957-58
By ship
By sledge
By airplane
By snow tractor
160° Longitude West of Greenwich
180°
Longitude East of Greenwich
160°

CARIBBEAN SEA
ATLANTIC OCEAN
PACIFIC
Equator
NETHERLANDS ANTILLES
Willemstad
WEST INDIES
BARBADOS
Bridgetown
GRENADA
TRINIDAD & TOBAGO
Port of Spain
Trinidad
Pta. Gallinas
G. de Venezuela
VENEZUELA
Caracas
Valencia
Barquisimeto
Maracaibo
Mérida
San Cristóbal
Pto. Cabello
Coro
Cumaná
Carúpano
Ciudad Bolívar
Ciudad Guayana
Puerto Ayacucho
Orinoco
GUYANA
Georgetown
SURINAM
Paramaribo
Nieuw Nickerie
FR. GUIANA
Cayenne
Oyapock
COLOMBIA
Bogotá
Barranquilla
Cartagena
Sta. Marta
Medellín
Montería
Manizales
Cali
Popayán
Pasto
Neiva
Tunja
Bucaramanga
Cúcuta
Pto. Carreño
Buenaventura
Magdalena
PANAMA
Panamá
Colón
CANAL ZONE (U. S.)
G. of Panamá
Malpelo I. (Col.)
ECUADOR
Quito
Guayaquil
Riobamba
Cuenca
G. of Guayaquil
PERU
Lima
Callao
Iquitos
Trujillo
Chiclayo
Piura
Sullana
Pta. de Aguja
Cajamarca
Huánuco
Huacho
Cerro de Pasco
Huancayo
Ayacucho
Pisco
Ica
Pta. Huacas
Arequipa
Mollendo
Puno
Tacna
Arica
Iquique
Ucayali
Marañon
Putumayo
Pto. Córdoba
BOLIVIA
La Paz
Sucre
Potosí
Oruro
L. Poopó
Cochabamba
Santa Cruz
Pto. Suárez
Trinidad
Cobija
Uyuni
Lake Titicaca
Beni
Mamoré
Guaporé
BRAZIL
Brasília
Belém
Mocajuba
Manaus
Itacoatiara
Santarém
Óbidos
Parintins
Macapá
I. de Maracá
I. de Marajó
São Luís
Parnaíba
Fortaleza
Mocoró
Natal
Recife
João Pessoa
Campina Grande
Caruarú
Maceió
Aracajú
Franca
Alagoinhas
Salvador (Bahia)
Jequié
Ilhéus
Juàzeiro do Norte
Teresina
Caxias
Belo Horizonte
Montes Claros
Carinhanha
Goiânia
Anápolis
Ipamerí
Araguarí
Goiás
Uberaba
Cuiabá
Corumbá
Aquidauana
Mato Grosso
Pôrto Nacional
Conceição do Araguaia
Pôrto Velho
Rio Branco
Benjamin Constant
Boa Vista
Amazon
Negro
Japurá
Madeira
Tapajós
Xingu
Tocantins
Araguaia
Parnaíba
São Francisco
Purus
Juruá
Branco
Roosevelt

SOUTH AMERICA
LAMBERT AZIMUTHAL
EQUAL-AREA PROJECTION
MILES
0 200 400 600
KILOMETRES
0 200 400 600
Capitals of Countries
International Boundaries
Copyright by C.S. HAMMOND & Co., N.Y.
Longitude West of Greenwich
OCEAN
ATLANTIC OCEAN
PARAGUAY
URUGUAY
Tropic of Capricorn
Tocopilla
Antofagasta
Taltal
Copiapó
La Serena
Ovalle
Valparaíso
Santiago
Rancagua
Talca
Concepción
Chillán
Angol
Temuco
Valdivia
Puerto Montt
Ancud
I. de Chiloé
ARCHIPIÉLAGO de los CHONOS
Aysén
I. Campana
I. Wellington
I. Madre de Dios
Santa Inés
I. Hoste
Cape Horn
Tierra del Fuego
Ushuaia
I. de los Estados
Punta Arenas
Río Gallegos
Bahía Grande
San Julián
Buenos Aires
C. Tres Puntas
G. San Jorge
Comodoro Rivadavia
Chubut
Pto. Madryn
San Carlos de Bariloche
Viedma
G. San Matías
Negro
Colorado
Curacó
Neuquén
Bahía Blanca
Mar del Plata
Tandil
Azul
Santa Rosa
La Plata
Buenos Aires
Avellaneda
C. San Antonio
Río de la Plata
Montevideo
Minas
Durazno
Melo
Paysandú
Salto
Mercedes
Paraná
Concordia
Rosario
Santa Fe
Córdoba
Villa María
Río Cuarto
Mercedes
San Juan
Aconcagua
Mendoza
S. Rafael
Salado
La Rioja
Catamarca
Santiago del Estero
Tucumán
Salta
Jujuy
Formosa
Resistencia
Corrientes
Goya
Posadas
Encarnación
Asunción
Concepción
Pto. Casado
Villarrica
San Pedro
Rosario
Uruguay
Paraná
Pelotas
L. dos Patos
Bagé
Santana do Livramento
Uruguaiana
Santa Maria
Pôrto Alegre
Caxias do Sul
Passo Fundo
Florianópolis
Paranaguá
Joinville
Curitiba
Londrina
Marília
São Paulo
Santos
Campinas
Juiz de Fora
Campos
C. de São Tomé
Niterói
Cabo Frio
Rio de Janeiro
FALKLAND ISLANDS
(IS. MALVINAS)
(Br. - claimed by Arg.)
Stanley
I. San Félix (Chile)
I. San Ambrosio (Chile)
I. Alejandro Selkirk
I. Robinson Crusoe
Juan Fernández Is. (Chile)
N

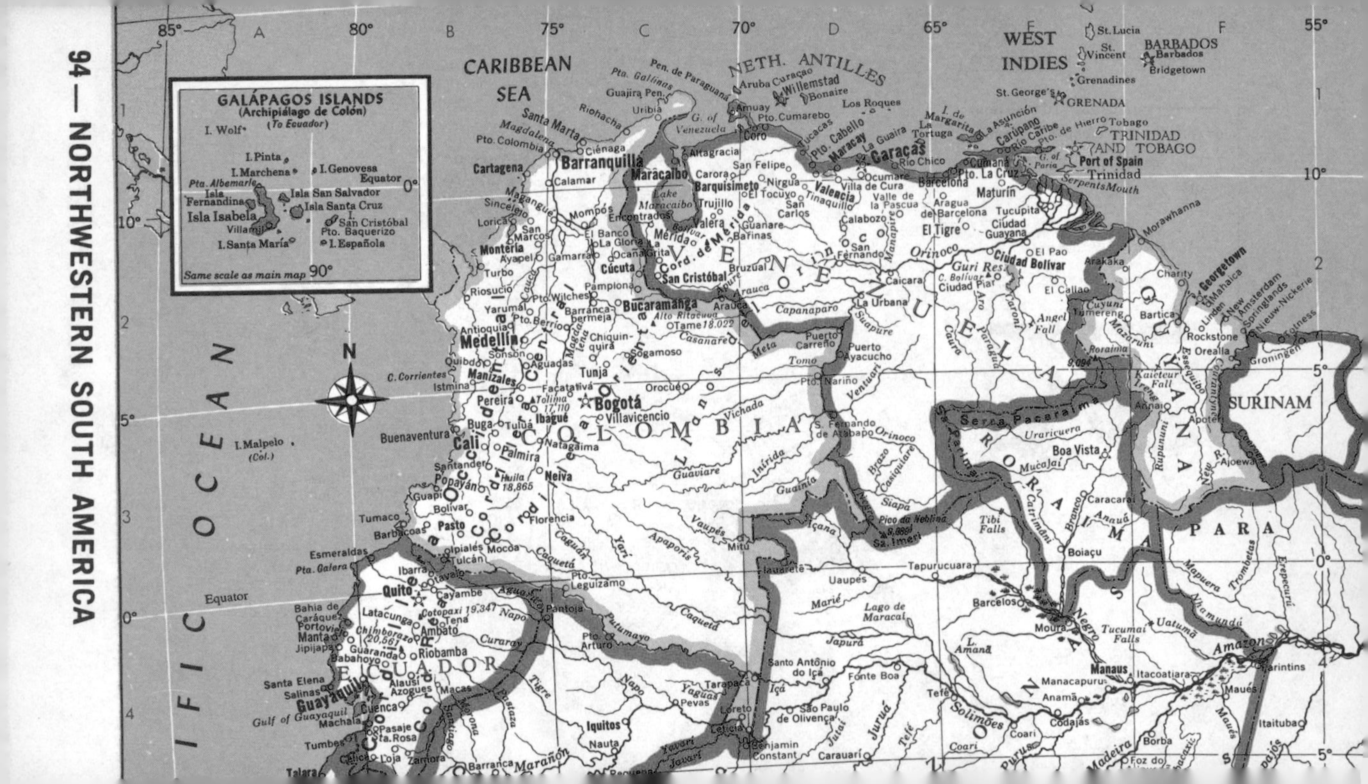

94 — NORTHWESTERN SOUTH AMERICA

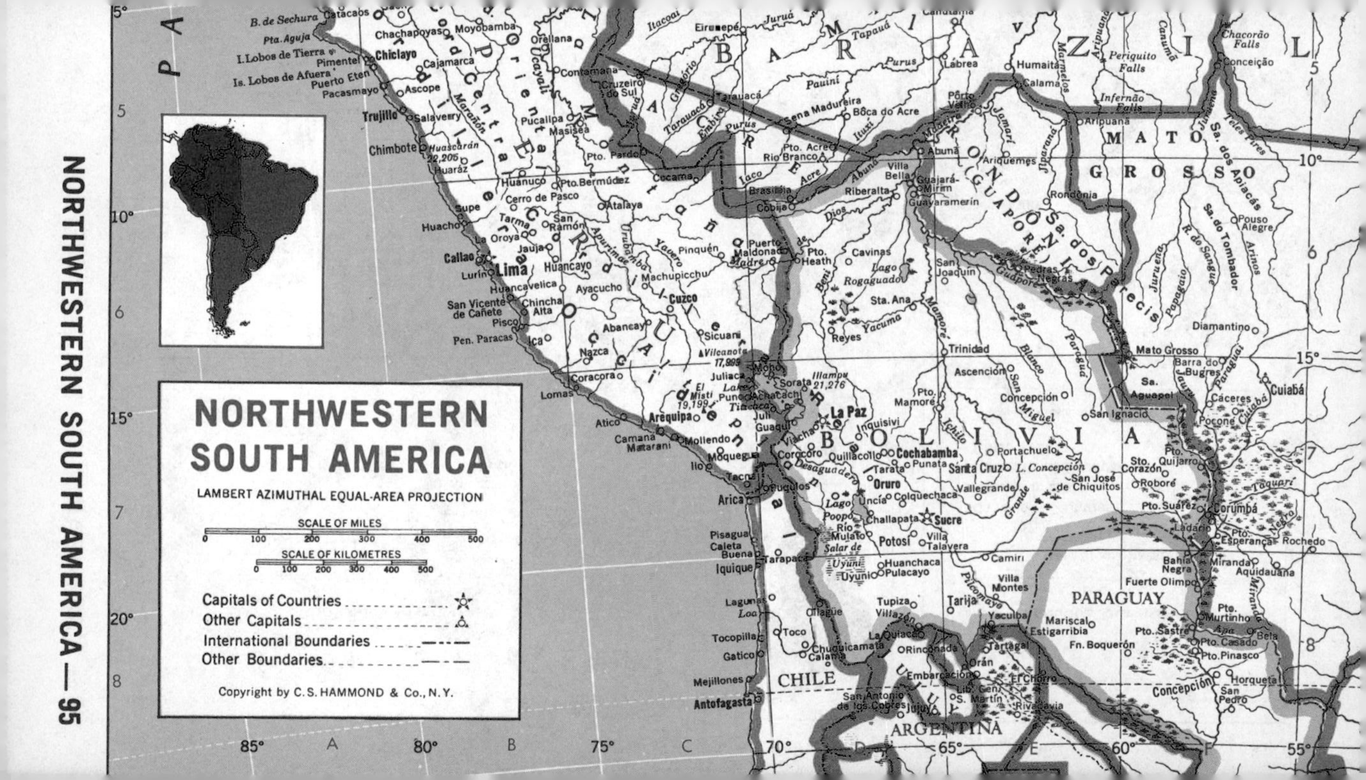

NORTHWESTERN SOUTH AMERICA
LAMBERT AZIMUTHAL EQUAL-AREA PROJECTION
SCALE OF MILES
0 100 200 300 400 500
SCALE OF KILOMETRES
0 100 200 300 400 500
Capitals of Countries
Other Capitals
International Boundaries
Other Boundaries
Copyright by C. S. HAMMOND & Co., N.Y.
BRAZIL
MATO GROSSO
ACRE
RONDÔNIA (GUAPORÉ)
BOLIVIA
PARAGUAY
CHILE
ARGENTINA
Lima
Callao
Cuzco
Arequipa
La Paz
Sucre
Cochabamba
Oruro
Potosí
Santa Cruz
Trinidad
Arica
Antofagasta
Iquique
Corumbá
Cuiabá
Concepción
Chiclayo
Trujillo
85° 80° 75° 70° 65° 60° 55°
5° 10° 15° 20°

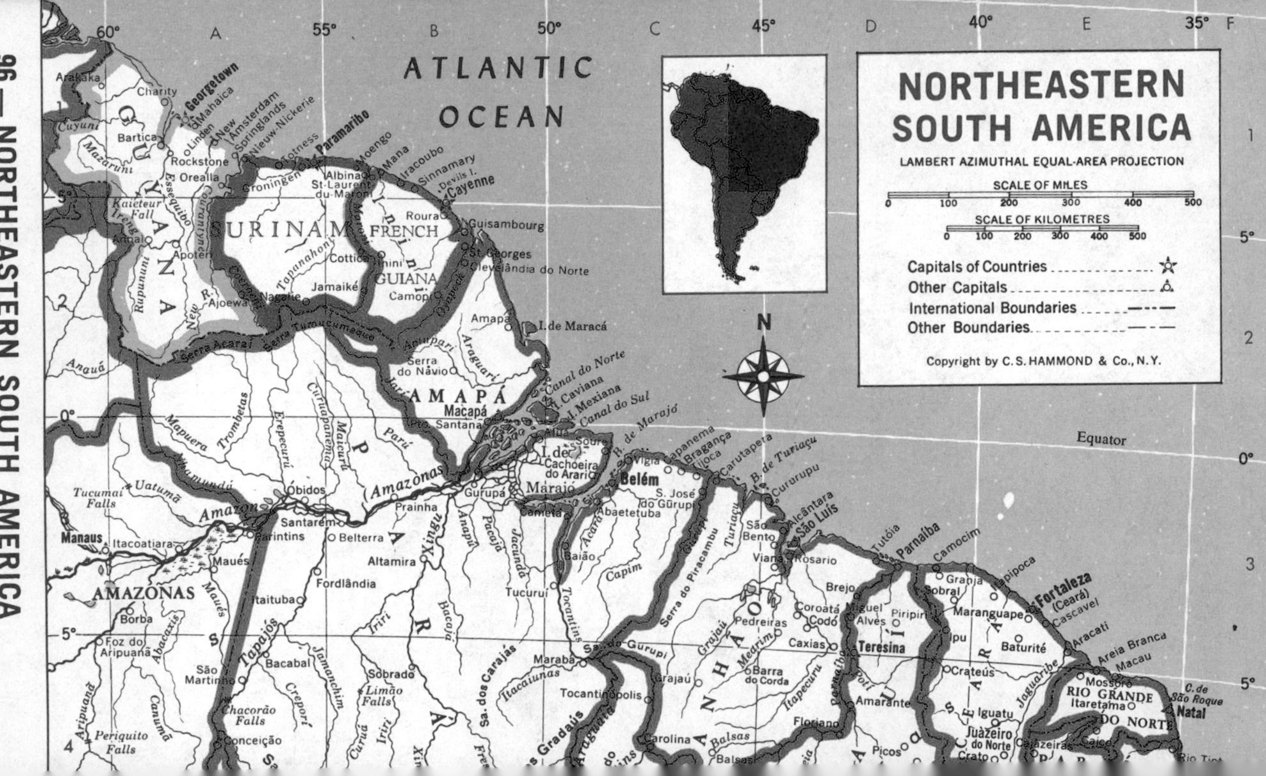
NORTHEASTERN SOUTH AMERICA
LAMBERT AZIMUTHAL EQUAL-AREA PROJECTION
SCALE OF MILES
0 100 200 300 400 500
SCALE OF KILOMETRES
0 100 200 300 400 500
Capitals of Countries
Other Capitals
International Boundaries
Other Boundaries
Copyright by C.S. HAMMOND & Co., N.Y.
ATLANTIC OCEAN
Equator
GUYANA
SURINAM
FRENCH GUIANA
AMAPÁ
PARÁ
AMAZONAS
MARANHÃO
PIAUÍ
CEARÁ
RIO GRANDE DO NORTE
Georgetown
Paramaribo
Cayenne
Macapá
Belém
São Luís
Teresina
Fortaleza
Natal
Manaus
Santarém
Amazonas
Tocantins
Xingu
Tapajós
I. de Marajó
I. de Maracá
Serra do Piracambu
Sa. dos Carajás
Serra Acaraí
Serra Tumucumaque

ATLANTIC OCEAN
Recife
(Pernambuco)
PERNAMBUCO
Maceió
ALAGOAS
SERGIPE
Aracaju
Salvador (Bahia)
B. de Todos-os-Santos
Ilhéus
Itabuna
Vitória da Conquista
Jequié
Feira de Santana
B A H I A
Juazeiro
Petrolina
Canavieiras
Belmonte
Pôrto Seguro
Caravelas
São Mateus
ESPÍRITO SANTO
Vitória
Campos
Cabo Frio
Niterói
RIO DE JANEIRO
Petrópolis
Juiz de Fora
Belo Horizonte
Gov. Valadares
Teófilo Otoni
Montes Claros
Diamantina
M I N A S   G E R A I S
São Francisco
Paracatu
Uberaba
Uberlândia
Ribeirão Prêto
Campinas
Jundiaí
SÃO PAULO
Santos
Sorocaba
S Ã O   P A U L O
Bauru
Marília
Araçatuba
Pres. Prudente
Londrina
Ponta Grossa
P A R A N Á
Castro
Tropic of Capricorn
DISTRITO FEDERAL
Brasília
Anápolis
Goiânia
G O I Á S
Catalão
Formosa
Planalto de Mato Grosso
Cuiabá
Corumbá
Campo Grande
Aquidauana
M A T O   G R O S S O
Sa. do Roncador
Sa. dos Xavantes
Ilha do Bananal
Xingu
Araguaia
Tocantins
Sa. dos Apiacás
Sa. do Tombador
Sa. Formosa
BOLIVIA
Concepción
Asunción
P A R A G U A Y
ARG.
Longitude West of 35° Greenwich

98 — CENTRAL AND SOUTHERN SOUTH AMERICA

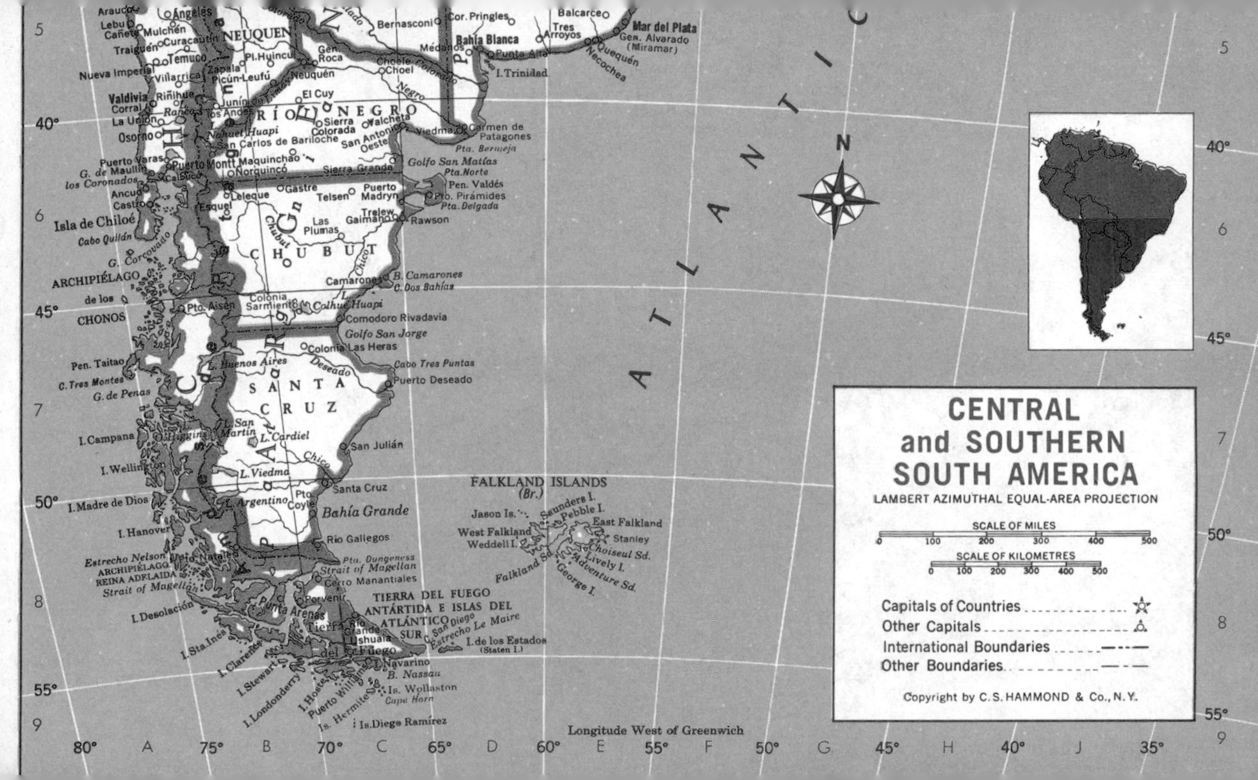
CENTRAL
and SOUTHERN
SOUTH AMERICA
LAMBERT AZIMUTHAL EQUAL-AREA PROJECTION
SCALE OF MILES
0 100 200 300 400 500
SCALE OF KILOMETRES
0 100 200 300 400 500
Capitals of Countries
Other Capitals
International Boundaries
Other Boundaries
Copyright by C.S. HAMMOND & Co., N.Y.
ATLANTIC
N
NEUQUEN
RÍO NEGRO
CHUBUT
SANTA CRUZ
FALKLAND ISLANDS
(Br.)
TIERRA DEL FUEGO
ANTÁRTIDA E ISLAS DEL ATLÁNTICO SUR
Mar del Plata
Bahía Blanca
Isla de Chiloé
Bahía Grande
Strait of Magellan
Cape Horn
Longitude West of Greenwich

ARCTIC OCEAN
North Pole
GREENLAND SEA
GREENLAND (Den.)
KING CHRISTIAN LAND
KING CHRISTIAN LAND
FREDERIK VI COAST
PEARY LAND
PRUDHOE LAND
LINCOLN SEA
Jan Mayen (Nor.)
Str. of Denmark
ICELAND
Reykjavik
Angmagssalik
C. Farewell
Julianehåb
Godthåb
Thule
Disko I.
Davis Strait
Baffin Bay
Baffin I.
Cumberland Sd.
Pangnirtung
Foxe Basin
Ellesmere I.
Devon I.
Lancaster Sd.
QUEEN ELIZABETH IS.
Axel Heiberg I.
SVERDRUP IS.
Prince Patrick I.
Melville I.
M'Clure Str.
Banks I.
Victoria I.
M'Clintock Chan.
Parry Channel
G. of Boothia
Boothia Pen.
Melville Pen.
Southampton I.
Coats I.
Mansel I.
Hudson Str.
Ungava Bay
C. Chidley
Labrador
Goose Bay
Hamilton Inlet
Battle Harbour
Str. of Belle Isle
Newfoundland
St. John's
C. Race
C. Breton I.
Sydney
G. of St. Lawrence
Charlottetown
Halifax
C. Sable
Saint John
St. Lawrence
Québec
Montréal
Ottawa
Toronto
L. Ontario
L. Huron
Portland
Boston
C. Cod
La Grande-Rivière
Nottaway
James Bay
BELCHER IS.
Hudson Bay
Moosonee
Timmins
Albany
L. Nipigon
Sault Ste. Marie
Thunder Bay
L. Superior
Duluth
St. Paul
L. Winnipeg
Nelson
Churchill
Chesterfield Inlet
Wager Bay
Winnipeg
Fargo
Bismarck
L. Manitoba
Prince Albert
Saskatoon
Saskatchewan
Regina
Medicine Hat
Reindeer L.
L. Athabasca
Athabasca
Fort McMurray
Ft. Smith
Great Slave L.
Yellowknife
Great Bear L.
Port Radium
Dease Str.
Amundsen Gulf
Edmonton
Calgary
Lethbridge
Great Falls
Yellowstone
Butte
Spokane
Seattle
Tacoma
Portland
Columbia
Vancouver
Victoria
Vancouver I.
Juan de Fuca Str.
Fraser
Peace
Liard
Prince Rupert
Mackenzie
Norman Wells
Inuvik
Aklavik
Ft. Simpson
Whitehorse
Dawson
Juneau
Ketchikan
Dixon Entr.
QUEEN CHARLOTTE IS.
ALEXANDER ARCH.
G. of Alaska
BEAUFORT SEA
Pt. Barrow
Wrangel I.
Bering Str.
Arctic Circle
ASIA
Anadyr'
St. Lawrence I.
BERING SEA
Norton Sd.
Seward Pen.
Nome
Koyukuk
Yukon
Fairbanks
UNITED STATES
ALASKA
Anchorage
Kuskokwim
Bristol Bay
Kodiak I.
Alaska Pen.
PACIFIC OCEAN
CANADA

NORTH AMERICA
LAMBERT AZIMUTHAL EQUAL-AREA PROJECTION
SCALE OF MILES
0 200 400 600 800 1000
SCALE OF KILOMETERS
0 200 400 600 800 1000
Capitals of Countries
International Boundaries
Canals
Copyright by C.S. HAMMOND & Co., N.Y.
UNITED STATES
MEXICO
CENTRAL AMERICA
SOUTH AMERICA
PACIFIC OCEAN
ATLANTIC
Gulf of Mexico
CARIBBEAN SEA
Gulf of California
Lower California
BAHAMAS
CUBA
JAMAICA
HAITI
DOMINICAN REP.
PUERTO RICO (U.S.)
GUATEMALA
EL SALVADOR
NICARAGUA
COSTA RICA
PANAMA
BELIZE
CANAL ZONE (U.S.)
GALÁPAGOS IS. (Ecuador)
Tropic of Cancer
Equator
Longitude 100° West of Greenwich 90°
Bermuda (Br.) Hamilton
Philadelphia
Chicago
Los Angeles
San Francisco
Mexico City
Havana

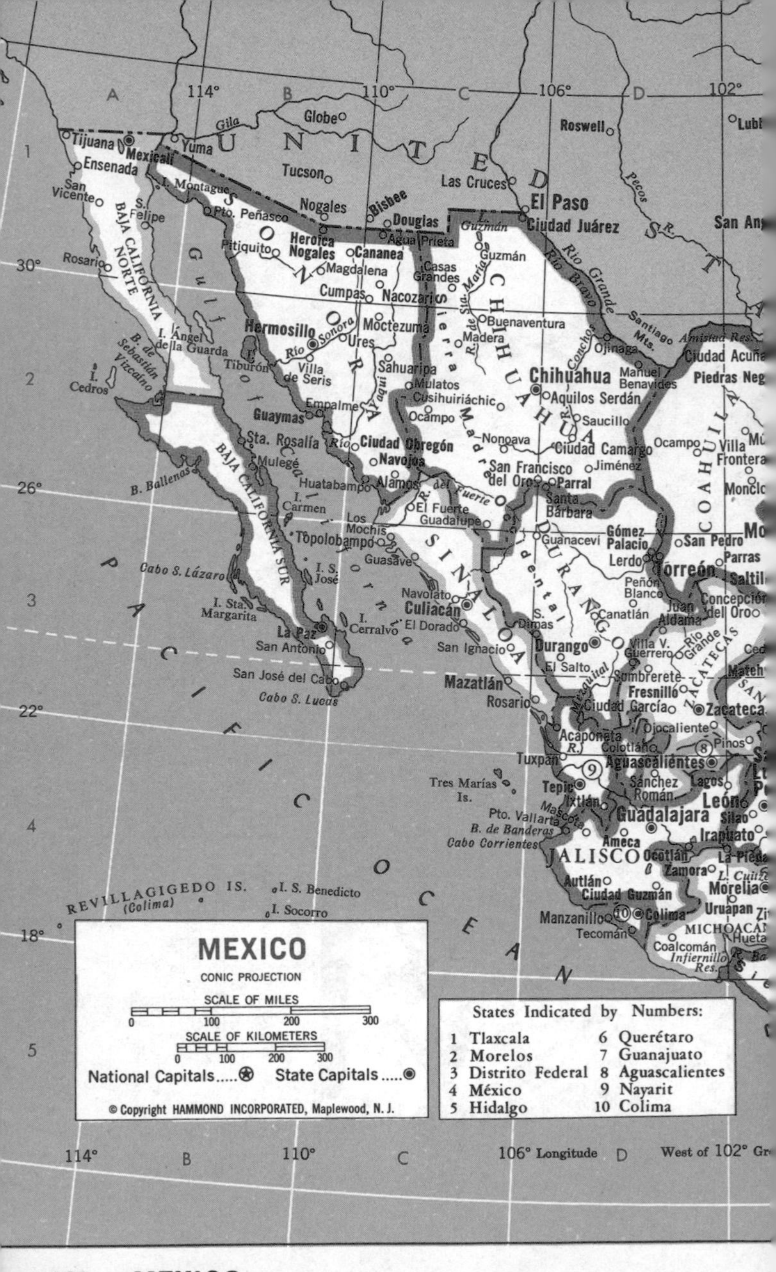

MEXICO
CONIC PROJECTION
SCALE OF MILES
0 100 200 300
SCALE OF KILOMETERS
0 100 200 300
National Capitals.....
State Capitals.....
© Copyright HAMMOND INCORPORATED, Maplewood, N. J.
States Indicated by Numbers:
1 Tlaxcala
2 Morelos
3 Distrito Federal
4 México
5 Hidalgo
6 Querétaro
7 Guanajuato
8 Aguascalientes
9 Nayarit
10 Colima
UNITED STATES
PACIFIC OCEAN
Gulf of California
BAJA CALIFORNIA NORTE
BAJA CALIFORNIA SUR
SONORA
CHIHUAHUA
SINALOA
DURANGO
COAHUILA
ZACATECAS
JALISCO
MICHOACAN
Sierra Madre Occidental
REVILLAGIGEDO IS. (Colima)
I. S. Benedicto
I. Socorro
Tres Marías Is.
Tijuana
Mexicali
Ensenada
San Vicente
S. Felipe
Rosario
Yuma
Gila
Globe
Tucson
Nogales
Bisbee
Douglas
Agua Prieta
Heroica Nogales
Cananea
Pitiquito
Magdalena
Cumpas
Nacozari
Moctezuma
Hermosillo
Río Sonora
Ures
Villa de Seris
Sahuaripa
Tiburón
I. Ángel de la Guarda
B. de Sebastián Vizcaíno
I. Cedros
Guaymas
Empalme
Yaqui
Sta. Rosalía
Mulegé
B. Ballenas
Ciudad Obregón
Navojoa
Álamos
Huatabampo
I. Carmen
Los Mochis
Topolobampo
Guasave
I. S. José
Cabo S. Lázaro
I. Sta. Margarita
La Paz
San Antonio
San José del Cabo
Cabo S. Lucas
I. Cerralvo
Navolato
Culiacán
El Dorado
San Ignacio
Mazatlán
Rosario
El Fuerte
Guadalupe
R. del Fuerte
Roswell
Las Cruces
El Paso
Ciudad Juárez
Pecos
Río Grande
Río Bravo
Guzmán
Casas Grandes
Sta. María
Buenaventura
Madera
Santiago Mts.
Conchos
Ojinaga
Chihuahua
Aquilos Serdán
Mulatos
Cusihuiriáchic
Ocampo
Saucillo
Manuel Benavides
Nonoava
Ciudad Camargo
Jiménez
San Francisco del Oro
Parral
Santa Bárbara
Gómez Palacio
Lerdo
Guanaceví
Torreón
San Pedro
Peñón Blanco
Canatlán
Juan Aldama
S. Dimas
Durango
El Salto
Villa V. Guerrero
Sombrerete
Fresnillo
Ciudad García
Zacatecas
Ojocaliente
Pinos
Acaponeta
Colotlán
Aguascalientes
Tuxpan
Tepic
Ixtlán
Sánchez Román
Lagos
León
Silao
Irapuato
Guadalajara
Mascota
Pto. Vallarta
B. de Banderas
Cabo Corrientes
Ameca
Ocotlán
Zamora
La Piedad
Autlán
Ciudad Guzmán
Morelia
Uruapan
Manzanillo
Colima
Tecomán
Coalcomán
Infiernillo Res.
Amistad Res.
Ciudad Acuña
Piedras Negras
Ocampo
Villa Frontera
Monclova
Parras
Saltillo
Concepción del Oro
Lubbock
San Angelo
114°
110°
106°
102°
30°
26°
22°
18°
106° Longitude
West of 102° Gr

GULF OF MEXICO
Bay of Campeche
Tropic of Cancer
Campeche Bank
Gulf of Tehuantepec
Gulf of Honduras
MEXICO CITY
Puebla
Veracruz
Jalapa
Orizaba
Tampico
Mérida
Campeche
Oaxaca
Guatemala
Belize City
Brownsville
Matamoros
Corpus Christi
Laredo
Ciudad Victoria
YUCATAN
CAMPECHE
TABASCO
CHIAPAS
OAXACA
QUINTANA ROO
GUATEMALA
BELIZE
HONDURAS
TLAXCALA
MORELOS
HIDALGO
GULF OF MEXICO
MILES

CENTRAL AMERICA
CONIC PROJECTION
SCALE OF MILES
0 25 50 100 150
SCALE OF KILOMETERS
0 25 50 100 150
Capitals of Countries
International Boundaries
Canals
Copyright by C.S. HAMMOND & Co., N.Y.
MEXICO
GUATEMALA
HONDURAS
BELIZE
EL SALVADOR
NICARAGUA
PACIFIC OCEAN
Palenque
Tenosique
Ciudad de las Casas (San Cristóbal)
Comitán
Usumacinta
Jataté
Grijalva
Sierra Madre
L. Petén-Itzá
San Andres
Flores
Maya Mts.
Belmopan
Belize City
Cayo
Stann Creek
Corozal
Orange Walk
Hondo
Ambergris Cay
Turneffe I.
Punta Gorda
San Luis
Gulf of Honduras
Chinaja
Pto. Barrios
Jacaltenango
Huehuetenango
San Pedro Carchá
Cobán
L. de Izabal
Motagua
Totonicapán
Comalapa
Zacapa
Tapachula
Guatemala
Antigua
Chiquimula
Jalapa
Quezaltenango
Retalhuleu
Mazatenango
L. de Atitlán
Escuintla
Jutiapa
Ahuachapán
San José
Sta. Rosa de Copán
Nueva Ocotepeque
Gracias
San Pedro Sula
Ulúa
El Progreso
Tela
La Ceiba
Aguán
Yoro
Sta. Bárbara
L. de Yojoa
Comayagua
La Paz
Tegucigalpa
Yuscarán
Danlí
Juticalpa
Catacamas
Trujillo
Iriona
Pto. Cortes
Utila
Roatán
Islas de la Bahía
C. de Honduras
Sa. de Esperanza
Sico
Patuca
Brus Laguna
Lempa
Sta. Ana
Sonsonate
Sta. Tecla (Nva. S. Salvador)
San Salvador
Cojutepeque
San Vicente
San Miguel
Amapala
G. de Fonseca
Nacaome
Choluteca
Ocotal
Jalapa
Somoto
Cord. Isabella
Jinotega
Estelí
Matagalpa
Grande
Chinandega
León
L. de Managua
Boaco
Mts. de Huapí
Managua
Juigalpa
Masaya
Granada
Jinotepe
Lake Nicaragua
I. de Ometepe
Rivas
San Juan del Sur
G. de Papagayo
Liberia
Canas
C. Velas
Sta. Cruz
Puntarenas
Alajuela
C. Blanco
G. de Nicoya
Guardian Bank
A 92° B 88° C
Longitu
16°
12°
8°
1
2
3
4

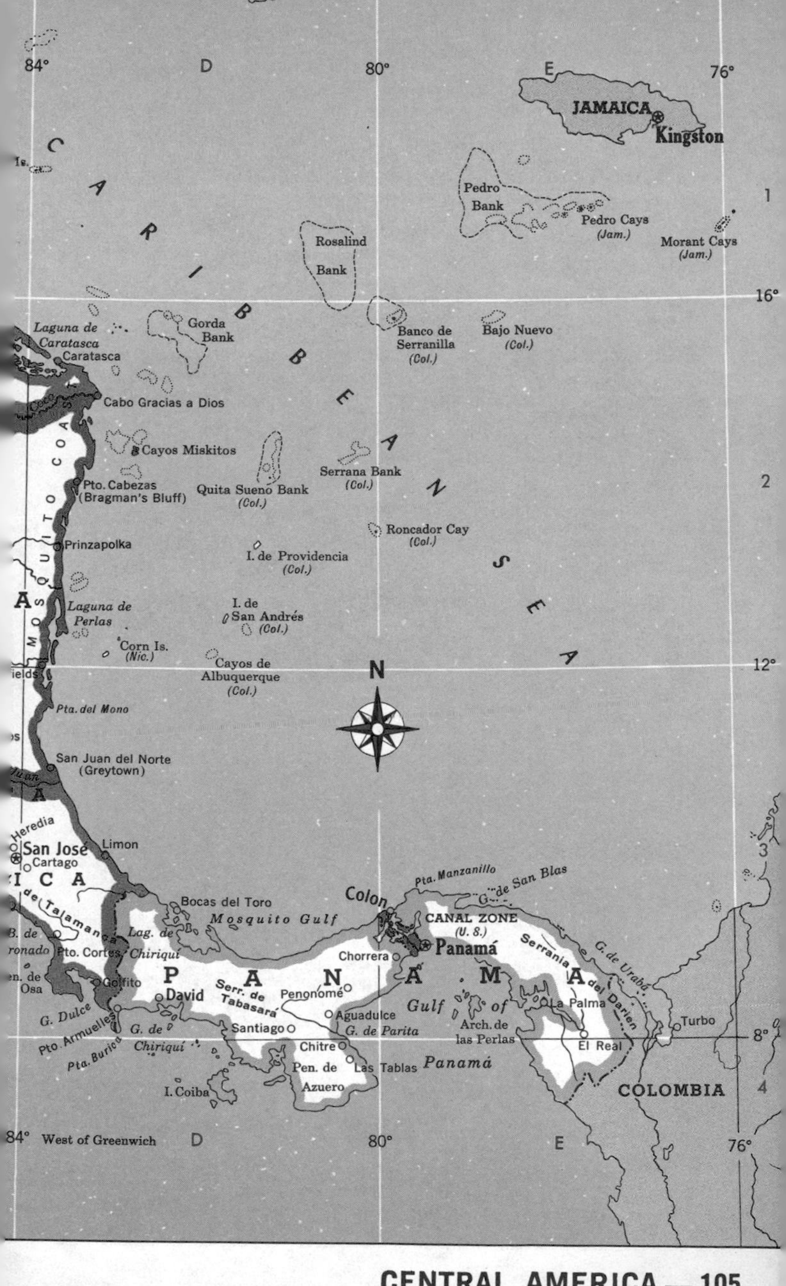

JAMAICA
Kingston
Pedro Bank
Pedro Cays
(Jam.)
Morant Cays
(Jam.)
Rosalind Bank
CARIBBEAN SEA
Laguna de Caratasca
Caratasca
Gorda Bank
Banco de Serranilla
(Col.)
Bajo Nuevo
(Col.)
Cabo Gracias a Dios
Cayos Miskitos
MOSQUITO COAST
Pto. Cabezas
(Bragman's Bluff)
Quita Sueno Bank
(Col.)
Serrana Bank
(Col.)
Roncador Cay
(Col.)
Prinzapolka
I. de Providencia
(Col.)
Laguna de Perlas
I. de San Andrés
(Col.)
Corn Is.
(Nic.)
Cayos de Albuquerque
(Col.)
Pta. del Mono
San Juan del Norte
(Greytown)
Heredia
San José
Cartago
Limon
Pta. Manzanillo
G. de San Blas
Colon
Bocas del Toro
Mosquito Gulf
CANAL ZONE
(U. S.)
Panamá
Lag. de Chiriquí
Chorrera
Serrania del Darien
G. de Urabá
PANAMA
Pto. Cortes
Golfito
David
Serr. de Tabasará
Penonomé
Gulf of Panamá
Arch. de las Perlas
La Palma
Aguadulce
G. de Parita
Turbo
G. Dulce
Pto. Armuelles
Pta. Burica
G. de Chiriquí
Santiago
Chitre
Las Tablas
El Real
Pen. de Azuero
I. Coiba
COLOMBIA
84°
80°
76°
16°
12°
8°
West of Greenwich

GULF OF MEXICO
UNITED STATES
L. Okeechobee
W. Palm Beach
Miami
Cape Sable
Dry Tortugas
Key West
FLORIDA KEYS
Straits of Florida
Grand Bahama
The Biminis
Berry Is.
Great Abaco
N. E. Providence Chan.
Harbour I.
Nassau
Eleuthera
New Providence I.
Andros I.
Tongue of the Ocean
Exuma Sd.
Great Exuma
Cat I.
San Salvador (Watling I.)
Long I.
Samana
Crooked I.
Acklins I.
Mayaguana
Ragged I.
Caicos Pass
Caicos Is.
Grand Caicos
TURKS AN
Turks Is
Little Inagua
Cockburn Har.
Great Inagua
BAHAMA
ATLANTIC
Tropic of Cance
Great Bahama Bank
Santarén Chan.
HAVANA
Mariana
Guanabacoa
Matanzas
Cárdenas
Guanajay
Artemisa
Pinar del Río
Güines
Colón
Sagua la Grande
Caibarién
Cayo Romano
Santa Clara
Morón
Sancti-Spíritus
Cienfuegos
Trinidad
Guane
G. de Batabanó
Nueva Gerona
Cabo de San Antonio
I. of Pines
Cayo Largo
CUBA
Camagüey
Nuevitas
Gibara
Holguín
Antilla
Guantánamo
Jardines de la Reina
G. de Guacanayabo
Manzanillo
Bayamo
San Luis
Baracoa
C. Maisí
I. de la Tortuga
Santiago de Cuba
(U.S. Base)
Windward Pass
Little Cayman
Grand Cayman
Cayman Brac
Georgetown
CAYMAN IS. (Br.)
C. Cruz
GREATER
Cap-Haïtien
Port-de-Paix
Gonaïves
Île de la Gonâve
HAITI
Monte-cristi
La Vega
DOMINIC REPUBL
Puer Plat
Port-au-Prince
C. Dame Marie
Navassa I. (U.S.)
Jérémie
Les Cayes
Petit-Goâve
Jacmel
Neiba
Baní
Barahona
C. Beata
HISPA
ANTI
Jamaica Chan.
Montego Bay
St. Anns Bay
Port Antonio
Savanna la Mar
Spanish Town
Kingston
Morant Point
Portland Point
JAMAICA
Swan Is. (Hond.)
Pedro Bank
Rosalind Bank
HONDURAS
C. Gracias a Dios
CARIBBEA
NICARAGUA
Puerto Cabezas (Bragmans Bluff)
Prinzapolca
Quita Sueño Bank
I. de Providencia (Col.)
I. San Andrés
Corn Is.
Cayos de Albuquerque (Col.)
Bluefields
N
NET
Aruba
Wille
Pta. Gallinas
Amuay
Ríohacha
G. of Venezuela
Santa Marta
Ciénaga
Barranquilla
Cartagena
Calamar
Maracaibo
Altagracia
L. Maracaibo
La Ceiba
Barqu
Santa Bárbara
COLOMBIA
Mompós
Magangué
Encontrados
Lorica
Montería
G. de Urabá
Limón
COSTA RICA
L. de Chiriquí
CANAL ZONE (U.S.)
Colón
PANAMA
Belén
Balboa
Panamá
David
VE
Longitude
A B C D
82° 78° 74°
24° 20° 16° 12°
1 3 4 5

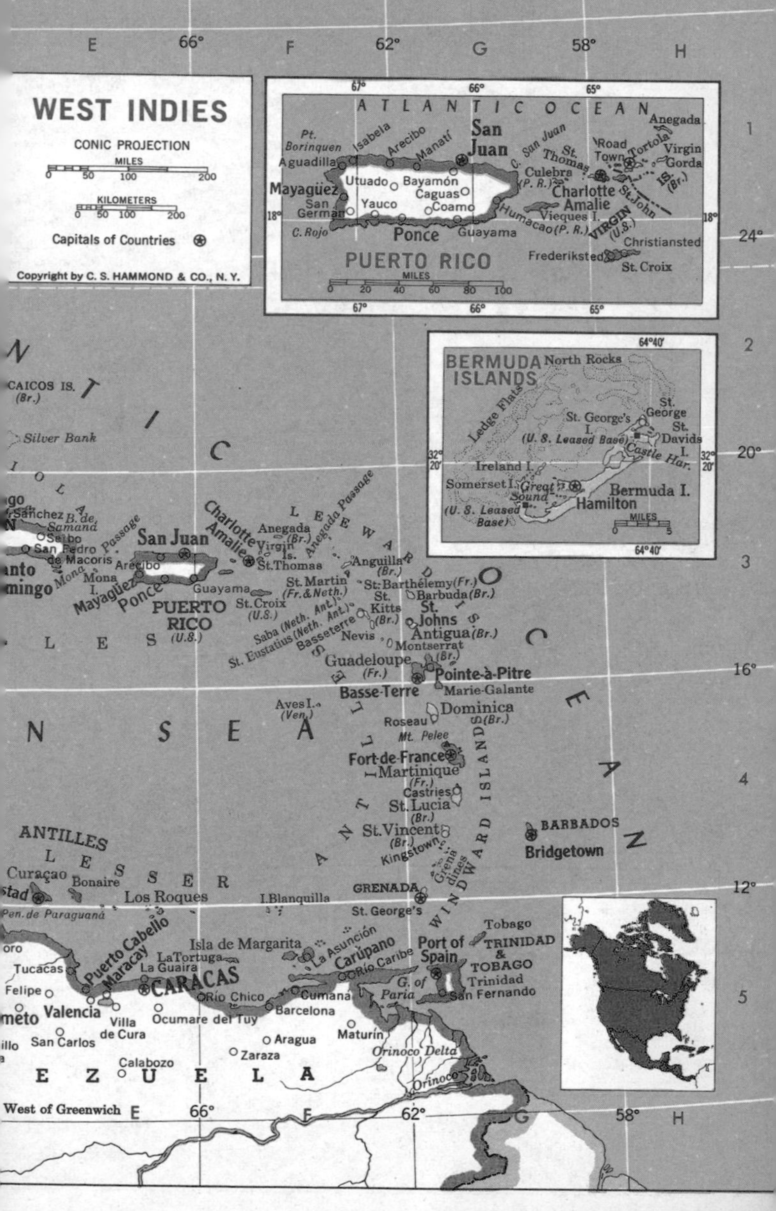

WEST INDIES
CONIC PROJECTION
MILES
KILOMETERS
Capitals of Countries
Copyright by C. S. HAMMOND & CO., N. Y.
ATLANTIC OCEAN
PUERTO RICO
San Juan
Mayagüez
Ponce
Aguadilla
Arecibo
Manatí
Isabela
Utuado
Bayamón
Caguas
Yauco
Coamo
San German
Guayama
Humacao (P. R.)
Culebra (P. R.)
Vieques I.
Charlotte Amalie
St. Thomas
St. John
Road Town
Tortola
Virgin Gorda
Anegada
VIRGIN IS.
Christiansted
Frederiksted
St. Croix
BERMUDA ISLANDS
North Rocks
Ledge Flats
St. George's I.
St. George
St. Davids I.
Castle Har.
(U. S. Leased Base)
Ireland I.
Somerset I.
Great Sound
Bermuda I.
Hamilton
CAICOS IS. (Br.)
Silver Bank
Samaná
Seibo
San Pedro de Macoris
Mona Passage
Mona I.
Anegada Passage
LEEWARD ISLANDS
Anguilla (Br.)
St. Martin (Fr. & Neth.)
St. Barthélemy (Fr.)
Barbuda (Br.)
St. Kitts (Br.)
Nevis
St. Johns
Antigua (Br.)
Saba (Neth. Ant.)
St. Eustatius (Neth. Ant.)
Basseterre
Montserrat (Br.)
Guadeloupe (Fr.)
Pointe-à-Pitre
Basse-Terre
Marie-Galante
Aves I. (Ven.)
Dominica (Br.)
Roseau
Mt. Pelee
Fort-de-France
Martinique (Fr.)
Castries
St. Lucia (Br.)
St. Vincent (Br.)
Kingstown
Grenadines
WINDWARD ISLANDS
BARBADOS
Bridgetown
GRENADA
St. George's
LESSER ANTILLES
Curaçao
Bonaire
Los Roques
I. Blanquilla
Pen. de Paraguaná
Puerto Cabello
Maracay
La Guaira
CARACAS
Isla de Margarita
La Asunción
La Tortuga
Carúpano
Río Caribe
Port of Spain
Tobago
TRINIDAD & TOBAGO
Trinidad
San Fernando
G. of Paria
Cumaná
Barcelona
Río Chico
Tucacas
Valencia
Villa de Cura
Ocumare del Tuy
San Carlos
Calabozo
Aragua
Zaraza
Maturín
Orinoco Delta
Orinoco
VENEZUELA
West of Greenwich

A
B
C
D
E
F
G
150°
140°
130°
120°
110°
100°
60°
50°
40°
3
4
5
6
7
8
BEAUFORT SEA
Banks Island
Melville I.
Bathurst I.
Cornwallis I.
PARRY
N. MAGNETIC POLE
Vis. Melville Sd.
Somerset I.
Prince of Wales I.
McClintock Channel
Amundsen Gulf
Holman I.
DISTRICT
Victoria I.
Boothia Pen.
Read I.
Cambridge Bay
Dease Str.
King William I.
Gjoa Haven
Victoria St.
Yukon
UNITED STATES
ALASKA
Fairbanks
Tanana
Circle
Porcupine
Fort Yukon
Arctic Circle
Mackenzie Bay
Aklavik
Tuktoyaktuk
Inuvik
Arctic Red River
Paulatuk
Peel
Stewart River
Dawson
Mayo
YUKON
Fort Selkirk
Pelly
Mt. St. Elias 18,008
Mt. Logan 19,850
Mt. Fairweather 15,300
Whitehorse
Skagway
Teslin L.
Watson Lake
Liard
Juneau
Sitka
Mackenzie
Norman Wells
Ft. Norman
Great Bear Lake
Coppermine
Bathurst Inlet
NORTHWEST
DISTRICT OF MACKENZIE
Port Radium
Wrigley
Arctic Circle
Garry L.
Baker Lake
Lac la Martre
Rae
Yellowknife
Ft. Simpson
Great Slave Lake
Ft. Providence
Ft. Reliance
Dubawnt L.
Chesterfield Inlet
Padlei
Eskimo
Trout L.
Ft. Liard
Fort Nelson
Hay River
Ft. Resolution
Ft. Smith
Churchill Pk. 10,500
Hay
Ft. Vermilion
L. Athabasca
Uranium City
Kasba L.
Nueltin L.
COAST
ROCKY
BRITISH
COLUMBIA
Prince of Wales I.
Prince Rupert
Ketchikan
Stewart
Queen Charlotte Is.
Hecate St.
Queen Charlotte Sd.
Smithers
Terrace
Kitimat
Burns Lake
Vanderhoof
Prince George
Quesnel
Ft. St. John
Peace
Dawson Cr.
Grande Prairie
Peace River
Ft. Chipewyan
Wollaston L.
Cree L.
Fort McMurray
Athabasca
Reindeer L.
Brochet
Lynn Lake
Churchill
Amery
Nelson
Church
ALBERTA
Edmonton
St. Paul
Jasper
Edson
Vermilion
Wetaskiwin
Leduc
Camrose
Lloydminster
Meadow Lake
SASKATCHEWAN
Flin Flon
Sherridon
MANITOBA
Mt. Waddington 13,260
Vancouver I.
Courtenay
Nanaimo
Juan de Fuca Str.
Kamloops
Lake Louise
Red Deer
North Battleford
Prince Albert
The Pas
Melfort
Humboldt
Winnipegosis
Winnipeg
Kelowna
Vernon
Revelstoke
Banff
Calgary
Drumheller
Biggar
Saskatoon
Berens River
Vancouver
New Westminster
Victoria
Cranbrook
Nelson
Medicine Hat
Swift Current
Yorkton
Melville
Kamsack
Seattle
WASH.
Spokane
IDA.
MONT.
Lethbridge
Cardston
Moose Jaw
Regina
Brandon
Dauphin
L. Manitoba
Red Lake
Portland
Columbia
Shaunavon
Weyburn
Estevan
Souris
St. Boniface
Morden
Kenora
L. of the Woods
Steep Rock
Rainy River
Fort Frances
N. DAK.
Bismarck
Fargo
Duluth
MINN.
S. DAK.
Pierre
Minneapolis
St. P
UNITED S
Missouri
Niobrara
NEBR.
IOWA
Longitude
West
82°
80°
43°
Walkerton
Newmarket
Orangeville
Cobourg
Lake Huron
Listowel
Fergus
Oshawa
Goderich
Toronto
Kitchener
Guelph
Brampton
Ontario
Galt
Hamilton
St. Catharines
MICH.
Stratford
Niagara Falls
Port Huron
Woodstock
Brantford
Sarnia
London
Simcoe
Welland
Niagara Falls
St. Thomas
Welland Canal
Buffalo
N.Y.
Wallaceburg
Detroit
Windsor
Chatham
Long Pt.
Miles
0 20 40
Sandwich
Leamington
Lake Erie
Erie
Pt. Pelee

CANADA
CONIC PROJECTION
SCALE OF MILES
0 100 200 300 400 500
SCALE OF KILOMETRES
0 100 200 300 400 500
Capitals of Countries
Provincial & Territorial Capitals
Canals
Copyright by C. S. Hammond & Co., N. Y.
BAFFIN BAY
DAVIS STRAIT
GREENLAND
BAFFIN ISLAND
FOXE BASIN
ATLANTIC OCEAN
HUDSON BAY
QUÉBEC
NEWFOUNDLAND
NEW BRUNSWICK
NOVA SCOTIA
PRINCE EDWARD I.
ST. PIERRE & MIQUELON (Fr.)
Gulf of St. Lawrence
Montreal
Quebec
Ottawa
Toronto
Halifax
Sherbrooke
Laval
Valleyfield
Granby
Magog

PACIFIC OCEAN
CANADA
MEXICO
WASHINGTON
OREGON
CALIFORNIA
IDAHO
NEVADA
MONTANA
WYOMING
UTAH
ARIZONA
COLORADO
NEW MEXICO
NORTH DAKOTA
SOUTH DAKOTA
NEBRASKA
ALASKA
HAWAII
Seattle
Portland
San Francisco
Los Angeles
San Diego
Salt Lake City
Denver
Phoenix
Tucson
Albuquerque
El Paso
Las Vegas
Reno
Sacramento
Boise
Helena
Cheyenne
Santa Fe
Carson City
Honolulu
Anchorage
Fairbanks
Juneau

UNITED STATES
POLYCONIC PROJECTION
SCALE OF MILES
0 100 200 300 400
SCALE OF KILOMETERS
0 100 200 300 400
Capitals of Countries
State Capitals
International Boundaries
State Boundaries
Copyright by C. S. HAMMOND & Co., N. Y.
C A N A D A
Moosonee
Albany
Missinaibi
Timmins
Lake Nipigon
Thunder Bay
Lake Superior
Sault Ste. Marie
Sudbury
Ottawa
Chicoutimi
Québec
Trois-Rivières
Montréal
Sherbrooke
St. Lawrence
Fredericton
Houlton
L. Champlain
Toronto
Hamilton
Niagara Falls
L. Ontario
L. Huron
L. Michigan
L. Erie
MAINE
Bangor
Augusta
Lewiston
Portland
VT.
Burlington
Montpelier
Rutland
N. H.
Berlin
Concord
Manchester
MASS.
Boston
Worcester
Springfield
CONN.
Hartford
New Haven
R.I.
Providence
Long I.
NEW YORK
Watertown
Rochester
Syracuse
Utica
Albany
Buffalo
Elmira
Binghamton
Jamestown
New York
PENNSYLVANIA
Erie
Scranton
Reading
Harrisburg
Pittsburgh
Philadelphia
Newark
Trenton
Wilmington
Dover
DEL.
MD.
Baltimore
Washington
Annapolis
Delaware Bay
Chesapeake Bay
VIRGINIA
Richmond
Lynchburg
Danville
Roanoke
Norfolk
Portsmouth
WEST VIRGINIA
Clarksburg
Charleston
Huntington
Wheeling
OHIO
Toledo
Akron
Cleveland
Youngstown
Canton
Springfield
Columbus
Dayton
Cincinnati
Covington
MICHIGAN
Marquette
Traverse City
Saginaw
Flint
Port Huron
Pontiac
Detroit
Lansing
Grand Rapids
Kalamazoo
Jackson
INDIANA
South Bend
Gary
Fort Wayne
Indianapolis
Terre Haute
Evansville
KENTUCKY
Louisville
Frankfort
Lexington
Owensboro
Paducah
Middlesboro
ILLINOIS
Chicago
Rockford
Rock Island
Peoria
Decatur
Springfield
East St. Louis
Cairo
WISCONSIN
Superior
Wausau
Eau Claire
Green Bay
Oshkosh
Sheboygan
Milwaukee
Madison
Racine
Beloit
La Crosse
MINNESOTA
Virginia
Hibbing
Duluth
Brainerd
St. Cloud
St. Paul
Minneapolis
Mankato
Winona
Rochester
IOWA
Mason City
Dubuque
Waterloo
Cedar Rapids
Davenport
Des Moines
Council Bluffs
Ottumwa
Sioux City
MISSOURI
Quincy
Hannibal
St. Joseph
Kansas City
Missouri
Sedalia
Jefferson City
St. Louis
Springfield
Cape Girardeau
Joplin
Tulsa
Muskogee
Fort Smith
ARKANSAS
Jonesboro
Hot Springs Nat'l Park
Little Rock
Pine Bluff
Texarkana
El Dorado
TENNESSEE
Nashville
Oak Ridge
Knoxville
Jackson
Memphis
Chattanooga
NORTH CAROLINA
Asheville
Winston-Salem
Durham
Raleigh
Charlotte
Wilmington
C. Hatteras
SOUTH CAROLINA
Greenville
Columbia
Charleston
Savannah
GEORGIA
Rome
Atlanta
Augusta
Macon
Columbus
Waycross
Savannah
ALABAMA
Gadsden
Birmingham
Tuscaloosa
Montgomery
Dothan
Mobile
MISSISSIPPI
Clarksdale
Greenville
Jackson
Meridian
Vicksburg
Hattiesburg
Biloxi
LOUISIANA
Monroe
Shreveport
Alexandria
Baton Rouge
New Orleans
Marshall
Tyler
Beaumont
Port Arthur
Galveston
FLORIDA
Pensacola
Panama City
Tallahassee
Jacksonville
St. Augustine
Daytona Beach
Orlando
C. Canaveral
Tampa
St. Petersburg
W. Palm Beach
Ft. Myers
Miami
Key West
Straits of Florida
Gulf of Mexico
ATLANTIC OCEAN
BAHAMAS
Gd. Bahama
Nassau
Eleuthera
Cat I.
Andros I.
Tropic of Cancer
Gt. Exuma
Long I.
CUBA
Havana
Longitude 95° West H of 90° Greenwich J
85° K 80° 75°
95° H 90° J 85° K 80° L 75° M 70° N
45° 40° 35° 30° 25°

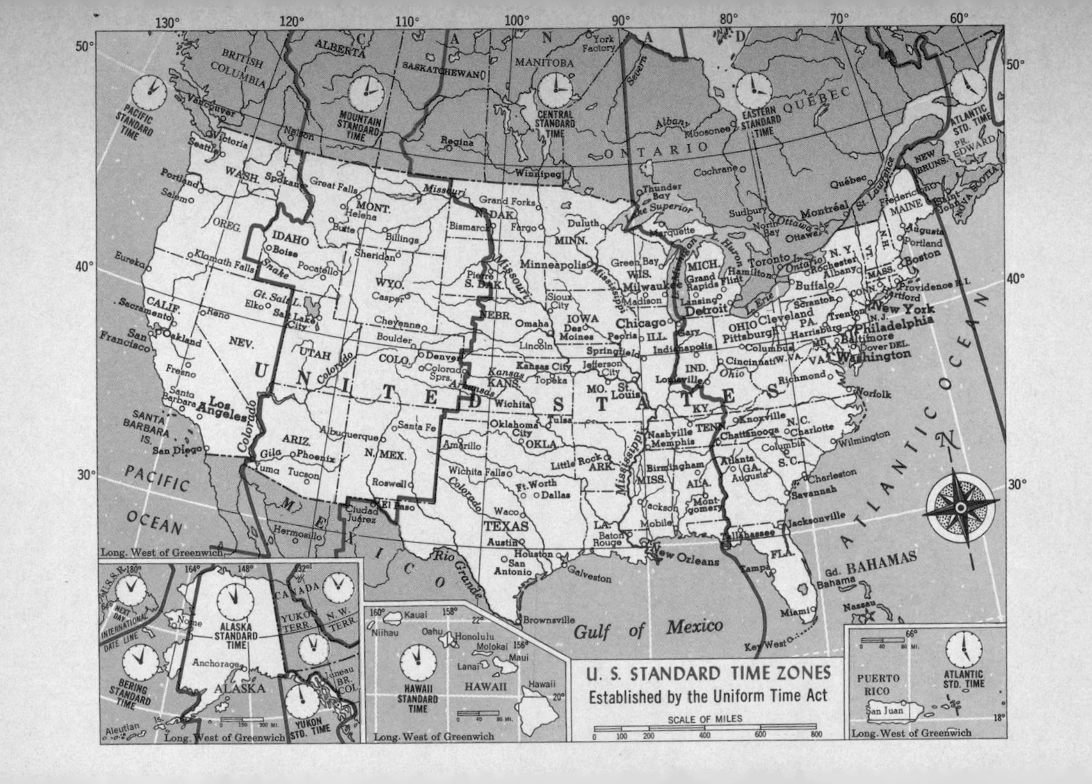
U. S. STANDARD TIME ZONES
Established by the Uniform Time Act
SCALE OF MILES
PACIFIC STANDARD TIME
MOUNTAIN STANDARD TIME
CENTRAL STANDARD TIME
EASTERN STANDARD TIME
ATLANTIC STD. TIME
ALASKA STANDARD TIME
BERING STANDARD TIME
YUKON STD. TIME
HAWAII STANDARD TIME
ATLANTIC OCEAN
PACIFIC OCEAN
Gulf of Mexico
UNITED STATES
CANADA
MEXICO
HAWAII
ALASKA
PUERTO RICO
Gd. BAHAMAS
INTERNATIONAL DATE LINE
Long. West of Greenwich

FACTS ABOUT THE STATES AND PROVINCES

U.S. States	Admitted to Union	Date Settled	State Nickname	State Flower
Alabama	1819	1702	Yellowhammer	Camellia
Alaska	1959	1784	The Great Land	Forget-me-not
Arizona	1912	1580	Grand Canyon	Saguaro Cactus
Arkansas	1836	1686	Land of Opportunity	Apple Blossom
California	1850	1769	Golden	Golden Poppy
Colorado	1876	1858	Centennial	Rocky Mtn. Columbine
Connecticut	1788	1633	Constitution	Mountain Laurel
Delaware	1787	1638	Diamond	Peach Blossom
Florida	1845	1565	Sunshine	Orange Blossom
Georgia	1788	1733	Peach	Cherokee Rose
Hawaii	1959	——	Aloha	Red Hibiscus
Idaho	1890	1842	Gem	Syringa
Illinois	1818	1699	Prairie	Native Violet
Indiana	1816	1732	Hoosier	Peony
Iowa	1846	1788	Hawkeye	Wild Rose
Kansas	1861	1827	Sunflower	Sunflower
Kentucky	1792	1774	Bluegrass	Goldenrod
Louisiana	1812	1699	Pelican	Magnolia
Maine	1820	1624	Pine Tree	Pine Cone & Tassel
Maryland	1788	1634	Old Line	Blackeyed Susan
Massachusetts	1788	1620	Bay	Mayflower
Michigan	1837	1668	Wolverine	Apple Blossom
Minnesota	1858	1805	North Star	Lady-slipper
Mississippi	1817	1699	Magnolia	Magnolia
Missouri	1821	1735	Show Me	Hawthorn
Montana	1889	1809	Treasure	Bitterroot
Nebraska	1867	1847	Cornhusker	Goldenrod
Nevada	1864	1851	Silver	Sagebrush
New Hampshire	1788	1623	Granite	Purple Lilac
New Jersey	1787	1617	Garden	Purple Violet
New Mexico	1912	1595	Land of Enchantment	Yucca
New York	1788	1614	Empire	Rose
North Carolina	1789	1650	Tarheel	Dogwood
North Dakota	1889	1780	Flickertail	Wild Prairie Rose
Ohio	1803	1788	Buckeye	Scarlet Carnation
Oklahoma	1907	1889	Sooner	Mistletoe
Oregon	1859	1811	Beaver	Oregon Grape
Pennsylvania	1787	1643	Keystone	Mountain Laurel
Rhode Island	1790	1636	Little Rhody	Violet
South Carolina	1788	1670	Palmetto	Yellow Jessamine
South Dakota	1889	1817	Coyote	Pasqueflower
Tennessee	1796	1757	Volunteer	Iris
Texas	1845	1682	Lone Star	Bluebonnet
Utah	1896	1847	Beehive	Sego Lily
Vermont	1791	1724	Green Mountain	Red Clover
Virginia	1788	1607	Old Dominion	American Dogwood
Washington	1889	1811	Evergreen	Western Rhododendron
West Virginia	1863	1727	Mountain	Big Rhododendron
Wisconsin	1848	1701	Badger	Wood Violet
Wyoming	1890	1834	Equality	Indian Paintbrush

Canadian Provinces	Date of Admission	Date Settled	Provincial Flower
Alberta	1905	1795	Wild Rose
British Columbia	1871	1843	Dogwood
Manitoba	1870	1812	Prairie Crocus
New Brunswick	1867	1611	Purple Violet
Newfoundland	1949	1610	Pitcher Plant
Nova Scotia	1867	1605	Trailing Arbutus
Ontario	1867	1671	White Trillium
Prince Edward Island	1873	1713	Lady's Slipper
Quebec	1867	1608	Madonna Lily
Saskatchewan	1905	1774	Wild Wood Lily

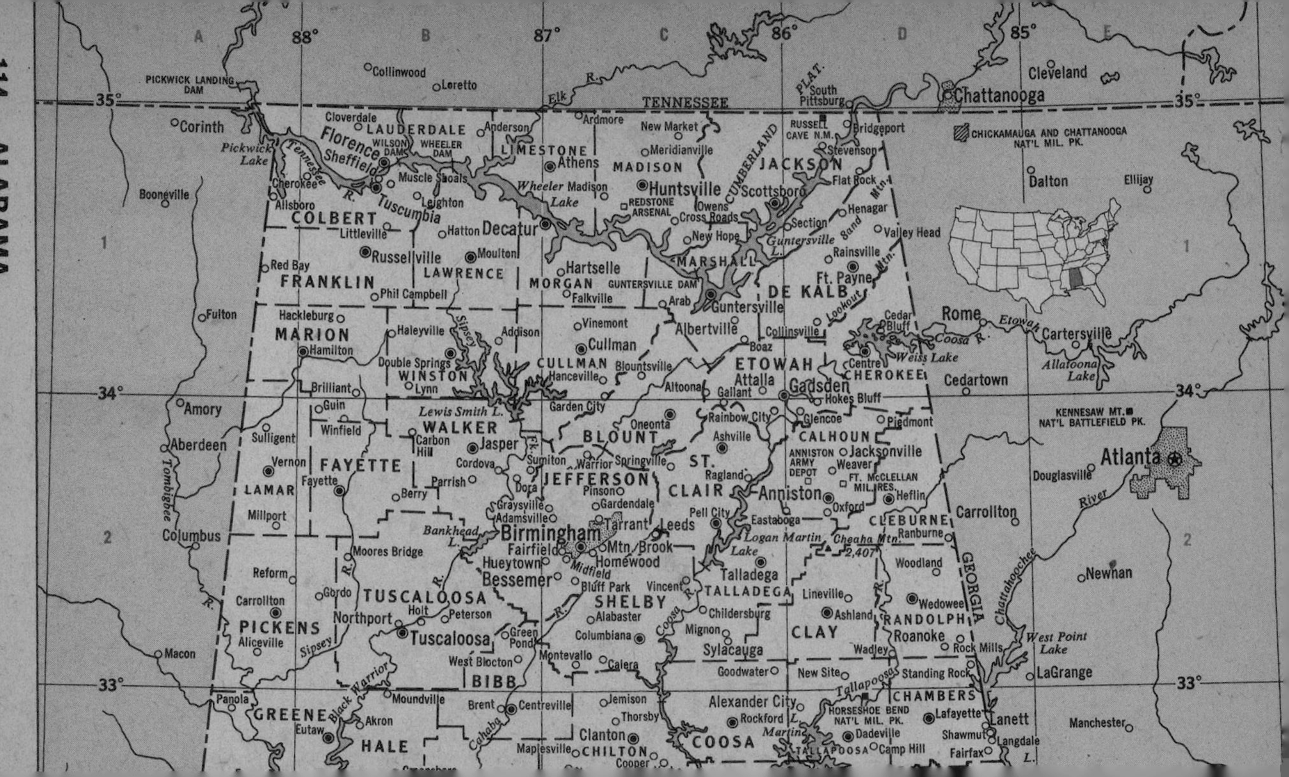

A
B
C
D
E
88°
87°
86°
85°
35°
34°
33°
1
2
TENNESSEE
GEORGIA
PICKWICK LANDING DAM
Pickwick Lake
Collinwood
Loretto
Corinth
Booneville
Fulton
Amory
Aberdeen
Tombigbee
Columbus
Macon
Elk R.
Tennessee R.
Cloverdale
LAUDERDALE
Florence
Sheffield
WILSON DAM
WHEELER DAM
Muscle Shoals
Leighton
Tuscumbia
Cherokee
Allsboro
COLBERT
Littleville
Anderson
Ardmore
LIMESTONE
Athens
Wheeler Lake
Madison
MADISON
Huntsville
REDSTONE ARSENAL
Meridianville
New Market
Owens Cross Roads
New Hope
South Pittsburg
CUMBERLAND PLAT.
RUSSELL CAVE N.M.
Bridgeport
Stevenson
JACKSON
Scottsboro
Flat Rock
Section
Henagar
Sand Mtn.
Valley Head
Guntersville L.
Rainsville
Ft. Payne
DE KALB
Lookout Mtn.
Chattanooga
Cleveland
CHICKAMAUGA AND CHATTANOOGA NAT'L MIL. PK.
Dalton
Ellijay
Hatton
Decatur
Moulton
Red Bay
FRANKLIN
Russellville
LAWRENCE
Phil Campbell
Hartselle
MORGAN
GUNTERSVILLE DAM
Falkville
MARSHALL
Arab
Guntersville
Albertville
Boaz
Collinsville
Cedar Bluff
Rome
Etowah
Cartersville
Coosa R.
Weiss Lake
Allatoona Lake
Centre
CHEROKEE
Cedartown
Hackleburg
MARION
Hamilton
Haleyville
Sipsey
Addison
Vinemont
Cullman
CULLMAN
Blountsville
Double Springs
WINSTON
Lynn
Hanceville
ETOWAH
Attalla
Altoona
Gallant
Gadsden
Hokes Bluff
Brilliant
Guin
Winfield
Lewis Smith L.
Garden City
Rainbow City
Glencoe
Piedmont
KENNESAW MT. NAT'L BATTLEFIELD PK.
Sulligent
WALKER
Carbon Hill
Jasper
Oneonta
BLOUNT
Ashville
CALHOUN
Jacksonville
ANNISTON ARMY DEPOT
Weaver
FT. McCLELLAN MIL. RES.
Atlanta
Douglasville
Vernon
FAYETTE
Cordova
Sumiton
Warrior
Springville
ST. CLAIR
Ragland
Anniston
Heflin
LAMAR
Fayette
Parrish
Dora
JEFFERSON
Pinson
Gardendale
Berry
Graysville
Adamsville
Millport
Bankhead L.
Birmingham
Tarrant
Leeds
Pell City
Eastaboga
Oxford
CLEBURNE
Carrollton
River
Columbus
Fairfield
Mtn. Brook
Homewood
Logan Martin Lake
Cheaha Mtn. 2,407
Ranburne
Moores Bridge
Hueytown
Midfield
Bessemer
Talladega
Woodland
Chattahoochee
Newnan
Reform
Bluff Park
Vincent
TALLADEGA
Lineville
Wedowee
Gordo
TUSCALOOSA
Holt
Peterson
SHELBY
Childersburg
Ashland
RANDOLPH
Carrollton
PICKENS
Northport
Alabaster
Mignon
CLAY
Roanoke
Aliceville
Tuscaloosa
Green Pond
Columbiana
Sylacauga
Wadley
Rock Mills
West Point Lake
Sipsey
West Blocton
Montevallo
Calera
Goodwater
New Site
Standing Rock
LaGrange
BIBB
Black Warrior
Panola
Moundville
Brent
Centreville
Jemison
Thorsby
Alexander City
Tallapoosa
CHAMBERS
GREENE
Akron
Cahaba
Clanton
Rockford
HORSESHOE BEND NAT'L MIL. PK.
Lafayette
Lanett
Manchester
Eutaw
HALE
Maplesville
CHILTON
COOSA
Martin L.
Dadeville
Shawmut
Langdale
Cooper
TALLAPOOSA
Camp Hill
Fairfax

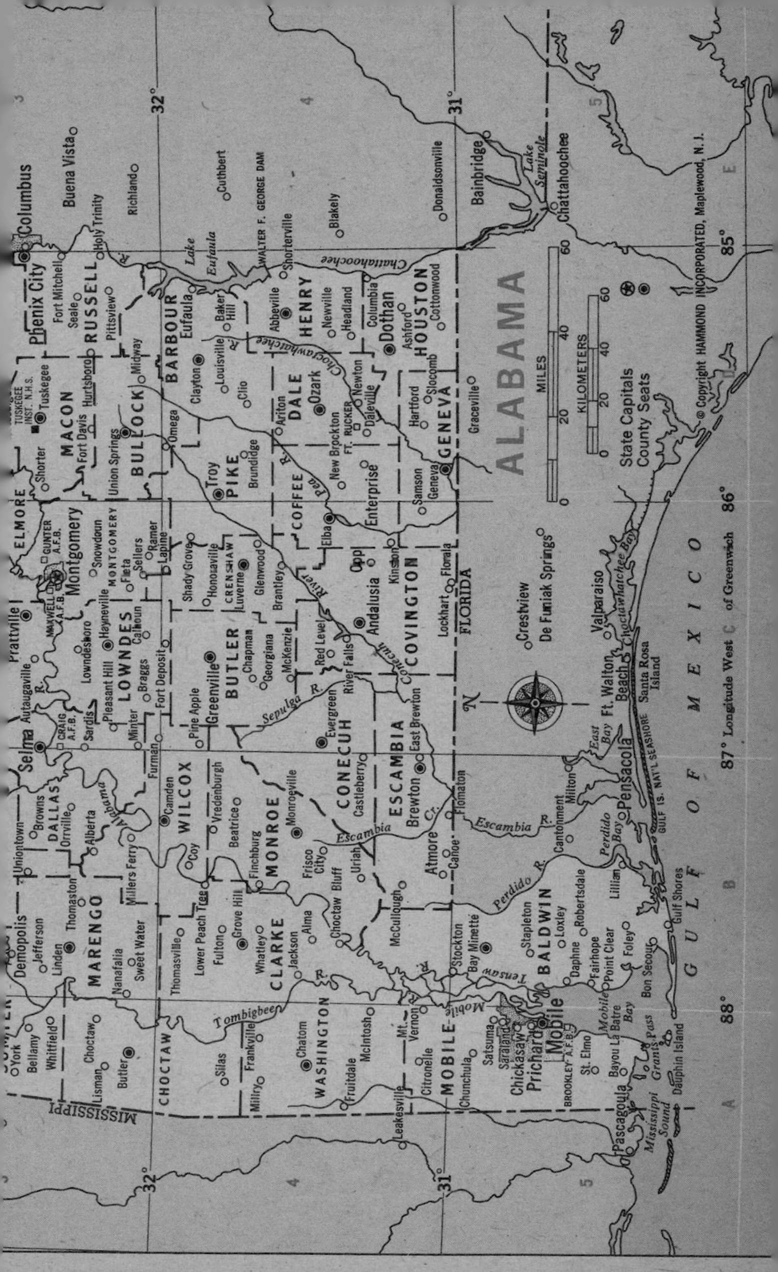
Columbus
Phenix City
Buena Vista
Richland
Cuthbert
Fort Mitchell
Seale
RUSSELL
Holy Trinity
Pittsview
Hurtsboro
MACON
Tuskegee
TUSKEGEE INST. N.H.S.
Fort Davis
Shorter
ELMORE
Montgomery
GUNTER A.F.B.
MAXWELL A.F.B.
MONTGOMERY
Snowdoun
Prattville
Autaugaville
CRAIG A.F.B.
Selma
Sardis
Pleasant Hill
Lowndesboro
Hayneville
LOWNDES
Fleta
Calhoun
Braggs
Sellers
Ramer
Lapine
Fort Deposit
Minter
Furman
Union Springs
BULLOCK
Midway
BARBOUR
Eufaula
Clayton
Lake Eufaula
WALTER F. GEORGE DAM
Shorterville
Abbeville
Baker Hill
Louisville
Clio
Omega
Troy
PIKE
Brundidge
Ariton
DALE
Ozark
FT. RUCKER
HENRY
Newville
Headland
Columbia
Dotham
Ashford
HOUSTON
Cottonwood
Newton
Daleville
New Brockton
Enterprise
COFFEE
Elba
Pea R.
Choctawhatchee R.
Chattahoochee
Blakely
Donaldsonville
Bainbridge
Lake Seminole
Chattahoochee
Hartford
Slocomb
GENEVA
Geneva
Samson
Graceville
ALABAMA
MILES
0 20 40 60
KILOMETERS
0 20 40 60
State Capitals
County Seats
Shady Grove
Honoraville
CRENSHAW
Luverne
Glenwood
Brantley
Opp
Kinston
Andalusia
Red Level
River Falls
COVINGTON
Florala
Lockhart
FLORIDA
Crestview
De Funiak Springs
Valparaiso
Choctawhatchee Bay
Ft. Walton Beach
Santa Rosa Island
Greenville
BUTLER
Chapman
Georgiana
McKenzie
Pine Apple
Sepulga R.
Evergreen
CONECUH
Castleberry
East Brewton
ESCAMBIA
Brewton
Flomaton
Atmore
Canoe
Escambia R.
Milton
East Bay
Pensacola
Cantonment
Perdido Bay
GULF IS. NAT'L SEASHORE
Lillian
Gulf Shores
Dallas
Orrville
Browns
Uniontown
Alberta
Alabama R.
Millers Ferry
Camden
WILCOX
Vredenburgh
Beatrice
Coy
MONROE
Monroeville
Finchburg
Frisco City
Uriah
Lower Peach Tree
Grove Hill
Fulton
Thomasville
CLARKE
Whatley
Jackson
Alma
Choctaw Bluff
McCullough
Stockton
Bay Minette
BALDWIN
Stapleton
Loxley
Robertsdale
Perdido R.
Daphne
Fairhope
Point Clear
Foley
Bon Secour
Tensaw R.
Demopolis
Jefferson
Linden
MARENGO
Thomaston
Nanafalia
Sweet Water
York
Bellamy
Whitfield
Choctaw
Lisman
Butler
CHOCTAW
Silas
Tombigbee R.
Frankville
Millry
Chatom
WASHINGTON
Fruitdale
McIntosh
Mt. Vernon
Citronelle
MOBILE
Chunchula
Satsuma
Saraland
Chickasaw
Prichard
Mobile
Mobile Bay
BROOKLEY A.F.B.
St. Elmo
Bayou La Batre
Grants Pass
Dauphin Island
Leakesville
MISSISSIPPI
Pascagoula
Mississippi Sound
GULF OF MEXICO
32°
31°
88°
87° Longitude West of Greenwich
86°
85°
© Copyright HAMMOND INCORPORATED, Maplewood, N.J.

ALASKA
MILES
KILOMETERS
State and Provincial Capitals
Court Houses
ARCTIC
Chukchi Sea
U.S.S.R.
BERING SEA
PACIFIC OCEAN
ALEUTIAN ISLANDS
FOX IS.
PRIBILOF IS.
Bristol Bay
Norton Sound
Kotzebue Sound
Seward Pen.
Alaska Pen.
Nome
Bethel
Barrow
Kotzebue
Dillingham
Nunivak I.
St. Lawrence I.
St. Matthew I.
St. Paul I.
St. George I.
Unimak I.
Shumagin Is.
Kodiak
Yukon
© Copyright HAMMOND INCORPORATED, Maplewood, N. J.
Longitude West of Greenwich

OCEAN
Prudhoe Bay
Kaktovik
Mt. Michelson 8,855
BRITISH MTS.
Arctic Village
Venetie
Porcupine R.
Ft. Yukon
Beaver
Stevens Village
Circle
YUKON TERR.
College
Fairbanks
North Pole
Nenana
Eagle
Tanana R.
Delta Jct.
Tanacross
Tetlin
Northway
Cantwell
Talkeetna
Glennallen
Copper Center
WRANGELL MTS.
Palmer
Anchorage
Spenard
Valdez
Whittier
Cordova
Seward
Montague I.
Mt. St. Elias 18,008
Mt. Logan 19,850
ST. ELIAS MTS.
Icy B.
Yakutat Bay
Yakutat
Dry B.
KLONDIKE REGION
Dawson
Yukon R.
Ft. Selkirk
Haines Jct.
Whitehorse
BRIT. COL.
CANADA
Skagway
Juneau
COAST MTS.
GLACIER BAY NAT'L MON.
Pelican
SOUTH-EASTERN ALEXANDER ARCHIPELAGO
Sitka
C. Ommaney
Angoon
Petersburg
Wrangell
Stewart
Ketchikan
Craig
Metlakatla
Prince Rupert
Kitimat
Swanson
Dixon Entr.
Hecate Str.
Graham I.
QUEEN CHARLOTTE IS.
Moresby I.
STIKINE MTS.
Stikine R.
GULF OF ALASKA
Skagway
Atlin
Haines
Mt. Fairweather 15,300
GLACIER BAY NAT'L MON. Gustavus
Lynn Canal
Juneau
BRIT. COLUMBIA
COAST MOUNTAINS
Cross Sd.
Pelican
Hoonah
Chichagof I.
Admiralty I.
Chatham Strait
Angoon
Telegraph Creek
ALEXANDER
Sitka
SITKA N.H.P.
Frederick Sd.
Kake
Petersburg
Baranof I.
Kuiu I.
Kupreanof I.
Wrangell
SOUTHEASTERN ARCHIPELAGO
Edna Bay
Prince of Wales I.
Klawock
Craig
Revillagigedo I.
Ketchikan
Hydaburg
Metlakatia
Dall I.
C. Muzon
Dixon Entrance
Prince Rupert
CANADA
0 50 100 mi.
0 50 100 km.
136°
132°
58°
56°
54°
BERING SEA
ALEUTIAN ISLANDS
SOUTH CENTRAL
Attu I.
Kiska I.
RAT IS.
Amchitka I.
Semisopochnoi I.
Kanaga I.
Atka I.
Amlia I.
Tanaga I.
Adak I.
ANDREANOF IS.
PACIFIC OCEAN
Same scale as main map
175°
180°
175°
140°
130°
120°
65°
60°
55°
E
F
G
H

114°
112°
110°
36°
Cedar City
CEDAR BREAKS NAT'L MON.
BRYCE CANYON NAT'L PARK
CAPITOL REEF NAT'L MON.
Lake Powell
NATURAL BRIDGES NAT'L MON.
HOVENWEEP N. M.
GLEN CANYON NAT'L REC. AREA
ZION NAT'L PARK
St. George
Hurricane
Kanab
UTAH
NEVADA
COLO.
Juan R.
Only point in U.S. common to 4 states.
RAINBOW BRIDGE NAT'L MON.
GLEN CANYON DAM
Page
KAIBAB IND. RES.
Fredonia
Littlefield
Virgin R.
Meadow Valley Wash
PIPE SPRING NAT'L MON.
NAVAJO NAT'L MON.
Kayenta
Pastora Pk. 9,420
KAIBAB PLAT.
Colorado R.
COLORADO PLATEAU
NAVAJO IND. RES.
BLACK MESA
CANYON DE CHELLY NAT'L MON.
Chinle
Overton
Lake Mead
LAKE MEAD NAT'L REC. AREA
HOOVER DAM
GRAND CANYON NATIONAL PARK
North Rim
Tuba City
HOPI IND. RES.
Grand Canyon
Little
PAINTED DESERT
Hotevilla
Polacca
Keams Canyon
APACHE
Fort Defiance
Window Rock
Oraibi
Shungopovy
Ganado
HUBBELL TRADING POST NAT'L HIST. SITE
Cornfields
St. Michaels
Klagetoh
COCONINO PLAT.
COCONINO
HUALAPAI IND. RES.
Peach Springs
Nelson
WUPATKI N. M.
Colorado
Lake Mohave
BLACK MTS.
Chloride
Valentine
MOHAVE
Kingman
DAVIS DAM
Davis Dam
Bullhead City
Hualapai Pk. 8,266
FT. MOHAVE IND. RES.
Seligman
Ash Fork
Williams
Humphreys Pk. 12,633
Bellemont
Flagstaff
SUNSET CRATER NAT'L MON.
WALNUT CANYON NAT'L MON.
METEOR CRATER
NAVAJO
Lupton
Houck
Sanders
Chambers
PAINTED DESERT SECTION
Winslow
Joseph City
Holbrook
PETRIFIED FOREST NAT'L PARK
NEW MEXICO
Needles
Topock
Lake Havasu
Lake Havasu City
CALIFORNIA
PARKER DAM
Sedona
Chino Valley
Clarkdale
TUZIGOOT NAT'L MON.
Cottonwood
Happy Jack
YAVAPAI
Cornville
MONTEZUMA CASTLE NAT'L MON.
Baker Butte 8,182
Bagdad
Prescott
Humboldt
Camp Verde
Verde
Skull Valley
Mayer
MOGOLLON RIM
Payson
Heber
Snowflake
Taylor
Clay Springs
Show Low
Concho
St. Johns

ARIZONA
GILA
GREENLEE
GRAHAM
PINAL
MARICOPA
YUMA
PIMA
COCHISE
SANTA CRUZ
MEXICO
SONORA
GREAT DESERT
Phoenix
Tucson
Yuma
Glendale
Scottsdale
Mesa
Tempe
Chandler
Gilbert
Avondale
Buckeye
Sun City
El Mirage
Wickenburg
Cave Creek
Globe
Miami
Superior
Florence
Coolidge
Casa Grande
Eloy
Marana
S. Tucson
Nogales
Heroica Nogales
Bisbee
Douglas
Agua Prieta
Tombstone
Benson
Willcox
Safford
Clifton
Morenci
Ajo
Sells
Gila Bend
Somerton
Wellton
Tacna
Quartzsite
Blythe
Puerto Peñasco
Cananea
Magdalena
Mt. Graham
10,713
PAPAGO IND. RES.
SAN CARLOS IND. RES.
ORGAN PIPE CACTUS NAT'L MON.
SAGUARO NAT'L MON.
CHIRICAHUA NAT'L MON.
CORONADO NAT'L MEM.
FT. BOWIE NAT'L HIST. SITE
TONTO NAT'L MON.
TUMACACORI NAT'L MON.
CASA GRANDE RUINS N. M.
YUMA PROVING GROUND
GILA BEND MTS.
PAINTED ROCK DAM
Gila River
Colorado
Santa Cruz R.
San Pedro R.
Salt
T. Roosevelt Lake
Imperial Res.
Laguna Res.
32°
110°
114°
Longitude West 112° of Greenwich
MILES
0 20 40 60 80
KILOMETERS
0 20 40 60 80
State Capital
County Seats
© Copyright HAMMOND INCORPORATED, Maplewood, N.J.

95°
94°
93°
A
B
C
1
2
3
4
36°
35°
34°
33°
Lake o' the Cherokees
Pryor
Fort Gibson Lake
Ft. Gibson
Stilwell
Sallisaw
Stigler
Panama
Poteau
Howe
Heavener
Broken Bow
Valliant
Idabel
Red
Clarksville
New Boston
Mt. Pleasant
Linden
Atlanta
Gravette
PEA RIDGE NAT'L MIL. PK.
Beaver
Eureka Springs
Bentonville
BENTON
Rogers
Gentry
Siloam Springs
Springdale
Berryville
Green Forest
CARROLL
BOONE
Harrison
Table Rock Lake
Bull Shoals Lake
Norfork Lake
BAXTER
Cotter
Yellville
MARION
Huntsville
Fayetteville
BUFFALO NAT'L RIVER
Jasper
Summers
Prairie Grove
Lincoln
WASHINGTON
White R.
MADISON
NEWTON
Marshall
SEARCY
Leslie
Saint Paul
Deer
O Z A R K
P L A T E A U
B O S T O N M T S.
Mountainburg
CRAWFORD
FRANKLIN
JOHNSON
Walker Mtn. 2,128
POPE
VAN BUREN
Clinton
Van Buren
Mulberry
Alma
Ozark
Coal Hill
Clarksville
Lamar
Dover
Ft. Smith
FT. SMITH NAT'L HIST. SITE
Arkansas
Barling
Charleston
Paris
Dardanelle
Res.
Russellville
CONWAY
FAULK
Damasc
FORT CHAFFEE
Jenny Lind
Greenwood
SEBASTIAN
Magazine Mtn. 2,753
LOGAN
Magazine
Dardanelle
Atkins
Plumerville
Morrilton
River
Huntington
Booneville
State Sanatorium
Mansfield
Hartford
Danville
Ola
Perryville
Conwa
Plainview
Waldron
SCOTT
YELL
PERRY
Roland
Sherwo
N Little Rock
Cammack Village
Little Rock
O U A C H I T A M T S.
Blue Mtn. 2,623
L. Ouachita
GARLAND
Mountain Pine
SALINE
Bryant
Mena
Mount Ida
Hot Springs National Park
HOT SPRINGS NAT'L PARK
Benton
Bauxite
OKLAHOMA
POLK
MONTGOMERY
Jones Mill
Saline
Sheridan
Wickes
Glenwood
HOT SPRING
Malvern
GRANT
HOWARD
L. Greeson
Amity
De Gray L.
SEVIER
Dierks
PIKE
Arkadelphia
CLARK
Ouachita
Carthage
River
De Queen
Murfreesboro
Curtis
DALLAS
Lockesburg
Nashville
Gurdon
Horatio
Millwood L.
Mineral Springs
Beirne
Sparkman
Fordyce
LITTLE RIVER
Foreman
Ashdown
HEMPSTEAD
Hope
Prescott
Laneburg
OUACHITA
Bearden
Thornton
CALHOUN
TEXAS
River
NEVADA
Camden
Hampton
Mandeville
Stephens
Texarkana
Texarkana
Stamps
Lewisville
Waldo
McNeil
Smackover
Calion
MILLER
LAFAYETTE
Magnolia
COLUMBIA
El Dorado
UNION
Urbana
L. Erling
Taylor
Doddridge
Bradley
Junction City
LOUISIANA
Haynesville
© Copyright HAMMOND INCORPORATED, Maplewood, N. J.

ARKANSAS
MILES
0 10 20 30 40 50 60
KILOMETERS
0 10 20 30 40 50 60
State Capital
County Seats
92° Longitude West of Greenwich 91°
90°
89°
36°
35°
34°
33°
D
E
F
1
2
3
4
KY.
FULTON
IZARD
SHARP
RANDOLPH
CLAY
GREENE
LAWRENCE
INDEPENDENCE
CRAIGHEAD
MISSISSIPPI
POINSETT
JACKSON
CROSS
CRITTENDEN
WHITE
WOODRUFF
ST. FRANCIS
MONROE
LEE
LONOKE
PRAIRIE
PHILLIPS
ARKANSAS
LINCOLN
DESHA
DREW
ASHLEY
CHICOT
Doniphan
Thayer
Mammoth Spring
Salem
Hardy
Corning
Piggott
Rector
Pocahontas
Ash Flat
Black Rock
Marmaduke
Melbourne
Evening Shade
Walnut Ridge
Powhatan
Hoxie
Paragould
Cave City
Jonesboro
Swifton
Batesville
Tuckerman
Newark
Lake City
Monette
Bay
Leachville
Manila
Blytheville
Armorel
Luxora
Caraway
Osceola
Keiser
Trumann
Weiner
Lepanto
Marked Tree
Harrisburg
Wilson
Joiner
Tyronza
Newport
Pleasant Plains
Heber Springs
Bradford
Turrell
Bald Knob
Augusta
Searcy
Judsonia
Kensett
McCrory
Parkin
Earle
Marion
Wynne
Crawfordsville
W. Memphis
Gregory
Beebe
Cotton Plant
Forrest City
Proctor
Hulbert
Madison
Memphis
Capleville
Collierville
Cabot
Des Arc
Palestine
Hughes
Brinkley
Jacksonville
Lonoke
Hazen
Carlisle
De Valls Bluff
Marianna
Aubrey
Clarendon
La Grange
Holly Grove
Lexa
W. Helena
Helena
England
Marvell
Stuttgart
Humphrey
Altheimer
DeWitt
Ethel
Elaine
Friars Point
Pine Bluff
Grady
Gillett
ARKANSAS POST NAT'L MEM.
Gould
Star City
Dumas
Rosedale
Wilmar
Monticello
McGehee
Arkansas City
Dermott
Lake Village
Greenville
Hamburg
Portland
Eudora
Crossett
Wilmot
Malden
Campbell
Portageville
Kennett
Hayti
Tiptonville
Caruthersville
Steele
Dyersburg
Ripley
Brownsville
Covington
Somerville
Hernando
Holly Springs
Senatobia
Sardis Lake
Sardis
Oxford
Marks
Clarksdale
Charleston
Bruce
Cleveland
Itta Bena
Greenwood
Moorhead
Indianola
Eupora
Belzoni
MISSISSIPPI
RIVER
TENN.
Arkansas
River
Black
White
St. Francis
Bartholomew
Bayou
Yazoo
Pine Bluff

SAN FRANCISCO AND VICINITY
LOS ANGELES AND VICINITY
Counties indicated by numbers:
1 ALAMEDA C4
2 AMADOR C3
3 CALAVERAS C3
4 CONTRA COSTA B4
5 LAKE B3
6 MARIN B3
7 MERCED C4
8 NAPA B3
9 ORANGE E6
10 SACRAMENTO C3
11 SAN FRANCISCO B4
12 SAN JOAQUIN C4
13 SOLANO C3
14 STANISLAUS C4
15 SUTTER C3
16 YUBA C3
PACIFIC OCEAN
NEVADA
CRATER LAKE NAT'L PK.
OREGON CAVES N.M.
LAVA BEDS N.M.
LASSEN VOLCANIC NAT'L PK.
YOSEMITE NAT'L PK.
REDWOOD NAT'L PK.
PT. REYES NAT'L SEASHORE
Medford
Reno
Carson City
Sacramento
Stockton
Oakland
Berkeley
San Jose
SAN FRANCISCO
LOS ANGELES
Pasadena
Long Beach
Santa Ana
Anaheim
Eureka
Redding
Chico
Lake Tahoe
Mt. Shasta 14,162
Lassen Pk. 10,457

CALIFORNIA
PACIFIC OCEAN
MEXICO
ARIZONA
BAJA CALIFORNIA
MILES
KILOMETERS
State Capitals
County Seats
Longitude West of 118° Greenwich
© Copyright HAMMOND INCORPORATED
LOS ANGELES
SAN DIEGO
Long Beach
Santa Barbara
Bakersfield
Las Vegas
MOJAVE DESERT
DEATH VALLEY
SAN BERNARDINO
Salton Sea

COLORADO
MILES
0 20 40 60
KILOMETERS
0 20 40 60
State Capitals
County Seats
WYOMING
UTAH
ARIZ.
NEW MEXICO
MOFFAT
ROUTT
JACKSON
GRAND
RIO BLANCO
GARFIELD
EAGLE
SUMMIT
MESA
PITKIN
LAKE
PARK
DELTA
GUNNISON
CHAFFEE
MONTROSE
OURAY
SAGUACHE
SAN MIGUEL
HINSDALE
DOLORES
SAN JUAN
MINERAL
MONTEZUMA
LA PLATA
RIO GRANDE
ALAMOSA
ARCHULETA
CONEJOS
DINOSAUR NAT'L MON.
ROCKY MOUNTAIN NAT'L PARK
SHADOW MTN. NAT'L REC. AREA
COLORADO NAT'L MON.
BLACK CANYON OF THE GUNNISON NAT'L MON.
CURECANTI NAT'L REC. AREA
GREAT SAND DUNES NAT'L MON.
MESA VERDE NAT'L PK.
HOVENWEEP N.M.
YUCCA HOUSE N.M.
UTE MTN. IND. RES.
SOUTHERN UTE IND. RES.
JICARILLA IND. RES.
NAVAJO I. R.
ROAN PLATEAU
UNCOMPAHGRE PLATEAU
CONTINENTAL DIVIDE
SAWATCH RANGE
SAN JUAN MTS.
Only point in U. S. common to 4 states
Green R.
Yampa R.
White R.
Colorado R.
Gunnison R.
Dolores River
Animas R.
Rio Grande
Arkansas
N. Platte R.
Blue Mesa Res.
Morrow Pt. Res.
Ruedi Res.
Green Mtn. Res.
Platoro Res.
Navajo Res.
Lake Granby
Antero Res.
Red Feather Lakes
Walden
Steamboat Springs
Craig
Hayden
Jensen
Dinosaur
Oak Creek
Yampa
Kremmling
Granby
Hot Sulphur Sprs.
Grand Lake
Estes Park
Longs Pk. 14,256
Rangely
Meeker
Burns
Central City
Berthoud Pass
Georgetown
Idaho Sprs.
Vail
Minturn
Dillon
Eagle
Gypsum
Rifle
New Castle
Silt
Glenwood Sprs.
Grand Valley
Mt. of the Holy Cross 14,005
Redcliff
Breckenridge
Grant
Shawnee
Carbondale
Basalt
Climax
Leadville
De Beque
Collbran
Fairplay
Fruita
Mesa
Grand Junction
Orchard Mesa
Redstone
Aspen
Mt. Massive 14,421
Mt. Elbert 14,433
Castle Pk. 14,265
Crested Butte
Mt. Harvard 14,420
Buena Vista
Cedaredge
Paonia
Austin
Hotchkiss
Delta
Gateway
Tincup
Olathe
Salida
Gunnison
Montrose
Paradox
Uravan
Nucla
Naturita
Ridgway
Uncompahgre Pk. 14,309
Norwood
Ouray
Lake City
San Luis Pk. 14,014
Saguache
Telluride
Creede
Crestone Pk. 14,294
Mt. Wilson 14,246
Silverton
Dove Creek
Center
Rico
Windom Pk. 14,084
Del Norte
Mosca
Blanca Pk. 14,345
Dolores
Monte Vista
Alamosa
Cortez
Mancos
Blanca
Durango
Pagosa Sprs.
Summit Pk. 13,272
La Jara
Sanford
Capulin
Towaoc
Ignacio
Manassa
Conejos
Antonito
Chama
San Luis
© Copyright HAMMOND INCORPORATED, Maplewood, N. J.
Longitude 107° West of Greenwich 106°

105°
F
104°
G
103°
H
102°
J
Cheyenne
NEBR.
41°
Ovid
Julesburg
Peetz
Grant
LARIMER
Grover
LOGAN
River
SEDGWICK
Wellington
WELD
Sterling Res.
Iliff
Fleming
Laporte
Ft. Collins
Pierce
Ault
Eaton
Briggsdale
New Raymer
Sterling
Haxtun
Holyoke
PHILLIPS
Windsor
Thompson
Greeley
Evans
La Salle
Loveland
Berthoud
Johnstown
South
Platte
Ft. Morgan
Brush
MORGAN
NEBR.
Platteville
Keenesburg
Akron
Otis
YUMA
Yuma
Eckley
Wray
Longmont
Boulder
Erie
Ft. Lupton
Lafayette
Broomfield
Brighton
Northglenn
Thornton
Westminster
Arvada
Wheat Ridge
Golden
Lakewood
Littleton
Commerce City
DENVER
Aurora
Denver
Englewood
ADAMS
Bennett
Strasburg
Byers
ARAPAHOE
Deer Trail
Kiowa
Beaver Cr.
Woodrow
WASHINGTON
Arikaree
R.
Vernon
40°
N
Cope
Joes
Kirk
Republican
R.
JEFFERSON
Platte
Louviers
Castle Rock
Kiowa
Agate
DOUGLAS
ELBERT
Limon
Flagler
S. Fork
Stratton
Burlington
Goodland
2
1
Cheesman L.
South
Simla
Big
Hugo
KIT CARSON
TELLER
Palmer Lake
Black Forest
Calhan
Woodland Park
AIR FORCE ACADEMY
39°
Manitou Sprs.
Pikes Pk. 14,110
Colorado Springs
Ivywild
Security
Fountain
Cripple Creek
Victor
Broadmoor
Stratton Meadows
EL PASO
Sandy
Wild Horse
Cr.
Kit Carson
Cheyenne Wells
LINCOLN
Karval
CHEYENNE
Horse
Cr.
Canon City
Fountain
Cr.
E. Canon
Lincoln Park
Florence
Royal Gorge
Eads
Sheridan Lake
KIOWA
3
KANSAS
CROWLEY
Pueblo
R.
Boone
Sugar City
Ordway
Vineland
Avondale
CUSTER
Westcliffe
BENT'S OLD FORT NAT'L HIST. SITE
Ft. Lyon
Wiley
Lamar
Bristol
Holly
Fowler
PUEBLO
R.
Rocky Ford
Swink
La Junta
Las Animas
John Martin Res.
Granada
Arkansas
R.
38°
MTS.
Gardner
Huerfano
Timpas
OTERO
R.
BENT
PROWERS
HUERFANO
Walsenburg
Two Buttes
4
La Veta
Tyrone
Aguilar
Purgatoire
Springfield
Walsh
Pritchett
Hoehne
Villegreen
Kim
BACA
Chama
Valdez
Trinidad
Sopris
LAS ANIMAS
Campo
R.
Sanchez Res.
Elkhart
37°
Raton
Cimarron
OKLA.
MTS.
105°
F
104°
G
103°
H
102°

CONNECTICUT
LITCHFIELD
NEW HAVEN
FAIRFIELD
HAR
MASSACHUSETTS
NEW YORK
TACONIC MTS.
Mt. Frissell 2,380
Twin Lakes
Canaan
E. Canaan
Salisbury
Lakeville
Falls Village
Norfolk
Colebrook
E. Hartland
Barkhamsted Res.
Granby
Tariffville
Simsbury
Canton Ctr.
Canton
Avon
Collinsville
Farmington
Nepaug Res.
West Branch
E. Branch
Winchester Ctr.
Winsted
New Hartford
Millerton
Sharon
Amenia
Millbrook
Wappinger Cr.
Cornwall
W. Goshen
Goshen
Cornwall Bridge
Mohawk Mtn. 1,682
Torrington
Harwinton
Burlington
Unionville
W. Hartfo
Farmington
Litchfield
Bantam
Bantam L.
Thomaston Res.
Housatonic
River
Kent
L. Waramaug
New Preston
Bear Hill 1,281
Shepaug R.
Washington
Bethlehem
Thomaston
Terryville
Plymouth
Bristol
Plainville
New Britain
Watertown
Oakville
Waterville
Waterbury
Southington
Plantsville
Meride
Sherman
Pawling
Fishkill Cr.
New Milford
Roxbury
Bridgewater
Woodbury
Middlebury
Lake Candlewood
Candlewood Isle
Southbury
Union City
Naugatuck
Cheshire
Beacon Falls
Walling
Quinnipiac R.
Bethany
Housatonic River
Newtown
Sandy Hook
Danbury
Bethel
Brewster
Croton Falls Res.
Seymour
Naugatuck River
Hamden
Ansonia
Derby
Gaillard L.
Montowese
Whitneyville
Monroe
Shelton
New Haven
Redding
Stepney
Trap Fall Res.
Allingtown
Orange
W. Haven
E. Haven
Ridgefield
Saugatuck Res.
Easton Res.
Trumbull
Branchville
Easton
New Croton Res.
Mt. Kisco
Milford
Woodmont
Morgan Pt.
Wilton
Bridgeport
Stratford
New Canaan
Westport
Stratford Pt.
Norwalk
Fairfield
Kensico Res.
S. Norwalk
E. Norwalk
Noroton Hts.
Darien
White Plains
Stamford
Norwalk Is.
Cos Cob
Riverside
Greenwich
LONG
A
B
30'
73°
42°
41°
N
MILES
0 5 10 15 20
KILOMETERS
0 5 10 15 20
State Capital
Longitude West

Longmeadow
30'
D
72°
E
Webster
Webster L.
ongamond Lakes
Thompsonville
42°
Suffield
Hazardville
Somersville
Enfield
Staffordville
N. Grosvenor Dale
Stafford Springs
Thompson
Quaddick Res.
South Woodstock
Windsor Locks
Warehouse Point
Scantic R.
Putnam
Broad Brook
Ellington
Shenipsit L.
Willimantic R.
Eastford
Pomfret
Tolland
Willington
Natchaug R.
Windsor
Rockville
Abington
E. Killingly
Bloomfield
Vernon
TOLLAND
Dayville
S. Windsor
WINDHAM
RHODE ISLAND
Wilson
Talcottville
Storrs
E. Brooklyn
Danielson
artford
Manchester
Hampton
Brooklyn
E. Hartford
Coventry
Mansfield Ctr.
T F O R D
Andover
Willimantic
River
Central Village
mwood
Glastonbury
Moosup
Wethersfield
Columbia
Windham
Canterbury
Sterling
Newington
Plainfield
Rocky Hill
S. Glastonbury
Shetucket
Quinebaug
Hebron
Lebanon
Marlborough
Baltic
Jewett City
lin
Versailles
Cromwell
Portland
R.
Voluntown
E. Hampton
Colchester
Yantic
Pachaug Pd.
Middletown
Connecticut
Preston City
Gardner L.
Norwich
Middlefield
Moodus
R.
30'
Durham
Haddam
MIDDLESEX
NEW LONDON
Montville
E. Haddam
N. Stonington
Thames
Chesterfield
Uncasville
Pawcatuck
Chester
River
NAVAL SUBMARINE BASE
Quaker Hill
R.
Deep River
COAST GUARD ACADEMY
Groton
W. Mystic
Pawcatuck
Westerly
N. Guilford
Mystic
Essex
New London
Stonington
Ivoryton
Poquonock Bridge
Niantic
Morningside Pk.
Old Lyme
Waterford
Old Saybrook
Watch Hill Pt.
Stony Creek
Guilford
Clinton
Madison
Fishers I.
Hammonasset Pt.
Cornfield Pt.
Sachem Head
2
The Race
I S L A N D
S O U N D
Plum I.
Orient Pt.
Greenport
Gardiners Bay
Gardiners I.
Southold
Montauk Pt.
L O N G I S L A N D
ATLANTIC OCEAN
41°
Sag Harbor
of Greenwich
30'
D
72°
E

DELAWARE
MILES
0 5 10 15 20
KILOMETERS
0 5 10 15 20
State Capital
County Seats
NEW
CASTLE
NEW JERSEY
PENNSYLVANIA
Delaware River
CHESAPEAKE BAY
Chesapeake & Delaware Canal
Chester
Claymont
Bellefonte
Wilmington
Penns Grove
Swedesboro
Arden
Holly Oak
Centerville
442
Winterthur
Montchanin
Westover Hills
Elsmere
Elmhurst
Minquadale
Yorklyn
Marshallton
Newport
Wilmington Manor
Christiana
New Castle
Newark
Brookside Park
Christina R.
Bear
Red Lion
Delaware City
St. Georges
Reedy I.
Port Penn
Odessa
Middletown
Noxontown Lake
Townsend
Blackbird
Smyrna R.
Clayton
Smyrna
Leipsic R.
Kenton
Leipsic
Cheswold
Dupont Manor
Dover
Kent Acres
Hartly
Little Creek
Bombay Hook Island
Goose Pt.
Kent I.
Kelly I.
Deepwater Pt.
Salem R.
Salem
Alloways Cr.
Monroeville
Elmer
Newfield
Vineland
Bridgeton
Union L.
Millville
Cohansey R.
Maurice R.
Cedarville
Port Norris
Egg Island Pt.
N
Elkton
North East
Northeast R.
Elk River
Cecilton
Sassafras R.
Chester R.
Chestertown
Susquehanna R.
Rising Sun
Oxford
West Grove
Pine Grove Res.
Octoraro Cr.
76°
75°
30'
A
B
C
D
1
2

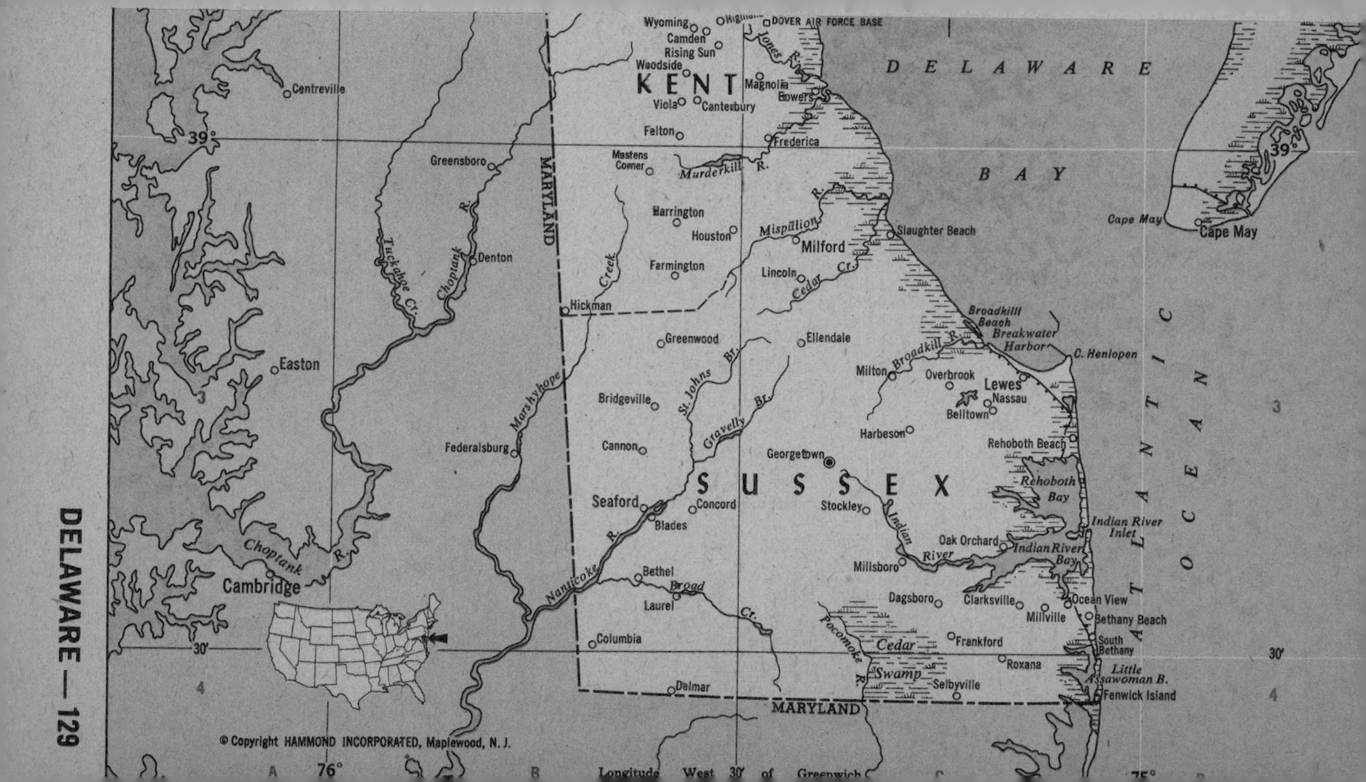
DELAWARE BAY
ATLANTIC OCEAN
KENT
SUSSEX
MARYLAND
MARYLAND
DOVER AIR FORCE BASE
Cape May
Cape May
C. Henlopen
Broadkill Beach
Breakwater Harbor
Slaughter Beach
Lewes
Nassau
Belltown
Overbrook
Rehoboth Beach
Rehoboth Bay
Indian River Inlet
Indian River Bay
Oak Orchard
Indian River
Ocean View
Bethany Beach
South Bethany
Little Assawoman B.
Fenwick Island
Millville
Clarksville
Roxana
Frankford
Selbyville
Dagsboro
Millsboro
Cedar Swamp
Pocomoke R.
Harbeson
Milton
Broadkill R.
Stockley
Georgetown
Ellendale
Milford
Lincoln
Cedar Cr.
Mispillion R.
Frederica
Bowers
Magnolia
Jones R.
Wyoming
Camden
Rising Sun
Woodside
Canterbury
Viola
Felton
Murderkill R.
Mastens Corner
Harrington
Houston
Farmington
Greenwood
Hickman
Creek
St. Johns Br.
Gravelly Br.
Bridgeville
Cannon
Concord
Blades
Seaford
Nanticoke R.
Bethel
Broad Cr.
Laurel
Columbia
Delmar
Marshyhope
Federalsburg
Denton
Greensboro
Choptank R.
Tuckahoe Cr.
Centreville
Easton
Cambridge
Choptank R.
39°
30'
76°
75°
Longitude West 30' of Greenwich
© Copyright HAMMOND INCORPORATED, Maplewood, N. J.

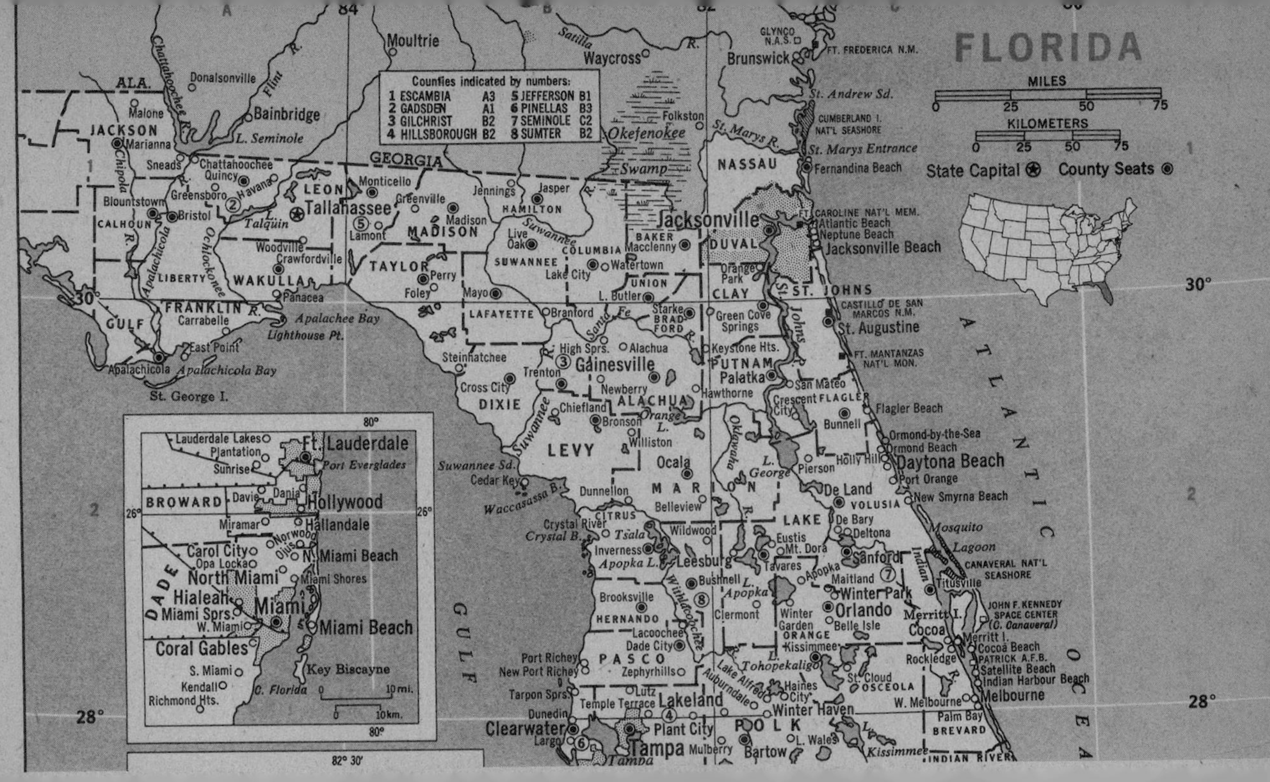
FLORIDA
MILES
0 25 50 75
KILOMETERS
0 25 50 75
State Capital
County Seats
Counties indicated by numbers:
1 ESCAMBIA A3
2 GADSDEN A1
3 GILCHRIST B2
4 HILLSBOROUGH B2
5 JEFFERSON B1
6 PINELLAS B3
7 SEMINOLE C2
8 SUMTER B2
ATLANTIC OCEAN
GULF
GEORGIA
ALA.
Moultrie
Waycross
Brunswick
Bainbridge
Donalsonville
Okefenokee Swamp
Tallahassee
Jacksonville
Jacksonville Beach
Fernandina Beach
St. Augustine
Palatka
Daytona Beach
De Land
Gainesville
Ocala
Leesburg
Sanford
Orlando
Winter Park
Titusville
Cocoa
Melbourne
Lakeland
Winter Haven
Plant City
Tampa
Clearwater
Apalachicola
Apalachee Bay
Ft. Lauderdale
Hollywood
Hallandale
N. Miami Beach
North Miami
Hialeah
Miami
Miami Beach
Coral Gables
Key Biscayne
30°
28°
26°
84°
80°
82° 30′

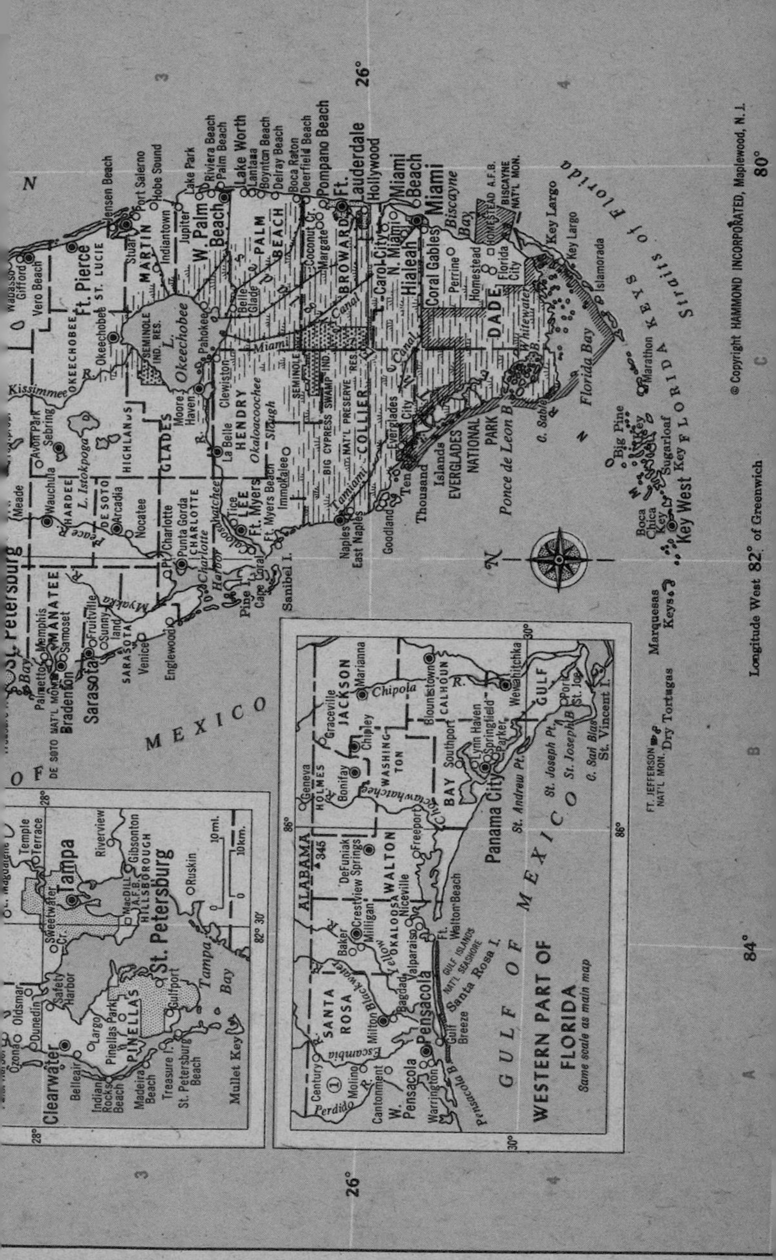

GULF OF MEXICO
St. Petersburg
Bradenton
Sarasota
Venice
Englewood
Punta Gorda
Pt. Charlotte
Charlotte Harbor
Pine I.
Cape Coral
Sanibel I.
Ft. Myers
Ft. Myers Beach
Naples
East Naples
Goodland
Immokalee
La Belle
Clewiston
Moore Haven
Belle Glade
Pahokee
L. Okeechobee
Okeechobee
Sebring
Avon Park
Wauchula
Arcadia
Nocatee
Ft. Pierce
Vero Beach
Stuart
Jensen Beach
Port Salerno
Hobe Sound
Jupiter
W. Palm Beach
Lake Worth
Riviera Beach
Palm Beach
Lantana
Boynton Beach
Delray Beach
Boca Raton
Deerfield Beach
Pompano Beach
Ft. Lauderdale
Hollywood
Miami Beach
Miami
N. Miami
Hialeah
Carol City
Coral Gables
Perrine
Homestead
Florida City
Biscayne Bay
Key Largo
Islamorada
Marathon
Big Pine Key
Sugarloaf Key
Key West
Boca Chica Key
Marquesas Keys
Dry Tortugas
FT. JEFFERSON NAT'L MON.
Florida Bay
C. Sable
Ponce de Leon B.
EVERGLADES NATIONAL PARK
Thousand Islands
Everglades City
Ten
BIG CYPRESS SWAMP
Tamiami Canal
Miami Canal
KEYS FLORIDA
Straits of Florida
MANATEE
SARASOTA
CHARLOTTE
DE SOTO
HARDEE
HIGHLANDS
GLADES
HENDRY
LEE
COLLIER
BROWARD
PALM BEACH
MARTIN
ST. LUCIE
OKEECHOBEE
DADE
MONROE
Longitude West 82° of Greenwich
26°
80°
84°
© Copyright HAMMOND INCORPORATED, Maplewood, N.J.
Clearwater
Dunedin
Oldsmar
Safety Harbor
Largo
Pinellas Park
Belleair
Indian Rocks Beach
Madeira Beach
Treasure I.
St. Petersburg Beach
Gulfport
Mullet Key
Tampa Bay
Tampa
Temple Terrace
Sweetwater Cr.
MacDILL A.F.B.
HILLSBOROUGH
Riverview
Gibsonton
Ruskin
PINELLAS
WESTERN PART OF FLORIDA
Same scale as main map
Pensacola
W. Pensacola
Warrington
Cantonment
Molino
Century
Milton
Bagdad
Gulf Breeze
Santa Rosa I.
GULF ISLANDS NAT'L SEASHORE
Valparaiso
Niceville
Crestview
Milligan
Baker
DeFuniak Springs
Freeport
Ft. Walton Beach
Geneva
Bonifay
Chipley
Graceville
Marianna
Blountstown
Wewahitchka
Port St. Joe
St. Vincent I.
C. San Blas
St. Joseph B.
Panama City
Lynn Haven
Springfield
Parker
Southport
St. Andrew Pt.
ALABAMA
SANTA ROSA
OKALOOSA
WALTON
HOLMES
WASHINGTON
JACKSON
CALHOUN
BAY
GULF

NORTH CAROLINA
SOUTH CAROLINA
ALABAMA
TENN.
Chattanooga
Cleveland
ATLANTA
Marietta
Decatur
Augusta
Athens
Macon
Rome
Dalton
Gainesville
Milledgeville
Griffin
La Grange
Carrollton
Cedartown
Elberton
Toccoa
Covington
Monroe
Hartwell Lake
Clark Hill Lake
Lake Sinclair
West Point Lake
Savannah R.
Ogeechee
Oconee
Ocmulgee
Chattahoochee
Tallapoosa
Salkehatchie
SAVANNAH RIVER ATOMIC ENERGY RESERVATION
GWINNETT
DE KALB
COBB
FULTON
Hapeville
Forest Park
College Park
E. Point
0 10 20 mi.
0 10 20 km.

GEORGIA
ATLANTIC OCEAN
GULF OF MEXICO
FLORIDA
ALABAMA
Savannah
Columbus
Phenix City
Albany
Americus
Cordele
Fitzgerald
Tifton
Moultrie
Thomasville
Valdosta
Waycross
Brunswick
Douglas
Jesup
Statesboro
Dublin
Vidalia
Bainbridge
Dothan
Jacksonville
Jacksonville Beach
Okefenokee Swamp
Counties indicated by numbers:
1 BALDWIN C3
2 BARROW C2
3 CATOOSA A2
4 CHEROKEE B2
5 CLAYTON B3
6 DADE A2
7 DE KALB B3
8 DOUGLAS B3
9 FAYETTE B3
10 GLASCOCK D3
11 MONTGOMERY D4
12 OCONEE C3
13 ROCKDALE B3
14 SCHLEY B4
15 TALIAFERRO D3
16 TOWNS C2
17 WALTON C3
18 WHITFIELD B2
MILES
KILOMETERS
State Capitals
County Seats
© Copyright HAMMOND INCORPORATED, Maplewood, N.J.
Longitude 84° West of Greenwich

KAUAI
Kilauea
Hanalei
KAUAI
Nohili Pt.
5,170
Kapaa
Wailua
Lehua
Kekaha
Lihue
Puuwai
Waimea
Kalaheo
Hanapepe
Koloa
NIIHAU
Makahuena Pt.
C. Kawaihoa
COUNTY
Kauai Channel
HONOLULU
Kahuku Pt.
Kahuku
Waialua
OAHU
Kaena Pt.
Wahiawa
Kaneohe
Waianae
Kailua
Aiea
Waipahu
Barbers Pt.
Honolulu
Pearl Har.
COUNTY
P A C I F I C
OAHU
Kahuku Pt.
Waialee
Kahuku
Mokuauia I.
Laie Pt.
Laie
Waimea
Kawailoa
Hauula
Haleiwa
Mokuleia
Keana Pt.
Waialua
DILLINGHAM A.F.B.
Kahana
Kahana Bay
Kaaawa
HONOLULU
Whitmore Vill.
Kaala 4,020
Waikane
Kaneohe Bay
Schofield Barracks
Wahiawa
WHEELER A.F.B.
Makaha
Kahaluu
Mokapu Pen.
Waianae
Waipio Acres
Mokapu
Kunia
Kailua Bay
COUNTY
Pacific Palisades
Kaneohe
Maili
Kailua
Pearl City
Waipahu
Aiea
Maunawili
Honouliuli
Pearl Harbor
Salt L.
Waimanalo Bay
Nanakuli
Waimanalo
Waimanalo Beach
Makakilo City
Ewa
Hickam Housing
Manana I.
Iroquois Pt.
Honolulu
Makapuu Pt.
Barbers Pt. Housing
Ewa Beach
Sand I.
Woodlawn
Kuapa Pd.
Mamala Bay
Barbers Pt.
BARBERS PT. N.A.S.
Aina Haina
Waikiki
Kahala
Koko Head
Diamond Head
Maunalua Bay

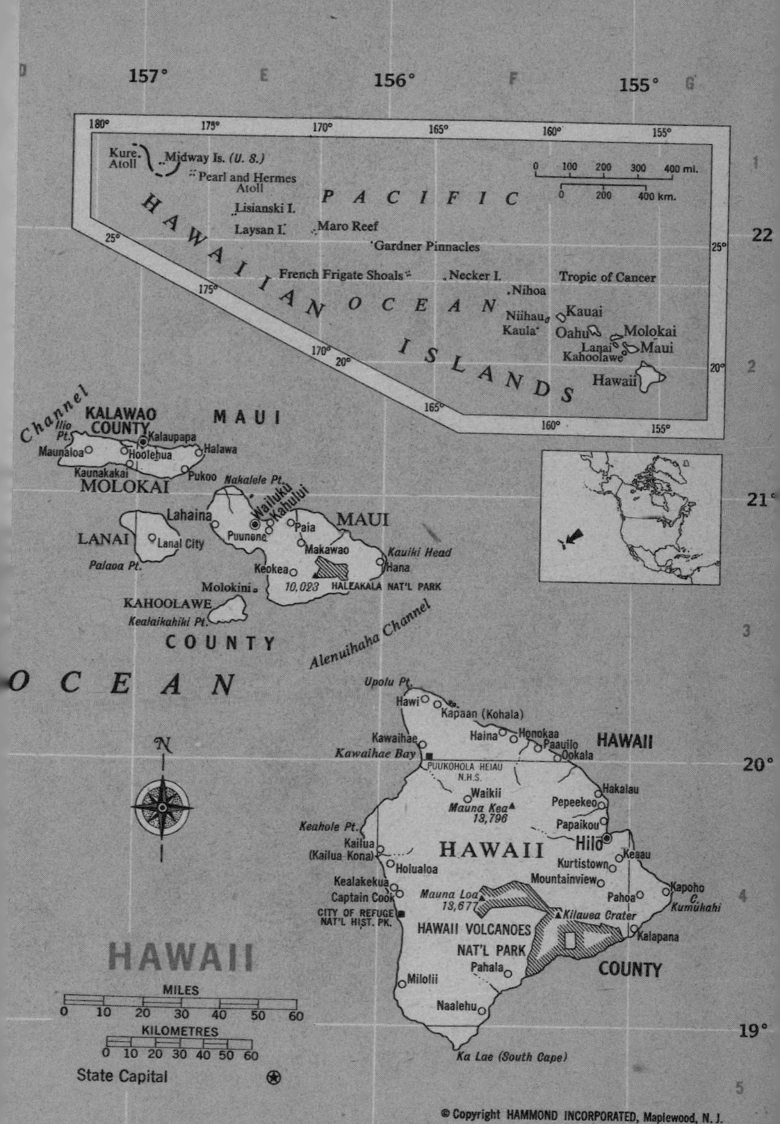

157°
156°
155°
180°
175°
170°
165°
160°
155°
Kure Atoll
Midway Is. (U. S.)
Pearl and Hermes Atoll
Lisianski I.
Laysan I.
Maro Reef
Gardner Pinnacles
French Frigate Shoals
Necker I.
Nihoa
PACIFIC OCEAN
HAWAIIAN ISLANDS
Tropic of Cancer
Niihau
Kaula
Kauai
Oahu
Molokai
Lanai
Maui
Kahoolawe
Hawaii
0 100 200 300 400 mi.
0 200 400 km.
25°
20°
22°
21°
20°
19°
Channel
Ilio Pt.
KALAWAO COUNTY
Kalaupapa
MAUI
Maunaloa
Hoolehua
Halawa
Kaunakakai
Pukoo
MOLOKAI
Nakalele Pt.
Wailuku
Kahului
Lahaina
MAUI
LANAI
Lanai City
Puunene
Paia
Makawao
Kauiki Head
Hana
Palaoa Pt.
Keokea
Molokini
10,023
HALEAKALA NAT'L PARK
KAHOOLAWE
Kealaikahiki Pt.
COUNTY
Alenuihaha Channel
OCEAN
Upolu Pt.
Hawi
Kapaan (Kohala)
Haina
Honokaa
Paauilo
Ookala
HAWAII
Kawaihae
Kawaihae Bay
PUUKOHOLA HEIAU N.H.S.
Waikii
Mauna Kea 13,796
Hakalau
Pepeekeo
Papaikou
Keahole Pt.
Hilo
Kailua (Kailua Kona)
HAWAII
Keaau
Kurtistown
Holualoa
Mountainview
Kealakekua
Captain Cook
Mauna Loa 13,677
Pahoa
Kapoho
C. Kumukahi
CITY OF REFUGE NAT'L HIST. PK.
Kilauea Crater
HAWAII VOLCANOES NAT'L PARK
Kalapana
Pahala
COUNTY
Milolii
Naalehu
Ka Lae (South Cape)
N
HAWAII
MILES
0 10 20 30 40 50 60
KILOMETRES
0 10 20 30 40 50 60
State Capital
Longitude 157° West of Greenwich 156° 155°

IDAHO
MILES
0 20 40 60 80
KILOMETRES
0 20 40 60 80
State Capitals
BRITISH COLUMBIA
ALBERTA
WATERTON-GLACIER
INT'L PEACE PK.
WATERTON LAKES NAT'L PARK
BLACKFEET INDIAN RESERVATION
GLACIER NATIONAL PARK
FLATHEAD INDIAN RESERVATION
CONTINENTAL DIVIDE
ROCKY MOUNTAINS
CABINET MTS.
BITTERROOT
CLEARWATER
IDAHO MOUNTAINS
MONTANA
WASHINGTON
Great Falls
Helena
Butte
Bozeman
Missoula
Kalispell
Libby
Lake Koocanusa
Kootenai R.
Flathead
Flathead L.
Hungry Horse Res.
Canyon Ferry Res.
Missouri
Smith R.
Jefferson R.
Clark Fork
Blackfoot R.
Bitterroot R.
Big Hole
Lost Trail Pass 6,990
Lolo Pass 5,187
E. Sister Pk. 6,866
BOUNDARY
BONNER
KOOTENAI
SHOSHONE
BENEWAH
LATAH
CLEARWATER
NEZ PERCE
LEWIS
Metaline Falls
Pend Oreille R.
KALISPEL IND. RES.
Priest L.
Bonners Ferry
Sandpoint
Clark Fork
Priest River
Spirit Lake
Lake Pend Oreille
Rathdrum
Hayden
Coeur d'Alene
Post Falls
Smelterville
Kellogg
Osburn
Wallace
Mullan
Coeur d'Alene L.
St. Maries
Avery
St. Joe R.
Potlatch
Colfax
Pullman
Troy
Moscow
Elk River
N. Fork
Genesee
Clearwater
Orofino
Headquarters
Pierce
Lochsa R.
NEZ PERCE NAT'L HIST PARK
Lapwai
Lewiston
Snake
Weippe
Kamiah
Craigmont
Nezperce
Kooskia
Cottonwood
Selway
Grangeville
High Mtn. Sheep Res.
White Bird
Elk City
River
Spokane
Spokane R.
48°
46°
112°
114°
116°

YELLOWSTONE
NATIONAL
PARK
Shoshone L.
Yellowstone
Jackson L.
GRAND TETON NAT'L PARK
Jackson
WYOMING
MONTANA
Hebgen L.
Madison
Dillon
CENTENNIAL MTS.
Island Park Res.
BEAVERHEAD MTS.
LEMHI RANGE
LOST RIVER RA.
Borah Pk. 12,662
FREMONT
Ashton
St. Anthony
Rexburg
TETON
Driggs
MADISON
Snake River
Rigby
Iona
Ammon
BONNEVILLE
Palisades Res.
Idaho Falls
JEFFERSON
CLARK
Dubois
Mud L.
NAT'L REACTOR TESTING STA.
U.S.A.E.C. RES.
Atomic City
7,659
BUTTE
Arco
Moore
Mackay
BINGHAM
Shelley
Blackfoot
FORT HALL IND. RES.
Grays L.
Blackfoot Res.
CARIBOU
Soda Sprs.
Georgetown
Grace
Lava Hot Sprs.
Chubbuck
Pocatello
Inkom
McCammon
BANNOCK
Downey
Montpelier
Paris
BEAR LAKE
Bear L.
Preston
FRANKLIN
Franklin
American Falls
American Falls Res.
Aberdeen
Springfield
POWER
Malad City
ONEIDA
UTAH
Walcott Res.
Rupert
Heyburn
MINIDOKA
Burley
CASSIA
Albion
Oakley
Cache Pk. 10,340
Almo
LEMHI
Salmon
Baker
Leadore
Cobalt
Challis
Twin Pks. 10,328
Clayton
CUSTER
Castle Pk. 11,820
SALMON RIVER MTS.
VALLEY
Mormon Mtn. 9,545
Middle Fork
Waugh Mtn. 8,882
SAWTOOTH RA.
Sun Valley
Ketchum
Hailey
Bellevue
BLAINE
Carey
CRATERS OF THE MOON NAT'L MON.
LINCOLN
Richfield
Shoshone
JEROME
Jerome
Paul
Kimberly
Shoshone Falls
Twin Falls
TWIN FALLS
Salmon Cr. Res.
Filer
Buhl
Wendell
Gooding
GOODING
Bliss
Fairfield
CAMAS
Atlanta
Anderson Ranch Res.
ELMORE
Mountain Home
Glenns Ferry
Hammett
Cape Horn Mtn. 9,600
SAWTOOTH NAT'L REC. AREA
Garden Valley
BOISE
Idaho City
Arrowrock Res.
Boise
Garden City
Meridian
ADA
Nampa
MOUNTAIN HOME A.F.B.
C. J. Strike Res.
SNAKE
OWYHEE
Murphy
Bruneau R.
Three Creek
NEVADA
Riddle
WESTERN SHOSHONE IND. RES.
CANYON
Caldwell
Parma
Homedale
GEM
Emmett
PAYETTE
Fruitland
New Plymouth
Payette
WASHINGTON
Weiser
Cambridge
Council
ADAMS
New Meadows
McCall
Cascade
Cascade Res.
Warm Lake
South Fork
Salmon
Riggins
He-Devil Mtn. 9,887
GRAND CANYON OF THE SNAKE
OXBOW DAM
BROWNLEE DAM
Enterprise
Vale
OREGON
Owyhee
Owyhee L.
44°
42°
112°
114°
116°
Longitude West of Greenwich
©C. S. HAMMOND & Co., Maplewood, N. J.

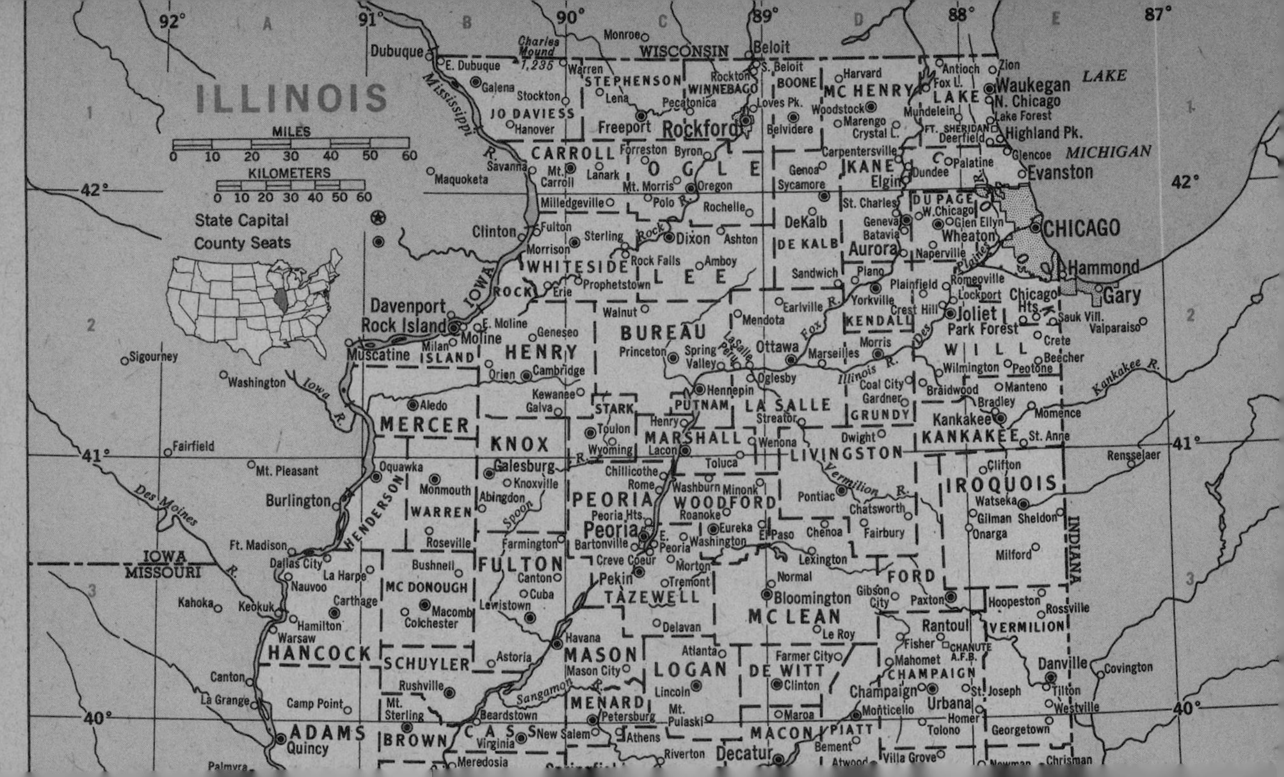

ILLINOIS
MILES
0 10 20 30 40 50 60
KILOMETERS
0 10 20 30 40 50 60
State Capital
County Seats
WISCONSIN
INDIANA
IOWA
MISSOURI
LAKE
MICHIGAN
Mississippi R.
Illinois R.
Rock R.
Fox R.
Des Plaines R.
Kankakee R.
Vermilion R.
Sangamon R.
Spoon R.
Iowa R.
Des Moines R.
Charles Mound 1,235
JO DAVIESS
STEPHENSON
WINNEBAGO
BOONE
MC HENRY
LAKE
CARROLL
OGLE
DE KALB
KANE
DU PAGE
COOK
WHITESIDE
LEE
ROCK ISLAND
HENRY
BUREAU
LA SALLE
KENDALL
WILL
GRUNDY
PUTNAM
STARK
MERCER
KNOX
MARSHALL
LIVINGSTON
KANKAKEE
IROQUOIS
WARREN
PEORIA
WOODFORD
HENDERSON
FULTON
TAZEWELL
MC LEAN
FORD
VERMILION
MC DONOUGH
HANCOCK
SCHUYLER
MASON
LOGAN
DE WITT
CHAMPAIGN
MENARD
ADAMS
BROWN
CASS
MACON
PIATT
Dubuque
E. Dubuque
Galena
Stockton
Hanover
Warren
Monroe
Lena
Freeport
Pecatonica
Rockton
Beloit
S. Beloit
Rockford
Loves Pk.
Belvidere
Harvard
Woodstock
Marengo
Crystal L.
Antioch
Zion
Fox L.
Waukegan
N. Chicago
Mundelein
Lake Forest
FT. SHERIDAN
Deerfield
Highland Pk.
Glencoe
Evanston
Savanna
Mt. Carroll
Lanark
Forreston
Byron
Mt. Morris
Oregon
Polo
Rochelle
Carpentersville
Genoa
Sycamore
Dundee
Palatine
Elgin
St. Charles
Geneva
Batavia
W. Chicago
Glen Ellyn
Wheaton
Naperville
Aurora
CHICAGO
Hammond
Gary
Maquoketa
Milledgeville
Clinton
Fulton
Morrison
Sterling
Dixon
Ashton
DeKalb
Rock Falls
Amboy
Sandwich
Plano
Plainfield
Romeoville
Lockport
Chicago Hts.
Sauk Vill.
Valparaiso
Erie
Prophetstown
Yorkville
Crest Hill
Joliet
Park Forest
Crete
Beecher
Davenport
Rock Island
Moline
E. Moline
Milan
Muscatine
Geneseo
Walnut
Mendota
Earlville
Ottawa
Marseilles
Morris
Wilmington
Peotone
Manteno
Sigourney
Washington
Orion
Cambridge
Princeton
Spring Valley
La Salle
Peru
Oglesby
Hennepin
Coal City
Gardner
Braidwood
Bradley
Momence
Kankakee
St. Anne
Kewanee
Galva
Aledo
Toulon
Henry
Streator
Wenona
Dwight
Fairfield
Lacon
Wyoming
Toluca
Rensselaer
Mt. Pleasant
Oquawka
Galesburg
Knoxville
Chillicothe
Rome
Washburn
Minonk
Pontiac
Clifton
Monmouth
Abingdon
Peoria Hts.
Roanoke
Chatsworth
Watseka
Gilman
Sheldon
Onarga
Burlington
Roseville
Farmington
Peoria
E. Peoria
Bartonville
Eureka
El Paso
Chenoa
Fairbury
Milford
Ft. Madison
Dallas City
La Harpe
Bushnell
Canton
Creve Coeur
Morton
Washington
Lexington
Nauvoo
Carthage
Cuba
Pekin
Tremont
Normal
Bloomington
Gibson City
Paxton
Hoopeston
Rossville
Kahoka
Keokuk
Hamilton
Warsaw
Macomb
Colchester
Lewistown
Delavan
Le Roy
Rantoul
Fisher
CHANUTE A.F.B.
Astoria
Havana
Mason City
Atlanta
Farmer City
Mahomet
Danville
Covington
Canton
La Grange
Camp Point
Rushville
Mt. Sterling
Beardstown
Lincoln
Clinton
Champaign
Urbana
St. Joseph
Tilton
Westville
Quincy
Virginia
New Salem
Petersburg
Athens
Mt. Pulaski
Maroa
Monticello
Homer
Tolono
Georgetown
Palmyra
Meredosia
Riverton
Decatur
Bement
Atwood
Villa Grove
Chrisman

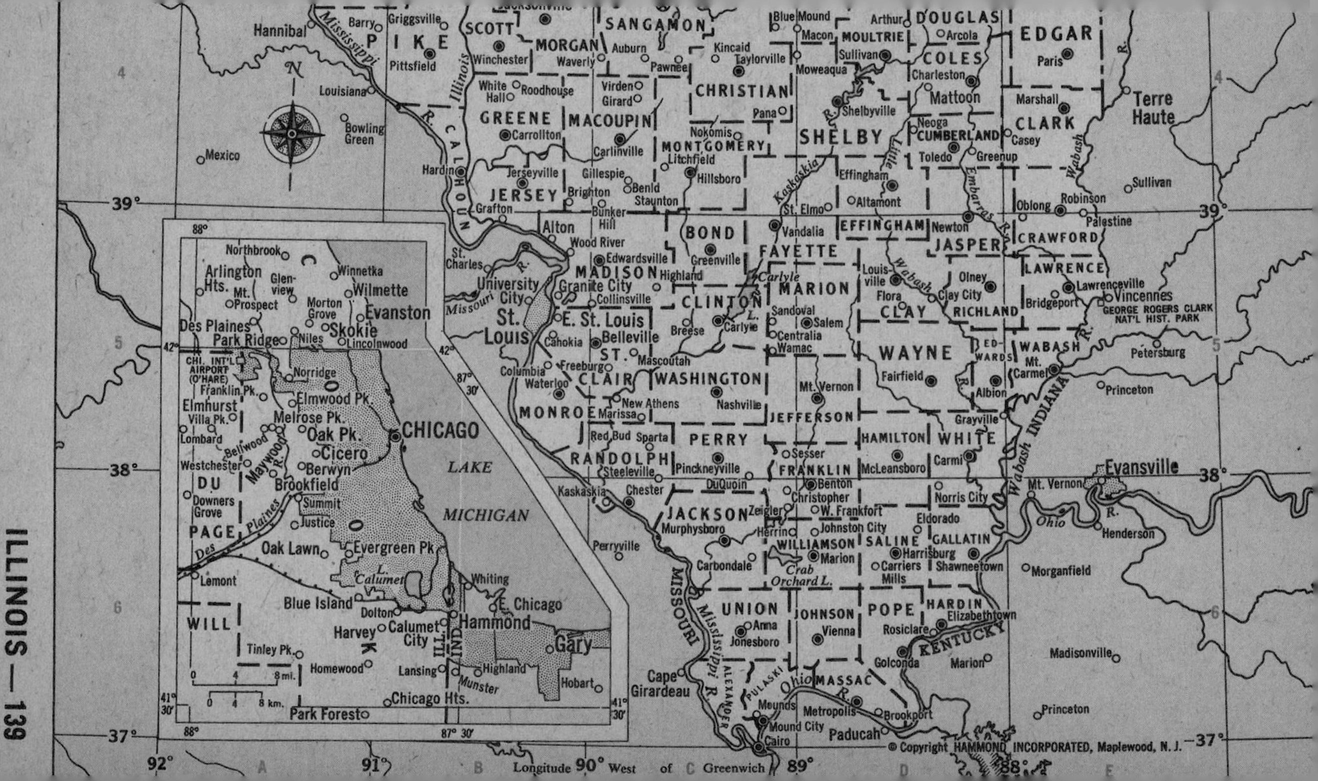

Hannibal
Barry
Griggsville
PIKE
Pittsfield
SCOTT
Winchester
MORGAN
Waverly
Auburn
SANGAMON
Pawnee
Kincaid
Taylorville
Blue Mound
Macon
Moweaqua
MOULTRIE
Sullivan
Arthur
DOUGLAS
Arcola
COLES
Charleston
Mattoon
EDGAR
Paris
Marshall
CLARK
Casey
Terre Haute
Louisiana
Bowling Green
Mexico
Illinois
White Hall
Roodhouse
Virden
Girard
CHRISTIAN
Pana
Shelbyville
SHELBY
Neoga
CUMBERLAND
Toledo
Greenup
GREENE
Carrollton
MACOUPIN
Carlinville
Nokomis
MONTGOMERY
Litchfield
Hillsboro
Kaskaskia
Effingham
Little Wabash
Embarras R.
Wabash
Sullivan
Hardin
CALHOUN
Jerseyville
Gillespie
Benld
Staunton
JERSEY
Brighton
Grafton
Bunker Hill
St. Elmo
Altamont
Oblong
Robinson
Palestine
Alton
Wood River
BOND
Vandalia
EFFINGHAM
Newton
JASPER
CRAWFORD
St. Charles
Edwardsville
Greenville
FAYETTE
MADISON
Highland
Carlyle
Louisville
Olney
LAWRENCE
Lawrenceville
Vincennes
GEORGE ROGERS CLARK NAT'L HIST. PARK
University City
Granite City
Collinsville
CLINTON
MARION
Flora
Clay City
Bridgeport
Missouri R.
St. Louis
E. St. Louis
Breese
Carlyle L.
Sandoval
Salem
Centralia
CLAY
RICHLAND
Cahokia
Belleville
Wamac
WAYNE
EDWARDS
WABASH
Mt. Carmel
Petersburg
Columbia
Freeburg
ST. CLAIR
Mascoutah
Waterloo
WASHINGTON
Fairfield
Albion
Princeton
New Athens
Nashville
Mt. Vernon
MONROE
Marissa
Red Bud
Sparta
JEFFERSON
Grayville
PERRY
HAMILTON
WHITE
RANDOLPH
Sesser
Pinckneyville
McLeansboro
Carmi
Steeleville
FRANKLIN
INDIANA
Evansville
Kaskaskia
Chester
Du Quoin
Benton
Norris City
Mt. Vernon
JACKSON
Zeigler
Christopher
W. Frankfort
Eldorado
Ohio R.
Henderson
Murphysboro
Herrin
Johnston City
SALINE
GALLATIN
Perryville
WILLIAMSON
Marion
Harrisburg
Carbondale
Carriers Mills
Shawneetown
Morganfield
MISSOURI
Crab Orchard L.
UNION
JOHNSON
POPE
HARDIN
Elizabethtown
Anna
Jonesboro
Vienna
Rosiclare
KENTUCKY
Golconda
Marion
Madisonville
Mississippi R.
Cape Girardeau
ALEXANDER
PULASKI
MASSAC
Mounds
Metropolis
Brookport
Princeton
Mound City
Paducah
Cairo
LAKE MICHIGAN
Northbrook
Winnetka
Arlington Hts.
Glenview
Mt. Prospect
Wilmette
Morton Grove
Evanston
Des Plaines
Park Ridge
Niles
Skokie
Lincolnwood
CHI. INT'L AIRPORT (O'HARE)
Norridge
Franklin Pk.
Elmwood Pk.
Elmhurst
Villa Pk.
Melrose Pk.
Lombard
Bellwood
Oak Pk.
CHICAGO
Westchester
Maywood
Cicero
Berwyn
DU PAGE
Brookfield
Downers Grove
Summit
Justice
Des Plaines R.
Oak Lawn
Evergreen Pk.
L. Calumet
Lemont
Whiting
E. Chicago
Blue Island
Dolton
Hammond
WILL
Harvey
Calumet City
ILL.
IND.
Gary
Tinley Pk.
COOK
Homewood
Lansing
Highland
Munster
Hobart
Chicago Hts.
Park Forest
0 4 8 mi.
0 4 8 km.
© Copyright HAMMOND INCORPORATED, Maplewood, N.J.
Longitude 90° West of Greenwich

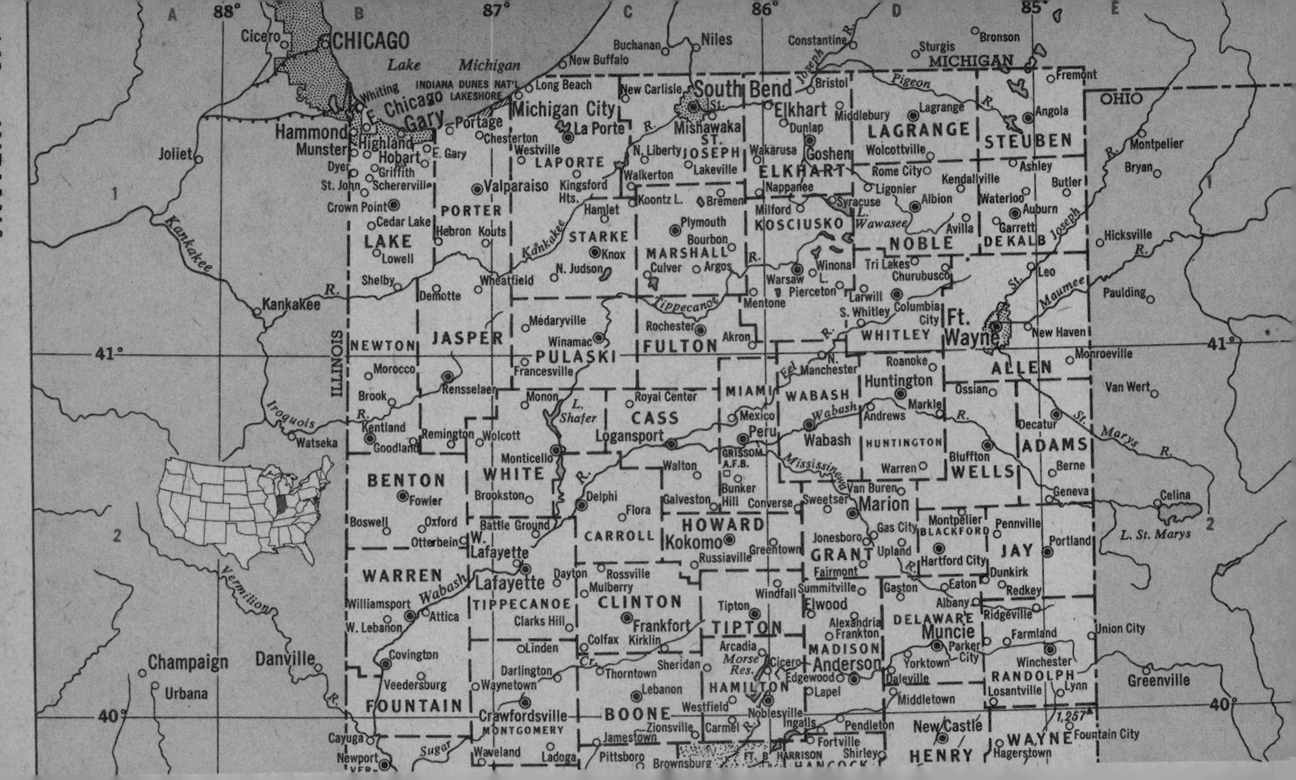
MICHIGAN
OHIO
ILLINOIS
Lake Michigan
INDIANA DUNES NAT'L LAKESHORE
CHICAGO
Cicero
Joliet
Kankakee
Watseka
Danville
Champaign
Urbana
Niles
Buchanan
New Buffalo
Constantine
Sturgis
Bronson
Fremont
Montpelier
Bryan
Hicksville
Paulding
Van Wert
Celina
L. St. Marys
Union City
Greenville
Fountain City
STEUBEN
LAGRANGE
ELKHART
ST. JOSEPH
LAPORTE
PORTER
LAKE
NEWTON
JASPER
STARKE
MARSHALL
KOSCIUSKO
NOBLE
DEKALB
ALLEN
WHITLEY
FULTON
PULASKI
WHITE
BENTON
WARREN
TIPPECANOE
CARROLL
CASS
MIAMI
WABASH
HUNTINGTON
WELLS
ADAMS
JAY
BLACKFORD
GRANT
HOWARD
CLINTON
TIPTON
MADISON
DELAWARE
RANDOLPH
HENRY
WAYNE
HAMILTON
BOONE
MONTGOMERY
FOUNTAIN
Hammond
Munster
Whiting
E. Chicago
Gary
E. Gary
Highland
Hobart
Griffith
Schererville
Dyer
St. John
Crown Point
Cedar Lake
Lowell
Shelby
Hebron
Kouts
Valparaiso
Chesterton
Portage
Michigan City
La Porte
Westville
Kingsford Hts.
Long Beach
New Carlisle
South Bend
Mishawaka
N. Liberty
Walkerton
Lakeville
Koontz L.
Hamlet
Knox
N. Judson
Wakarusa
Elkhart
Dunlap
Goshen
Nappanee
Bristol
Middlebury
Lagrange
Wolcottville
Rome City
Ligonier
Angola
Ashley
Auburn
Garrett
Butler
Waterloo
Kendallville
Albion
Avilla
Ft. Wayne
New Haven
Leo
Monroeville
Warsaw
Winona L.
Pierceton
Syracuse
L. Wawasee
Milford
Tri Lakes
Churubusco
Larwill
Columbia City
S. Whitley
Plymouth
Bremen
Bourbon
Argos
Culver
Rochester
Akron
Mentone
Huntington
Roanoke
Markle
Andrews
Wabash
N. Manchester
Mexico
Peru
Logansport
Royal Center
Winamac
Medaryville
Rensselaer
Remington
Wolcott
Monticello
Monon
L. Shafer
Francesville
Goodland
Kentland
Brook
Morocco
Demotte
Wheatfield
Fowler
Oxford
Boswell
Otterbein
Brookston
Battle Ground
Lafayette
W. Lafayette
Delphi
Flora
Kokomo
Russiaville
Greentown
Bunker Hill
GRISSOM A.F.B.
Converse
Galveston
Walton
Marion
Gas City
Jonesboro
Sweetser
Van Buren
Warren
Bluffton
Ossian
Decatur
Berne
Geneva
Portland
Pennville
Dunkirk
Redkey
Montpelier
Hartford City
Upland
Fairmount
Summitville
Gaston
Alexandria
Frankton
Elwood
Tipton
Windfall
Anderson
Muncie
Eaton
Albany
Yorktown
Parker City
Daleville
Middletown
New Castle
Winchester
Farmland
Ridgeville
Lynn
Losantville
Hagerstown
1,257
Pendleton
Ingalls
Fortville
Shirley
Noblesville
Cicero
Edgewood
Lapel
Arcadia
Morse Res.
Westfield
Carmel
Zionsville
Lebanon
Sheridan
Thorntown
Frankfort
Kirklin
Colfax
Rossville
Mulberry
Dayton
Clarks Hill
Linden
Darlington
Crawfordsville
Waynetown
Ladoga
Jamestown
Pittsboro
Brownsburg
Attica
Williamsport
W. Lebanon
Covington
Veedersburg
Newport
Cayuga
Waveland
Tippecanoe
Kankakee R.
Iroquois R.
Wabash R.
Eel R.
Mississinewa R.
St. Joseph R.
St. Marys R.
Maumee
Pigeon R.
Vermilion R.
Sugar Cr.
A
B
C
D
E
1
2
88°
87°
86°
85°
41°
40°

INDIANA
MILES
0 10 20 30 40 50
KILOMETERS
0 10 20 30 40 50
State Capital
County Seats
Longitude West of Greenwich
© C. S. HAMMOND & Co. Maplewood, N.J.

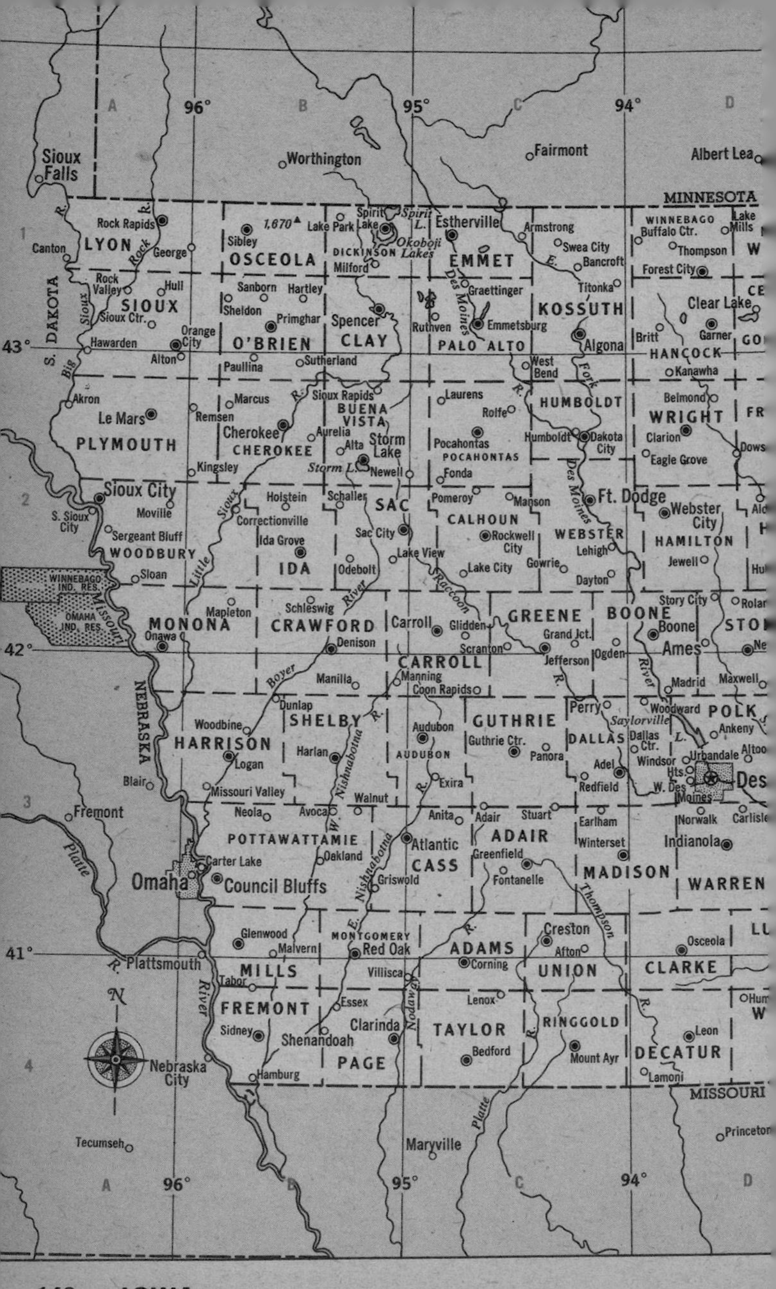
Sioux Falls
Worthington
Fairmont
Albert Lea
MINNESOTA
S. DAKOTA
LYON
OSCEOLA
DICKINSON
EMMET
KOSSUTH
WINNEBAGO
SIOUX
O'BRIEN
CLAY
PALO ALTO
HANCOCK
PLYMOUTH
CHEROKEE
BUENA VISTA
POCAHONTAS
HUMBOLDT
WRIGHT
WOODBURY
IDA
SAC
CALHOUN
WEBSTER
HAMILTON
MONONA
CRAWFORD
CARROLL
GREENE
BOONE
HARRISON
SHELBY
AUDUBON
GUTHRIE
DALLAS
POLK
POTTAWATTAMIE
CASS
ADAIR
MADISON
WARREN
MILLS
MONTGOMERY
ADAMS
UNION
CLARKE
FREMONT
PAGE
TAYLOR
RINGGOLD
DECATUR
NEBRASKA
MISSOURI
Sioux City
Le Mars
Cherokee
Storm Lake
Spencer
Estherville
Algona
Ft. Dodge
Webster City
Ames
Omaha
Council Bluffs
Atlantic
Red Oak
Creston
Clarinda
Shenandoah
Indianola
Nebraska City
Plattsmouth
Fremont
Maryville
96°
95°
94°
43°
42°
41°

IOWA
MILES
0
20
40
60
KILOMETERS
0
20
40
60
State Capital
County Seats
93°
92°
91°
90°
43°
42°
41°
E
F
G
1
2
3
4
Austin
L.
bert Lea
Caledonia
New Albin
Lansing
Waukon
MITCHELL
St. Ansgar
Riceville
Osage
HOWARD
Elma
Cresco
WINNESHIEK
Decorah
ALLAMAKEE
Calmar
Postville
EFFIGY MOUNDS NAT'L MON.
McGregor
Monona
Wisconsin R.
Nora Sprs.
FLOYD
Charles City
Rockford
New Hampton
CHICKASAW
Turkey
W. Union
CLAYTON
Elkader
Guttenberg
MISSISSIPPI
Nashua
Greene
Clarksville
Allison
BUTLER
Shell Rock
Parkersburg
Cedar
Tripoli
Waverly
BREMER
Denver
Sumner
Wapsipinicon
FAYETTE
Fayette
Oelwein
Strawberry Pt.
Hazleton
BUCHANAN
Edgewood
DELAWARE
Dyersville
Manchester
Farley
Dubuque
DUBUQUE
Hopkinton
Cascade
Lancaster
Platteville
Darlington
WISCONSIN
ILLINOIS
Galena
Bellevue
RIVER
Sheffield
Hampton
Ackley
Cedar Falls
GRUNDY
Waterloo
BLACK HAWK
Jesup
Evansdale
Independence
Grundy Ctr.
Hudson
Washburn
Reinbeck
La Porte City
River
Conrad
Iowa
Gladbrook
Traer
TAMA
Dysart
Vinton
BENTON
Center Point
Central City
LINN
Monticello
JONES
Anamosa
JACKSON
Maquoketa
Preston
Sabula
Wyoming
State Ctr.
Marshalltown
Hiawatha
Marion
Toledo
Tama
MARSHALL
SAC & FOX IND. RES.
River
Belle Plaine
Cedar Rapids
Lisbon
Mt. Vernon
Olin
CLINTON
Clinton
Mechanicsville
Clarence
De Witt
Camanche
Tipton
CEDAR
Amana
JASPER
Newton
Grinnell
Brooklyn
POWESHIEK
Marengo
Victor
IOWA
Williamsburg
Iowa City
Coralville
University Hts.
JOHNSON
W. Branch
H. HOOVER NAT'L HIST. SITE
Durant
SCOTT
Le Claire
Bettendorf
Davenport
Moline
Rock Island
Wilton
Skunk
Montezuma
N. English
W. Liberty
MUSCATINE
Muscatine
ILLINOIS
Red Rock L.
R.
New Sharon
Wellman
Kalona
WASHINGTON
What Cheer
Sigourney
Keota
Pella
MAHASKA
Washington
Columbus Jct.
MARION
Oskaloosa
KEOKUK
Skunk
LOUISA
Aledo
Melcher
Eddyville
Hedrick
Brighton
Wapello
Lovilia
WAPELLO
JEFFERSON
Winfield
Morning Sun
MONROE
Chariton
Albia
Ottumwa
Fairfield
HENRY
Mediapolis
RIVER
Des Moines
Eldon
Mt. Pleasant
New London
DES MOINES
Galesburg
Monmouth
Chariton
Rathbun L.
Moravia
DAVIS
VAN BUREN
W. Burlington
Mystic
Centerville
Keosauqua
Burlington
Bloomfield
Milton
APPANOOSE
Moulton
West Point
Farmington
Ft. Madison
LEE
MISSISSIPPI
Unionville
River
Kahoka
Keokuk
© Copyright HAMMOND INCORPORATED, Maplewood, N. J.
ngitude 93° West of Greenwich
92°
91°

KANSAS
CHEYENNE
RAWLINS
DECATUR
NORTON
PHILLIPS
SMITH
SHERMAN
THOMAS
SHERIDAN
GRAHAM
ROOKS
OSBORNE
WALLACE
LOGAN
GOVE
TREGO
ELLIS
RUSSELL
GREELEY
WICHITA
SCOTT
LANE
NESS
RUSH
BARTON
HAMILTON
KEARNY
FINNEY
HODGEMAN
PAWNEE
STAFFORD
STANTON
GRANT
HASKELL
GRAY
FORD
EDWARDS
KIOWA
PRATT
MORTON
STEVENS
SEWARD
MEADE
CLARK
COMANCHE
BARBER
COLORADO
NEBRASKA
TEXAS
SMOKY HILLS
Mt. Sunflower 4,039
McCook
Alma
Red Cloud
St. Francis
Bird City
Atwood
Oberlin
Norton
Almena
Phillipsburg
Kensington
Lebanon
Smith Ctr.
Lenora
Logan
Goodland
Colby
Hoxie
Hill City
Stockton
Osborne
Downs
Palco
Plainville
Natoma
Oakley
Quinter
WaKeeney
Lucas
Sharon Sprs.
Russell Sprs.
Gove
Ellis
Hays
Victoria
Russell
Tribune
Leoti
Scott City
Dighton
Ness City
La Crosse
Hoisington
Claflin
Great Bend
Ellinwood
Larned
FT. LARNED NAT'L HIST. SITE
Jetmore
Syracuse
Lakin
Garden City
Kinsley
Macksville
St. John
Stafford
Cimarron
Dodge City
Spearville
Ft. Dodge
Pratt
Johnson
Ulysses
Montezuma
Haviland
Greensburg
Sublette
Satanta
Bucklin
Minneola
Fowler
Richfield
Plains
Meade
Ashland
Coldwater
Protection
Medicine Lodge
Hugoton
Elkhart
Liberal
Kiowa
Alva
Swanson L.
Republican R.
Harlan Co. Lake
Beaver Cr.
Sappa Cr.
Norton Res.
Kirwin Res.
North Fork
Webster Res.
South Fork
North Fork
Saline R.
Cedar Bluff Res.
Smoky Hill River
Cheyenne Bottoms
Pawnee R.
L. McKinney
Arkansas R.
Crooked Cr.
Cimarron R.
Medicine Lodge
Beaver
Salt
River
102° 101° 100° 99°
40° 39° 38° 37°
A B C D
1 2 3 4 5
MILES
0 20 40 60 80
KILOMETERS
0 20 40 60 80
State Capital
County Seats
© Copyright HAMMOND INCORPORATED, Maplewood, N. J.
Counties indicated by nu
1 GEARY
2 JEFFERSON
3 LEAVENWORTH
4 SHAWNEE
5 WYANDOTTE

98°
E
97°
F
96°
G
95°
H
1
2
3
4
5
40°
39°
38°
37°
Big Blue R.
HOMESTEAD NAT'L MON.
Beatrice
Maryville
Little Blue
Fairbury
Superior
Falls City
MISSOURI
MISSOURI
Platte
REPUBLIC
WASHINGTON
Hanover
Sabetha
SAC-FOX IOWA IND. RES.
Belleville
Washington
Marysville
Seneca
Hiawatha
Highland
Troy
St. Joseph
MARSHALL
NEMAHA
BROWN
Wathena
Elwood
Scandia
Greenleaf
Centralia
Horton
DONIPHAN
Republican
Waterville
Blue Rapids
Frankfort
KICKAPOO I. R.
Atchison
Jewell
Clyde
Clifton
Effingham
Concordia
CLOUD
CLAY
Tuttle Creek Lake
Onaga
JACKSON
Holton
ATCHISON
Clay Ctr.
Westmoreland
Nortonville
FT. LEAVENWORTH
Solomon
Glasco
Miltonvale
Riley
POTAWATOMI IND. RES.
Valley Falls
Leavenworth
POTTAWATOMIE
Perry L.
Lansing
Kansas
Delphos
OTTAWA
Wakefield
RILEY
Wamego
St. Marys
Oskaloosa
Kansas City
City
Minneapolis
Milford L.
Ogden
Manhattan
Rossville
Tonganoxie
Bonner Sprs
Lincoln
Bennington
FT. RILEY
Kansas
Shawnee
Prairie Village
Chapman
Junction City
Alma
Topeka
Lawrence
Overland Pk.
Abilene
WABAUNSEE
FORBES A.F.B.
Eudora
Olathe
Solomon
Eskridge
DOUGLAS
Salina
Enterprise
Carbondale
Gardner
JOHNSON
SALINE
DICKINSON
MORRIS
Baldwin City
Spring Hill
Council Grove L.
Council Grove
Burlingame
Wellsville
Louisburg
Kanopolis
Gypsum
Pomona L.
Lyndon
des Cygnes
Kanopolis L.
Herington
Osage City
Ottawa
Paola
Lindsborg
MARION
LYON
Marais
Osawatomie
Marquette
Marion L.
Strong City
OSAGE
FRANKLIN
MIAMI
Geneseo
Emporia
Little River
MC PHERSON
Marion
COFFEY
Garnett
La Cygne
Lyons
McPherson
Canton
Hillsboro
Cottonwood
Cottonwood Falls
John Redmond Res.
LINN
Sterling
Inman
CHASE
Burlington
Pleasanton
Moundridge
Florence
Olpe
ANDERSON
Peabody
Mound City
Buhler
Hesston
FLINT HILLS
Madison
Lamont
Burrton
HARVEY
Le Roy
Neosho
Newton
BUTLER
Moran
BOURBON
Hutchinson
Halstead
Potwin
GREENWOOD
WOODSON
Haven
Sedgwick
Eureka
Yates Ctr.
Iola
Ft. Scott
Cheney
El Dorado
ALLEN
Valley Ctr.
Toronto L.
Toronto
Humboldt
Colwich
SEDGWICK
Towanda
Andover
Fall River L.
Chanute
NEOSHO
Arcadia
Kingman
Wichita
Augusta
WILSON
Erie
CRAWFORD
McCONNELL A.F.B.
Cheney
Fredonia
Girard
Arma
Haysville
Derby
Douglass
Howard
ELK
St. Paul
Frontenac
Clearwater
Mulvane
COWLEY
Neodesha
Pittsburg
Conway Sprs.
Belle Plaine
Burden
Moline
Cherryvale
Parsons
Weir
Wellington
LABETTE
Argonia
Elk City Lake
Altamont
CHEROKEE
Oxford
Winfield
Independence
Columbus
SUMNER
Sedan
MONTGOMERY
Oswego
Galena
Chikaskia
Arkansas City
Cedar Vale
CHAUTAUQUA
Coffeyville
Baxter Sprs.
Caldwell
Caney
Chetopa
OKLAHOMA
Hulah Lake
Miami
Great Salt Plains Lake
Bartlesville
Lake O' The Cherokees
Ponca City
Oologah Lake
Fork
Pawhuska
OSAGE IND. RES.
N
Arkansas R.
F3
G2
G2
G2
H2
Enid
Claremore
Keystone Lake
Tulsa
98° Longitude E West of 97° Greenwich F
96°
G
95°
H

WESTERN PART OF KENTUCKY
Same scale as main map
W. Frankfort
Marion
Harrisburg
Shawneetown
Uniontown
Morganfield
UNION
Sturgis
Clay
Anna
Vienna
Rosiclare
Golconda
CRITTENDEN
Marion
Salem
LIVINGSTON
Cape Girardeau
ILLINOIS
Metropolis
Paducah
Smithland
CALDWELL
Princeton
BARKLEY DAM
Eddyville
Calvert City
Mound City
BALLARD
La Center
Lone Oak
Barlow
Woodlawn
Tennessee R.
Cairo
Wickliffe
MC CRACKEN
KENTUCKY DAM
LYON
Charleston
Sikeston
E. Prairie
Mississippi
Bardwell
CARLISLE
Arlington
Benton
Kentucky Lake
Cadiz
MARSHALL
TRIGG
Clinton
Mayfield
LAND BETWEEN THE LAKES REC. AREA
New Madrid
MISSOURI
HICKMAN
Hickman
GRAVES
CALLOWAY
Murray
FULTON
Fulton
TENNESSEE
Barkley
88°
87°
86°
A
B
C
1
2
3
39°
38°
37°
Greenfield
Connersville
Rushville
Shelbyville
Flatrock
Greensburg
Batesv
Columbus
N. Vernon
Austin
Scottsburg
Madison
Carrollton
Bedford
TRIMBLE
Ohio
New Albany
St. Matthews
Middletown
LaGrange
OLDHAM
Crestwood
HENR
Louisville
Shively
Pleasure Ridge Pk.
Buechel
Okolona
Valley Sta.
SHELBY
Shelbyvil
Taylorsville
Salt R.
Mt. Washington
W. Point
Shepherdsville
BULLITT
Bloomfield
Wabash R.
Mt. Vernon
Evansville
Boonville
INDIANA
MEADE
Brandenburg
Tell City
Henderson
River
Hawesville
HANCOCK
Cloverport
Irvington
Muldraugh
Radcliff
Ft. Knox
Lebanon Jct.
Bardstown
NELSON
Vine Grove
Elizabethtown
WASHINGTON
Springfield
Uniontown
HENDERSON
Owensboro
Whitesville
Hardinsburg
BRECKINRIDGE
Rough R. L.
HARDIN
Morganfield
UNION
Sturgis
Sebree
WEBSTER
DAVIESS
Green
Calhoun
Fordsville
Rough
Hodgenville
New Haven
Rolling
Lebanon
Clay
Dixon
MC LEAN
Livermore
OHIO
Hartford
Leitchfield
GRAYSON
ABRAHAM LINCOLN BIRTHPLACE NAT'L HIST. SITE
Fork
MARION
Providence
Tradewater
HOPKINS
Madisonville
Earlington
Beaver Dam
Nolin L.
HART
Campbellsville
Greensburg
TAYLOR
Central City
Mortons Gap
CALDWELL
Dawson Sprs.
Nortonville
Greenville
Drakesboro
Morgantown
EDMONSON
Munfordville
Horse Cave
GREEN
Green R. Res.
Princeton
MUHLENBERG
BUTLER
Brownsville
MAMMOTH CAVE NAT'L PK.
Cave City
Columbia
Crofton
Lewisburg
Hadley
Barren
Smiths Grove
BARREN
ADAIR
CHRISTIAN
Bowling Green
Glasgow
Edmonton
METCALFE
Jamestown
Cadiz
Hopkinsville
LOGAN
Auburn
WARREN
Barren R. Lake
WOLF CR. DAM
Elkton
Burkesville
TRIGG
Lake
Pembroke
TODD
Russellville
Scottsville
MONROE
Tompkinsville
Cumberland R.
CLINTO
Albany
FT. CAMPBELL
Guthrie
Adairville
Franklin
SIMPSON
ALLEN
Barkley
TENNESSEE
Clarksville
Portland
Gamaliel
Springfield
Longitude

KENTUCKY
D
84°
E
83°
F
82°
MILES
0 5 10 20 30 40 50 60
KILOMETERS
0 5 10 20 30 40 50 60
State Capital
County Seats
1
2
3
4
39°
38°
37°
Counties indicated by numbers:
1 CAMPBELL D2
2 CUMBERLAND C4
3 GARRARD D3
4 JEFFERSON C2
5 JESSAMINE D3
6 LARUE C3
7 ROBERTSON D2
8 SPENCER C2
9 WOODFORD D2
INDIANA
OHIO
Eaton
Dayton
Middletown
Hamilton
Great Miami R.
Little Miami R.
Harrison
Greenfield
Hillsboro
Newport
Ft. Thomas
Ludlow
Covington
Florence
Park Hills
Alexandria
Independence
BOONE
KENTON
Walton
GRANT
Dry Ridge
Williamstown
PENDLETON
Augusta
Falmouth
BRACKEN
Brooksville
Georgetown
W. Union
Portsmouth
Fullerton
OHIO
Ohio
MASON
Maysville
Mt. Olivet
Licking
Vanceburg
LEWIS
GREENUP
Greenup
Ironton
Flatwoods
Ashland
Westwood
Catlettsburg
Huntington
Pomeroy
Point Pleasant
Gallipolis
River
HARRISON
Cynthiana
OWEN
SCOTT
NICHOLAS
Millersburg
Carlisle
Flemingsburg
FLEMING
Grayson
Olive Hill
CARTER
BOYD
WEST
VIRGINIA
Georgetown
BOURBON
Paris
Owingsville
Morehead
ROWAN
Grayson L.
Midway
FAYETTE
Versailles
Lexington
MONT-GOMERY
Mt. Sterling
BATH
ELLIOTT
Sandy Hook
LAWRENCE
Louisa
Big Sandy
Lawrenceburg
Winchester
CLARK
Clay City
Frenchburg
MENIFEE
W. Liberty
MORGAN
JOHNSON
Levisa Fk.
Tug Fk.
Inez
Beauty
Logan
Nicholasville
Wilmore
River
Stanton
POWELL
WOLFE
MAGOFFIN
Paintsville
MARTIN
Dewey L.
Williamson
Hardy
S. Williamson
Burgin
Richmond
Irvine
Campton
Salyersville
Auxier
Prestonsburg
Danville
MADISON
ESTILL
LEE
Lancaster
Berea
Beattyville
BREATHITT
Jackson
Martin
Betsy Layne
PIKE
McVeigh
Phelps
Stanford
LINCOLN
JACKSON
Booneville
FLOYD
Wayland
Pikeville
Fishtrap L.
Crab Orchard
ROCK-CASTLE
Mt. Vernon
McKee
OWSLEY
PLATEAU
KNOTT
Wheelwright
Virgie
Lookout
Sextons Cr.
Oneida
Combs
Hindman
Elkhorn City
Grundy
Hazard
McRoberts
PULASKI
Pittsburg
LAUREL
Manchester
Lothair
PERRY
Whitesburg
Jenkins
Clintwood
Somerset
London
Hima
Hyden
LESLIE
LETCHER
Laurel R.
CLAY
Leatherwood
Cumberland
CUMBERLAND
Burnside
Gray
KNOX
Lynch
Norton
Corbin
Barbourville
Benham
Black Mtn.
4,145
Evarts
Cumberland
Monticello
WHITLEY
Artemus
Loyall
Harlan
Big Stone Gap
Cawood
HARLAN
Pineville
Clinch R.
Holston R.
N. Fk.
Whitley City
Williamsburg
BELL
Stearns
MC CREARY
Pine Knot
Middlesboro
CUMBERLAND GAP NAT'L HIST. PK.
VIRGINIA
TENNESSEE
Bristol
Kingsport
Jellico
© Copyright HAMMOND INCORPORATED, Maplewood, N. J.
West of
Greenwich
D
84°
E
83°
F

A
94°
B
93°
C
92°
D
33°
ARKANSAS
Springhill
Cullen
Haynesville
Rodessa
Vivian
Plain Dealing
Bayou Bodcau Res.
CLAIBORNE
UNION
Marion
Ouachita
MOREHOUSE
Oak Grove
Caddo L.
Oil City
BOSSIER
Cotton Valley
Homer
Bernice
Bayou D'Arbonne
Farmerville
L.
Bastrop
WEST
EAST
Lake Provide
Benton
Red
Claiborne L.
Dubach
Mer Rouge
CARROLL
CARROL
Mooringsport
Bossier City
Minden
LINCOLN
Sterlington
Macon
CADDO
Ruston
W. Monroe
1
Marshall
Cross L.
Doyline
Gibsland
Arcadia
Grambling
Monroe
RICHLAND
Shreveport
BARKSDALE A.F.B.
WEBSTER
Driskill Mtn.
535
OUACHITA
Delhi
Rayville
Boeuf
Wallace L.
L. Bistineau
Ringgold
JACKSON
Hodge
Chatham
Archibald
Tallu
MADISO
Sabine
Gloster
BIENVILLE
Jonesboro
River
Winnsboro
Bayou
Carthage
DE SOTO
RED RIVER
Columbia
TENSAS
Mansfield
R.
Coushatta
Chestnut
WINN
Clarks
FRANKLIN
Newellton
32°
Logansport
Campti
Little
Joyce
CALDWELL
Wisner
St. Joseph
Saline L.
Winnfield
Olla
Center
Pleasant Hill
NATCHITOCHES
Urania
Waterproof
Natchitoches
Harrisonburg
N
Montgomery
CATAHOULA
Zwolle
SABINE
Natchez
L. Iatt
Jena
LA SALLE
Ferriday
Mississi
San Augustine
Many
Melrose
GRANT
Catahoula L.
Jonesville
Vidalia
Natche
2
Sam Rayburn Res.
Toledo Bend Res.
Cloutierville
Colfax
CON-CORDIA
Florien
Boyce
Red
MISS.
Pineville
ENGLAND A.F.B.
Homochitto
Vernon L.
Alexandria
Samtown
Effie
River
Anandale
Leesville
RAPIDES
Marksville
TOLEDO BEND DAM
L. Anacoco
FT. POLK
Lecompte
Mansura
31°
Cheneyville
Longleaf
Cottonport
Jasper
TEXAS
VERNON
Glenmora
Bunkie
Simmesport
WEST FELICIANA
AVOYELLES
Jackson
Elizabeth
Atchafalaya
De Ridder
Oakdale
EVANGELINE
ST. LANDRY
St. Francisvi
Merryville
Bundick L.
Ville Platte
Morganza
Calcasieu
Melville
POINTE
R.
BEAUREGARD
Oberlin
New Roads
EAST
Sabine
Mamou
Washington
3
Longville
ALLEN
Port Barre
COUPEE
Eunice
Opelousas
De Quincy
Kinder
Basile
Elton
Maringouin
Port Allen
6
Neches
CALCASIEU
Church Pt.
Sunset
Arnaudville
Baton Rouge
JEFFERSON
Iota
ACADIA
Breaux Bridge
Plaquemine
Starks
Hollywood
Welsh
Jennings
Rayne
LAFAYETTE
Sulphur
Lafayette
IBERVILLE
Vinton
Westlake
Iowa
DAVIS
Crowley
Broussard
ST. MARTIN
White Castle
Orange
Lake Charles
Lake Arthur
St. Martinville
Donalds
Beaumont
Intracoastal
New Iberia
Hackberry
Gueydan
Kaplan
Erath
30°
Paincourtvi
Abbeville
Grand L.
Port Arthur
Calcasieu L.
Waterway
IBERIA
Delcambre
Jeanerette
Sabine L.
Grand L.
Cameron
CAMERON
Baldwin
Weeks
White L.
Franklin
Louisiana Pt.
Grand Chenier
Vermilion B.
Bayou Vista
Patterson
Cote
VERMILION
Texas Pt.
Blanche
ST. MARY
Sabine Pass
B.
4
Marsh I.
Atchafalaya B.
GULF OF MEXICO
29°
© Copyright HAMMOND INCORPORATED, Maplewood, N. J.
A
94°
B
93°
C
92°
D
Longitud

LOUISIANA
Parishes indicated by numbers:
1 JEFFERSON E4
2 ORLEANS F3
3 ST. CHARLES E4
4 ST. JAMES E4
5 ST. JOHN THE BAPTIST E3
6 WEST BATON ROUGE D3
MILES
0 10 20 40 60
KILOMETERS
0 10 20 40 60
State Capital
County Seats
West of Greenwich
E F G H
90° 89° 88°
33° 32° 31° 30° 29°
1 2 3 4
EAST BATON ROUGE
Baton Rouge
Denham Sprs.
Walker
Livingston
LIVINGSTON
Natalbany
Hammond
Ponchatoula
TANGIPAHOA
Covington
ST. TAMMANY
Madisonville
Mandeville
Pearl River
Slidell
Pearl River
MISS.
French Settlement
Lake Maurepas
Pass Manchac
Lake Pontchartrain
Sunshine
Seymourville
Carville
Gonzales
Sorrento
IBERVILLE
White Castle
ASCENSION
ST. JOHN THE BAPTIST
ST. JAMES
Donaldsonville
Gramercy
Garyville
Reserve
La Place
Bonnet Carre Floodway
New Orleans
ORLEANS
Metairie
Convent
Lutcher
Edgard
Norco
Jefferson Hts.
Kenner
Arabi
CHALMETTE NAT'L HIST. PK.
Chalmette
L. Borgne
Paincourtville
Vacherie
Hahnville
Luling
Harahan
Westwego
Gretna
Mississippi R. Gulf Outlet
L. Verret
Napoleonville
Lac des Allemands
Paradis
Boutte
Marrero
Harvey
Terry Town
Violet
ASSUMPTION
Supreme
Des Allemands
ST. CHARLES
Belle Chasse
ST. BERNARD
Thibodaux
Labadieville
LAFOURCHE
JEFFERSON
Schriever
Bayou Lafourche
Raceland
L. Salvador
Barataria
PLAQUEMINES
Morgan City
Gibson
Berwick
Amelia
Lockport
Phoenix
TERREBONNE
Bayou Cane
Kentwood
WASHINGTON
Greensburg
Franklinton
ST. HELENA
TANGIPAHOA
Roseland
Amite
Bogalusa
Independence
ST. TAMMANY
MISSISSIPPI
Pearl R.
Gulfport
Biloxi
Pascagoula
Pascagoula R.
ALABAMA
Mobile
Mobile Bay
Grants Pass
Dauphin I.
Petit Bois I.
Horn I.
Ship I.
Mississippi Sound
Cat I.
L. Pontchartrain
L. Maurepas
Chandeleur Sound
Chandeleur Is.
Eloi B.
Breton Sound
Curlew I.
Breton Is.
Pointe a la Hache
Port Sulphur
Golden Meadow
Galliano
Larose
Chauvin
Houma
Theriot
Barataria B.
Buras
Triumph
Venice
Main Pass
Grand Isle
THE MISSISSIPPI DELTA
W. Bay
Southwest Pass
Terrebonne B.
Timbalier B.
Timbalier I.
Dernieres

CANADA
NEW BRUNSWICK
QUÉBEC
QUE.
N.B.
AROOSTOOK
PISCATAQUIS
SOMERSET
Edmundston
Madawaska
Frenchville
St. Agatha
Ft. Kent
Soldier Pond
Eagle Lake
Allagash
St. Francis
Grand Falls
Van Buren
Keegan
Stockholm
Long L.
Square L.
Eagle L.
Limestone
LORING A.F.B.
Caribou
Ft. Fairfield
Presque Isle
Mars Hill
Bridgewater
Monticello
Houlton
Woodstock
Washburn
Mapleton
Squa Pan L.
Ashland
Fish River L.
1,981
Oakfield
Island Falls
Patten
Sherman Mills
Mattawamkeag L.
Mattawamkeag
Millinocket L.
Millinocket
E. Millinocket
Danforth
Chiputneticook Lakes
Grand L.
Baskahegan L.
St. Croix
Magaguadavic L.
Oromocto L.
Saint John River
Tobique R.
Aroostook R.
Grand Lake Seboeis
East Branch
Mt. Katahdin 5,268
Seboeis L.
Schoodic
Pemadumcook L.
Brownville Jct.
Chesuncook L.
Chamberlain L.
Eagle L.
Chemquasabamticook L.
Allagash L.
Allagash R.
Saint John R.
L. de l'Est
St. Francis
Caucomgomoc L.
Seboomook L.
Moosehead Lake
Greenville
Big Squaw Mtn. 3,196
Rockwood
Jackman
Lac-Mégantic
L. Mégantic
L. St-François
Tring Jonction
Chaudière R.
Etchemin R.
Québec
Île d'Orléans
Montmagny
Baie-St-Paul
LAURENTIDES PROV. PK.
ST. LAWRENCE RIVER
67°
68°
69°
70°
71°
47°
46°

MAINE
MILES
0 10 20 30 40 50
KILOMETERS
0 10 20 30 40 50
State and Provincial Capitals
County Seats
© Copyright HAMMOND INCORPORATED, Maplewood, N. J.
Longitude 69° West of Greenwich
ATLANTIC OCEAN
FRANKLIN
OXFORD
ANDROSCOGGIN
CUMBERLAND
YORK
SAGADAHOC
KENNEBEC
LINCOLN
KNOX
WALDO
HANCOCK
WASHINGTON
NEW HAMPSHIRE
Augusta
Portland
Bangor
Lewiston
Auburn
Waterville
Biddeford
Saco
Portsmouth
Kittery
Sanford
Brunswick
Bath
Rockland
Belfast
Ellsworth
Machias
Calais
Eastport
Lubec
Skowhegan
Farmington
Rumford
Old Town
Orono
Brewer
Camden
Bar Harbor
Mt. Desert I.
ACADIA NAT'L PK.
Penobscot Bay
Casco Bay
Passamaquoddy Bay
Grand Manan I.
Monhegan I.
Matinicus I.
Isles of Shoals
Bigelow Mtn. 4,150
Saddleback Mtn. 4,116
Mooselookmeguntic L.
Umbagog L.
Sebago L.
Androscoggin R.
Kennebec R.
Saco R.
Salmon Falls R.
Narraguagus R.
Machias R.

PENNSYLVANIA
ALLEGANY
WASHINGTON
Hagerstown
FREDERICK
Frederick
CARROLL
HOWARD
MONTGOMERY
Rockville
VIRGINIA
W. VIRGINIA
APPALACHIAN MOUNTAINS
SOUTH MTN.
CATOCTIN MTN.
GETTYSBURG NAT'L MIL. PK.
ANTIETAM NAT'L BATTLEFIELD SITE
HARPERS FY. NAT'L HIST. PK.
CHESAPEAKE & OHIO CANAL NAT'L HIST. PARK
Potomac R.
Arlington
WASHINGTON
Alexandria
D.C.
CHARLE
MARYLAND
MILES
KILOMETERS
National Capital
State Capitals
County Seats
DISTRICT OF COLUMBIA
PRINCE GEORGES
WESTERN PART OF MARYLAND
Same scale as main map
GARRETT
ALLEGANY
Cumberland
Oakland
WEST VIRGINIA
BACKBONE MTN.
ALLEGHENY MTS.

PHILADELPHIA
Chester
NEW JERSEY
Wilmington
PENNSYLVANIA
Newark
Rising Sun
CECIL
Elkton
CONOWINGO DAM
Port Deposit
BAINBRIDGE NAVAL TR. CTR.
HARFORD
Bel Air
Havre de Grace
Charlestown
Aberdeen
Elk Neck
C. & D. Canal
Chesapeake City
Delaware R.
Vineland
Bridgeton
Millville
Middletown
Kingsville
ABERDEEN PROVING GD.
Edgewood
EDGEWOOD ARSENAL
Timonium
Towson
HAMPTON N.H.S.
Pikesville
Overlea
Middle River
Essex
FT. McHENRY N.M.
Dundalk
Lansdowne
Edgemere
Linthicum Hts.
FT. GEO. G. MEADE
Severna Pk.
Odenton
Crownsville
Bowie
Annapolis
U.S. NAVAL ACADEMY
Kennedyville
KENT
Chester R.
Chestertown
QUEEN ANNES
Rock Hall
Centreville
DELAWARE
Smyrna
N.J.
DEL.
Dover
DELAWARE BAY
Pretty boy Res.
Chester
Grasonville
Kent I.
Londontowne
ANNE ARUNDEL
Upper Marlboro
Churchton
Kent Pt.
Eastern B.
Greensboro
Ridgely
Denton
TALBOT
Easton
St. Michaels
Tilghman
Oxford
Choptank R.
CAROLINE
Federalsburg
Hurlock
Cambridge
Harrington
C. May
C. Henlopen
Lewes
Rehoboth Beach
Bridgeville
Georgetown
Seaford
Laurel
Owings
Chesapeake Beach
CALVERT
Prince Frederick
DORCHESTER
Nanticoke R.
Delmar
DELAWARE
Hebron
WICOMICO
Salisbury
Whaleysville
Berlin
Ocean City
Fruitland
ST. MARYS
Patuxent R.
Hollywood
Fishing Cr.
Wingate
Leonardtown
PATUXENT NAVAL AIR TEST CTR.
Lexington Park
Nanticoke
Princess Anne
WORCESTER
Snow Hill
Pocomoke R.
SOMERSET
Deal I.
Bloodsworth I.
Piney Pt.
Fairmount
Tangier Sd.
S. Marsh I.
Pocomoke City
Chincoteague Bay
Assateague
VIRGINIA
RIVER
Pt. Lookout
Smith I.
Crisfield
Pocomoke Sd.
CHESAPEAKE BAY
ATLANTIC OCEAN
© Copyright HAMMOND INCORPORATED, Maplewood, N. J.
West of 30′ Greenwich
30′
76°
30′
75°
39°
38°

VERMONT
NEW HAMPSHIRE
NEW YORK
CONNECTICUT
RHODE ISLAND
TACONIC MTS.
BERKSHIRE HILLS
BERKSHIRE
FRANKLIN
HAMPSHIRE
HAMPDEN
WORCESTER
MIDDLESEX
ESSEX
SUFFOLK
NORFOLK
Bennington
Brattleboro
Harriman Res.
Jaffrey
Williamstown
N. Adams
Mt. Greylock 3,491
Adams
Colrain
Northfield
Winchendon
Townsend
Greenfield
Turners Falls
Orange
Athol
Baldwinville
Gardner
Fitchburg
Buckland
Deerfield
Millers Falls
Templeton
Westminster
Leominster
Cheshire
Lanesboro
Montague
Dalton
Pittsfield
S. Deerfield
Quabbin Res.
Wachusett Mtn. 2,006
Barre
S. Lancaster
Williamsburg
Wachusett Res.
Lenox
Hatfield
Amherst
Rutland
Hadley
Holden
Lee
Stockbridge
Northampton
Belchertown
N. Brookfield
Shrewsbury
Chester
Easthampton
Ware
Worcester
Housatonic
Great Barrington
Mt. Tom 1,202
Spencer
Leicester
Blandford
Holyoke
Westover A.F.B.
Warren
Auburn
Chicopee
Palmer
Charlton City
Sheffield
Westfield
Springfield Armory N.H.S.
Three Rivers
Oxford
Whitinsville
W. Springfield
Springfield
Monson
Sturbridge
Southwick
Agawam
Longmeadow
Southbridge
E. Douglas
Webster
Deerfield R.
Millers R.
Westfield R.
Connecticut R.
Housatonic R.
Hoosic R.
Ware R.
Chicopee R.
N. Billerica
N. Reading
Manchester
Billerica
Danvers
Pinehurst
Wilmington
Beverly
Reading
Peabody
Salem Maritime Nat'l Hist. Site
Burlington
Salem
Marblehead
Bedford
Wakefield
Woburn
Stoneham
Saugus Iron Works Nat'l Hist. Site
S. Acton
Minute Man Nat'l Hist. Park
Swampscott
Concord
Winchester
Melrose
Saugus
Walden Pd.
Lexington
Lynn
Maynard
Lincoln Center
Mystic L.
Medford
Malden
Arlington
Everett
Revere
East Pt.
Cambridge Res.
Belmont
Chelsea
Boston
Sudbury
Waltham
Somerville
Winthrop
Watertown
Cambridge
Wayland
Logan Int'l Airport
Bay
Newton
L. Cochituate
John F. Kennedy N.H.S.
BOSTON
Brookline
Dorchester Hts. N.H.S.
Hull
Pt. Allerton
Wellesley
Charles R.
Sudbury R.
Concord R.
Dorchester
Natick
Needham
Adams Nat'l Hist. Site
Framingham
Milton
Dedham
Quincy
Hingham
Braintree
Weymouth
Holliston
Norwood
N. Scituate
Medfield
Neponset R.
Randolph
Walpole
Canton
0 2 4 6 mi.
0 2 4 6 km.
Westerly
Block Island
R.I.
N.Y.
73°
72°
71°
42°
41°

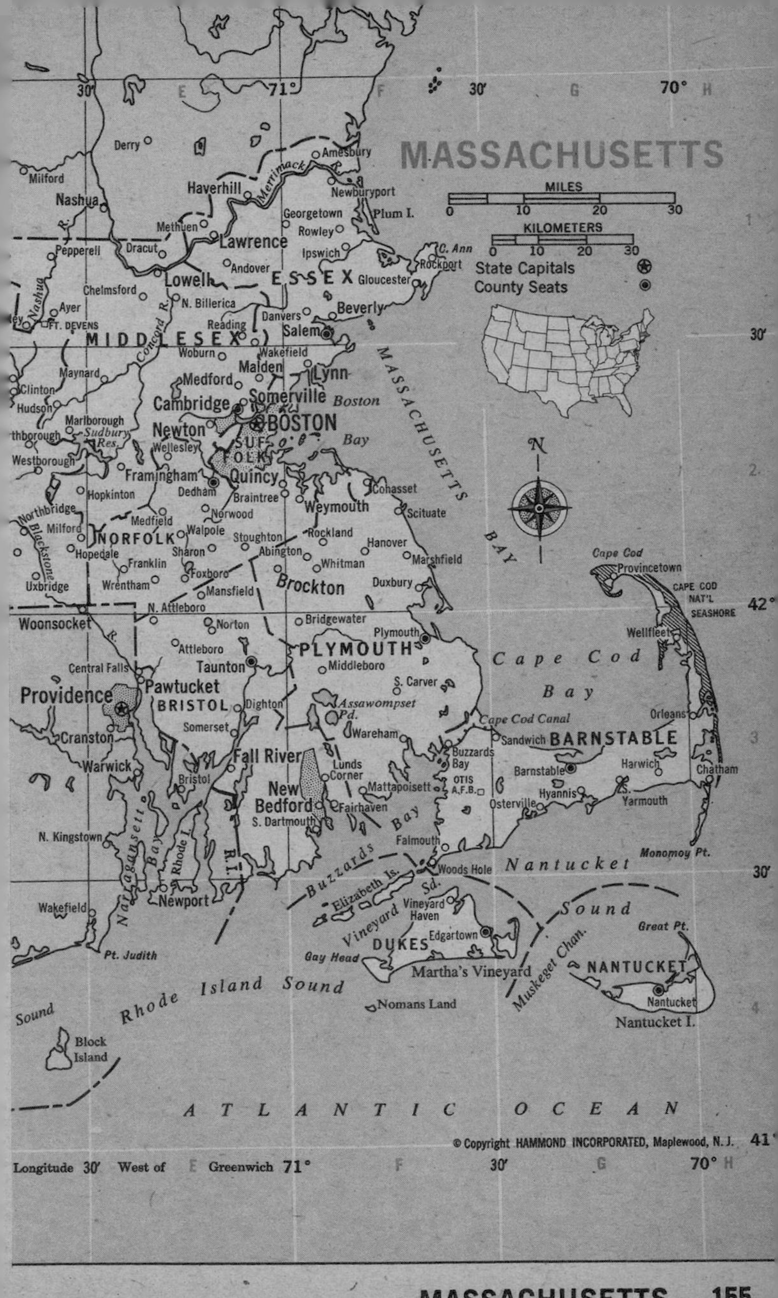
MASSACHUSETTS
MILES
KILOMETERS
State Capitals
County Seats
ESSEX
MIDDLESEX
SUFFOLK
NORFOLK
PLYMOUTH
BRISTOL
BARNSTABLE
DUKES
NANTUCKET
BOSTON
Providence
Cape Cod Bay
Massachusetts Bay
Nantucket Sound
Rhode Island Sound
ATLANTIC OCEAN
CAPE COD NAT'L SEASHORE
Martha's Vineyard
Nantucket I.
Block Island
© Copyright HAMMOND INCORPORATED, Maplewood, N.J.
Longitude 30' West of Greenwich 71° 30' 70°

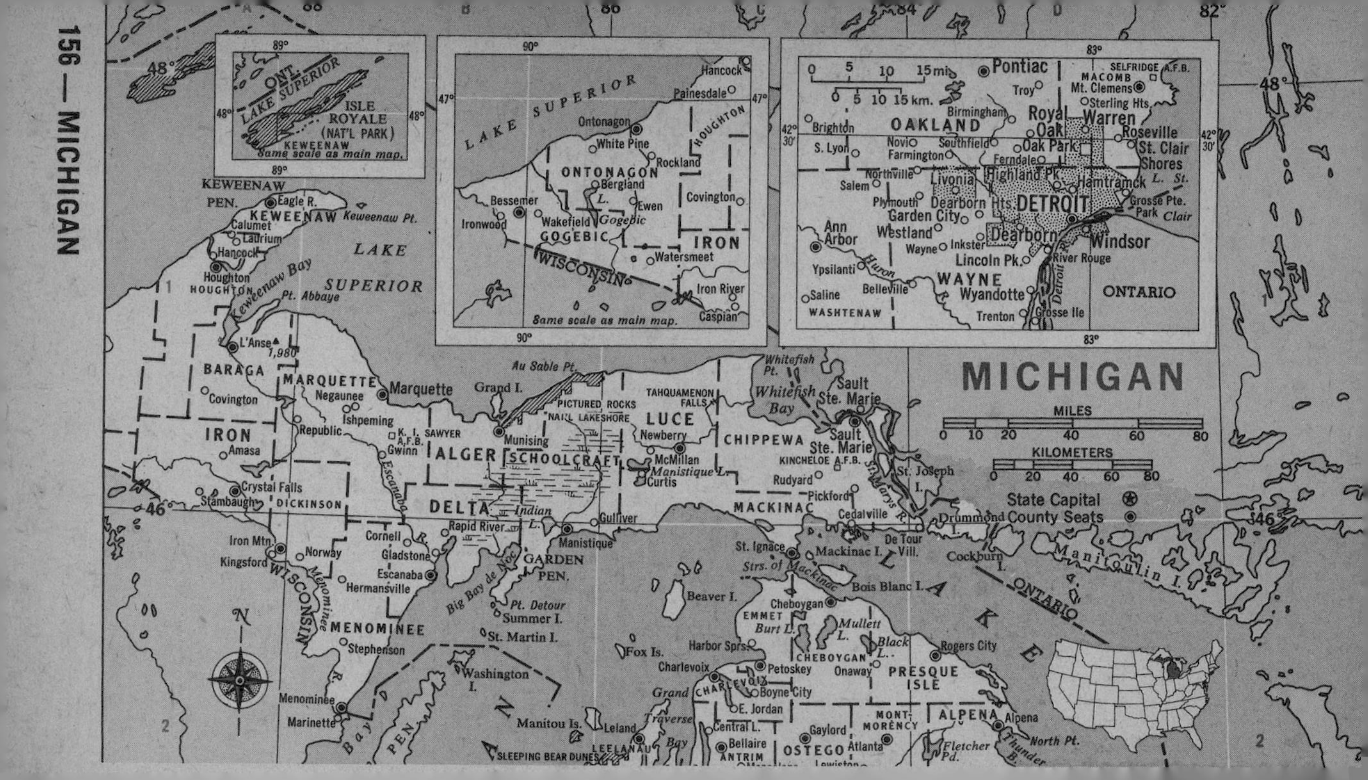
MICHIGAN
MILES
0 10 20 40 60 80
KILOMETERS
0 20 40 60 80
State Capital
County Seats
ISLE ROYALE (NAT'L PARK)
KEWEENAW
Same scale as main map.
LAKE SUPERIOR
ONT.
Hancock
Painesdale
HOUGHTON
Ontonagon
White Pine
Rockland
ONTONAGON
Bergland
Ewen
Covington
Bessemer
Ironwood
Wakefield
Gogebic L.
GOGEBIC
IRON
Watersmeet
WISCONSIN
Iron River
Caspian
Pontiac
Troy
MACOMB
SELFRIDGE A.F.B.
Mt. Clemens
Sterling Hts.
Royal Oak
Warren
Roseville
St. Clair Shores
Brighton
OAKLAND
Birmingham
S. Lyon
Novi
Southfield
Oak Park
Farmington
Ferndale
Northville
Livonia
Highland Pk.
Hamtramck
L. St. Clair
Salem
Plymouth
Dearborn Hts.
DETROIT
Grosse Pte. Park
Garden City
Ann Arbor
Westland
Dearborn
Windsor
Wayne
Inkster
Huron R.
Lincoln Pk.
River Rouge
Ypsilanti
WAYNE
Detroit R.
Belleville
Wyandotte
ONTARIO
Saline
WASHTENAW
Trenton
Grosse Ile
KEWEENAW PEN.
Eagle R.
KEWEENAW
Keweenaw Pt.
Calumet
Laurium
Houghton
HOUGHTON
Keweenaw Bay
Pt. Abbaye
L'Anse
1,980
BARAGA
MARQUETTE
Marquette
Negaunee
Ishpeming
Covington
Republic
K. I. SAWYER A.F.B.
Gwinn
IRON
Amasa
Crystal Falls
Stambaugh
DICKINSON
Escanaba R.
Au Sable Pt.
Grand I.
PICTURED ROCKS NAT'L LAKESHORE
Munising
ALGER
SCHOOLCRAFT
DELTA
Indian L.
Rapid River
Cornell
Gladstone
Escanaba
Hermansville
Iron Mtn.
Norway
Kingsford
WISCONSIN
Menominee R.
MENOMINEE
Stephenson
Menominee
Marinette
Big Bay de Noc
GARDEN PEN.
Manistique
Gulliver
Pt. Detour
Summer I.
St. Martin I.
Washington I.
TAHQUAMENON FALLS
LUCE
Newberry
McMillan
Manistique L.
Curtis
Whitefish Pt.
Whitefish Bay
Sault Ste. Marie
CHIPPEWA
KINCHELOE A.F.B.
Rudyard
Pickford
St. Joseph I.
St. Marys R.
MACKINAC
Cedarville
De Tour Vill.
Drummond I.
Cockburn I.
Manitoulin I.
ONTARIO
St. Ignace
Mackinac I.
Strs. of Mackinac
Bois Blanc I.
LAKE
Beaver I.
Cheboygan
EMMET
Burt L.
Mullett L.
Black L.
CHEBOYGAN
Onaway
Harbor Sprs.
Petoskey
Fox Is.
Charlevoix
CHARLEVOIX
Boyne City
Grand Traverse Bay
E. Jordan
Central L.
Gaylord
MONT-MORENCY
Atlanta
PRESQUE ISLE
Rogers City
ALPENA
Alpena
North Pt.
Thunder B.
Fletcher Pd.
Manitou Is.
Leland
LEELANAU
SLEEPING BEAR DUNES
Bellaire
ANTRIM
OSTEGO
N
48°
46°
82°
84°
86°
88°
89°
90°
83°
47°
42° 30'
1
2

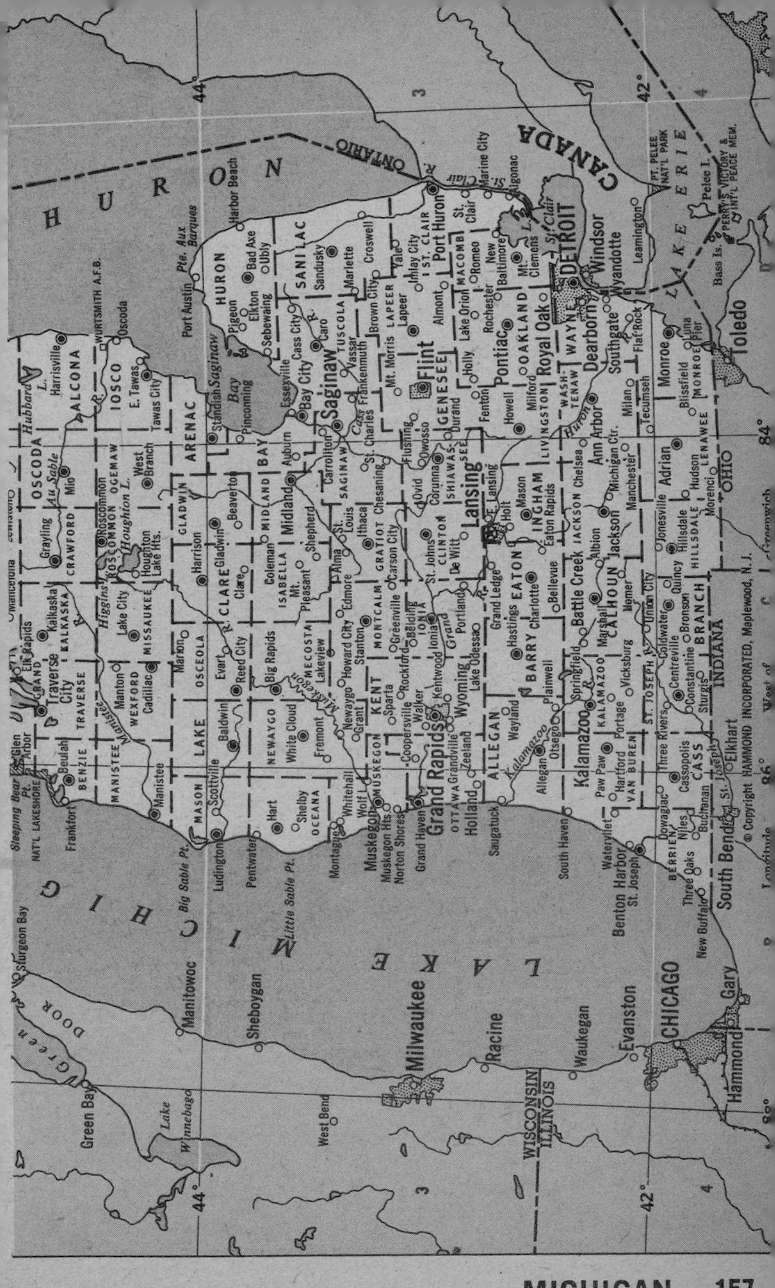
LAKE HURON
LAKE MICHIGAN
LAKE ERIE
CANADA
ONTARIO
DETROIT
Windsor
Flint
Lansing
Saginaw
Grand Rapids
Kalamazoo
Pontiac
Royal Oak
Dearborn
Ann Arbor
Jackson
Battle Creek
Port Huron
Bay City
Midland
Muskegon
Toledo
Monroe
Milwaukee
Racine
CHICAGO
Gary
Hammond
South Bend
Evanston
Waukegan
Sheboygan
Manitowoc
Green Bay
Benton Harbor
St. Joseph
Holland
Traverse City
Cadillac
WISCONSIN
ILLINOIS
INDIANA
OHIO
© Copyright HAMMOND INCORPORATED, Maplewood, N. J.

MINNESOTA
MILES
KILOMETERS
State Capital
County Seats
CANADA
MANITOBA
ONTARIO
NORTH DAKOTA
NORTHEASTERN MINNESOTA
Same scale as main map
Eagle Mtn. 2,301
MISQUAH HILLS
GRAND PORTAGE NAT'L MON.
Grand Portage
Hovland
Grand Marais
Tofte
LAKE SUPERIOR
QUETICO PROV. PK.
VOYAGEURS NAT'L PARK
KITTSON
ROSEAU
LAKE OF THE WOODS
MARSHALL
PENNINGTON
RED LAKE
POLK
BELTRAMI
KOOCHICHING
CLEARWATER
MAHNOMEN
NORMAN
CLAY
BECKER
HUBBARD
CASS
ITASCA
ST. LOUIS
LAKE
COOK
CARLTON
AITKIN
CROW WING
WADENA
Duluth
Superior
Bemidji
International Falls
Grand Rapids
Hibbing
Virginia
Moorhead
Fargo
Grand Forks
East Grand Forks
Crookston
Thief River Falls
Two Harbors
Cloquet
Ely
Lake of the Woods
Rainy Lake
Upper Red L.
Lower Red L.
Leech L.
L. Winnibigoshish
Mississippi R.
Red River of the North

WISC.
WISCONSIN
N.D.
S.D.
SOUTH DAKOTA
IOWA
Minneapolis
St. Paul
St. Cloud
Rochester
Bloomington
Mankato
Winona
La Crosse
Red Wing
Stillwater
Hastings
Northfield
Faribault
Owatonna
Austin
Albert Lea
New Ulm
St. Peter
Willmar
Brainerd
Little Falls
Fergus Falls
Sioux Falls
Brookings
Worthington
Fairmont
Mississippi
Minnesota
Mille Lacs Lake
Longitude West 94° of Greenwich
©Copyright HAMMOND INCORPORATED, Maplewood, N. J.

MISSISSIPPI
MILES
0 10 20 30 40 50 60
KILOMETERS
0 10 20 30 40 50 60
State Capitals
County Seats
TENNESSEE
ALABAMA
ARKANSAS
LOUISIANA
ALCORN
TIPPAH
BENTON
MARSHALL
DE SOTO
TATE
TUNICA
PRENTISS
TISHOMINGO
UNION
LEE
ITAWAMBA
PONTOTOC
LAFAYETTE
PANOLA
QUITMAN
COAHOMA
MONROE
CHICKASAW
CALHOUN
YALOBUSHA
TALLAHATCHIE
SUNFLOWER
BOLIVAR
CLAY
WEBSTER
MONTGOMERY
GRENADA
LEFLORE
CARROLL
LOWNDES
OKTIBBEHA
CHOCTAW
ATTALA
HOLMES
HUMPHREYS
WASHINGTON
NOXUBEE
WINSTON
LEAKE
NESHOBA
KEMPER
MADISON
YAZOO
SHARKEY
Memphis
W. Memphis
Corinth
Iuka
Woodall Mtn. 806
Booneville
Wheeler
Baldwyn
Belmont
Red Bay
BRICES CROSS ROADS NAT'L BATTLEFIELD
Tupelo
TUPELO NAT'L BATTLEFIELD
Fulton
Plantersville
Verona
Shannon
Nettleton
Amory
Hamilton
Sulligent
Columbus
COLUMBUS A.F.B.
Reform
Aliceville
Aberdeen
Okolona
Houlka
Houston
West Point
Starkville
State College
Mathiston
Brooksville
Macon
Shuqualak
Scooba
DeKalb
Louisville
Philadelphia
Noxapater
Ripley
Blue Mountain
New Albany
Pontotoc
Ashland
Holly Springs
Byhalia
Collierville
Olive Branch
Hernando
Horn Lake
Coldwater
Senatobia
Como
Sardis
Crenshaw
Oxford
Sardis Lake
Water Valley
Enid Lake
Batesville
Crowder
Marks
Lambert
Charleston
Coffeeville
Bruce
Pittsboro
Calhoun City
Vardaman
Maben
Walthall
Eupora
Ackerman
Weir
Ethel
Kosciusko
Carthage
Duck Hill
Winona
Kilmichael
Tie Plant
Grenada
Grenada Lake
Vaiden
Carrollton
Greenwood
Durant
Goodman
Pickens
Lexington
Tchula
Yazoo City
Bentonia
Tunica
Friars Point
Jonestown
Clarksdale
Tutwiler
Sumner
Webb
Drew
Ruleville
Itta Bena
Moorhead
Inverness
Sunflower
Indianola
Belzoni
Isola
Midnight
Shelby
Mound Bayou
Merigold
Cleveland
Boyle
Shaw
Benoit
Rosedale
Leland
Arcola
Hollandale
Greenville
Anguilla
Rolling Fork
Mayersville
Helena
Marianna
Forrest City
St. Francis R.
DeWitt
Dumas
McGehee
Dermott
Lake Village
Eudora
Lake Providence
Monticello
Star City
Warren
Crossett
Bastrop
Pine Bluff
Pickwick L.
Arkabutla L.
Tombigbee
Buttahatchee
Noxubee
Yockanookany
Tallahatchie
Yalobusha
Coldwater
Sunflower
Mississippi
White R.
Arkansas River
Bartholomew
Saline
Bayou
Macon
Demopolis Lake

ALABAMA
LOUISIANA
Jackson
Vicksburg
Meridian
Hattiesburg
Gulfport
Biloxi
Pascagoula
Mobile
New Orleans
Baton Rouge
Natchez
ISSAQUENA
WARREN
HINDS
RANKIN
SCOTT
NEWTON
LAUDERDALE
SMITH
JASPER
CLARKE
SIMPSON
COVINGTON
JONES
WAYNE
COPIAH
LAWRENCE
JEFFERSON DAVIS
MARION
LAMAR
FOR-REST
PERRY
GREENE
GEORGE
STONE
JACKSON
HARRISON
HANCOCK
PEARL RIVER
WALTHALL
PIKE
LINCOLN
FRANKLIN
AMITE
WILKINSON
ADAMS
JEFFERSON
CLAIBORNE
Canton
Madison
Ridgeland
Tougaloo
Clinton
Bolton
Flora
Raymond
Utica
Terry
Whitfield
Piney Woods
Crystal Springs
Hazlehurst
Wesson
Brookhaven
Summit
McComb
Magnolia
Osyka
Fernwood
Tylertown
Columbia
Foxworth
Morgantown
Prentiss
Monticello
Mendenhall
Magee
Brandon
Pelahatchie
Morton
Forest
Sebastopol
Lena
Union
Raleigh
Taylorsville
Mount Olive
Collins
Laurel
Ellisville
Moselle
Bay Springs
Paulding
Newton
Decatur
Hickory
Enterprise
Stonewall
Quitman
Marion
Lauderdale
Heidelberg
Sandersville
Shubuta
Waynesboro
Buckatunna
State Line
Richton
Beaumont
Leakesville
McLain
New Augusta
Petal
Sumrall
Purvis
Lumberton
Brooklyn
Wiggins
Perkinston
McHenry
Poplarville
McNeill
Carriere
Picayune
Lucedale
Bay St. Louis
Lakeshore
Waveland
Pass Christian
Long Beach
D'Iberville
Keesler A.F.B.
Ocean Sprs.
Moss Pt.
Escatawpa
Port Gibson
Alcorn College
Fayette
Roxie
Meadville
Bude
Gloster
Liberty
Crosby
Centreville
Woodville
Rayville
Delhi
Tallulah
Winnsboro
Vidalia
Ferriday
Clinton
Jackson
Kentwood
Amite
Franklinton
Bogalusa
Covington
Hammond
Slidell
Plaquemine
Livingston
York
Butler
Citronelle
Bayou La Batre
Mississippi
River
Yazoo
Big Black
Pearl
Strong
Bowie
Black
Leaf
Chickasawhay
Pascagoula
Tombigbee
Homochitto
Bogue Chitto
Ouachita
Tensas
Red
Ross Barnett Res.
Okatibbee Lake
Lake Pontchartrain
L. Maurepas
Mississippi Sound
Mobile Bay
Ship I.
Horn I.
Cat I.
Petit Bois I.
Dauphin I.
Gulf Islands Nat'l Seashore
Vicksburg Nat'l Mil. Park
32°
31°
92°
91°
90°
89° Longitude West of Greenwich 88°
© Copyright HAMMOND INCORPORATED, Maplewood, N.J.

96°
A
95°
B
94°
C
93°
D
92°
E
Bedford
Lamoni
Centerville
IOWA
Des Moines
ATCHISON
Tarkio
Auburn
Rock Port
NODAWAY
Grant City
WORTH
MERCER
Princeton
PUTNAM
Unionville
Lancaster
SCHUY-
LER
Memphis
Kahoka
NEBRASKA
Maryville
GENTRY
Albany
HARRISON
Bethany
SCOTLAND
CLARK
Missouri
Stanberry
Thompson
SULLIVAN
Milan
Chariton
Kirksville
Edina
LEWIS
Canton
Mound City
King City
GRUNDY
Trenton
Falls City
HOLT
ANDREW
DAVIESS
ADAIR
KNOX
Monticello
40°
Oregon
Savannah
Gallatin
LINN
La Plata
La Grange
DE KALB
Grand
Linneus
MACON
SHELBY
MARION
Maysville
Chillicothe
Brookfield
Shelbyville
Palmyra
St. Joseph
Cameron
Hamilton
Salt
Clarence
CLINTON
CALDWELL
Macon
Platte
Plattsburg
Kingston
LIVINGSTON
Marceline
Shelbina
Monroe City
Atchison
BUCHANAN
CHARITON
RANDOLPH
2
CARROLL
MONROE
PLATTE
Lathrop
Weston
RAY
Brunswick
Keytesville
Huntsville
Paris
Smithville
RALLS
Platte City
CLAY
Excelsior Sprs.
Carrollton
Salisbury
Moberly
Leavenworth
Liberty
Richmond
Slater
Glasgow
AUDRAIN
Kansas City
Lexington
Marshall
Fayette
Centralia
Mexico
Kansas
Independence
Higginsville
HOWARD
BOONE
39°
Topeka
Kansas City
LAFAYETTE
SALINE
New Franklin
Wellsville
Odessa
Concordia
Sweet Sprs.
Lawrence
JACKSON
Boonville
Columbia
Belton
PETTIS
COOPER
Fulton
Warrensburg
Knob Noster
Pleasant Hill
Holden
WHITEMAN A.F.B.
Sedalia
MONITEAU
CALLAWAY
KANSAS
Ottawa
des
Harrisonville
JOHNSON
Tipton
California
Missouri
Marais
Cygnes
3
CASS
Jefferson City
Windsor
Linn
N
Versailles
COLE
Adrian
Clinton
BENTON
MORGAN
OSAGE
GASCONADE
BATES
Eldon
Butler
HENRY
Warsaw
MARIES
Belle
Tuscumbia
L. of the Ozarks
Vienna
Appleton City
BAGNELL DAM
Rich Hill
Osage
ST.
R.
MILLER
38°
Osceola
CLAIR
HICKORY
St. James
Hermitage
CAMDEN
Camdenton
Dixon
Rolla
VERNON
Ft. Scott
PULASKI
Richland
PHELPS
Pomme de Terre L.
Nevada
El Dorado Sprs.
Waynesville
Neosho
CEDAR
Stockton
Lebanon
FT. LEONARD WOOD
Stockton Lake
Buffalo
BARTON
Bolivar
DALLAS
LACLEDE
Salem
4
Fall
Lamar
DADE
POLK
Pittsburg
DENT
Gasconade
WEBSTER
Parsons
Greenfield
Marshfield
GREENE
Houston
JASPER
WRIGHT
Independence
Carthage
OZARK NAT'L SCENIC RIVERWAY
Carl Jct.
Springfield
Hartville
TEXAS
LAWRENCE
Republic
Seymour
Joplin
Webb City
Mt. Vernon
Mountain Grove
Cabool
Coffeyville
WILSON'S CREEK NAT'L BATTLEFIELD
Ozark
37°
G. WASHINGTON CARVER N.M.
Granby
Aurora
Willow Sprs.
Miami
Neosho
Monett
STONE
CHRISTIAN
Ava
Lake O' The Cherokees
NEWTON
BARRY
Galena
DOUGLAS
West Plains
OKLAHOMA
Table Rock L.
Forsyth
Vinita
Cassville
Branson
OZARK
HOWELL
Oologah Lake
Pineville
Gainesville
5
MC DONALD
TANEY
Thayer
Norfork L.
Claremore
Rogers
Berryville
Bull Shoals Lake
Mountain Home
Verdigris
Springdale
Harrison
White
Ft. Gibson L.
Fayetteville
36°
A
95°
B
94°
C
93°
D
Longitude 92° West of

MISSOURI
St. Louis
Kansas City
Jefferson Nat'l Expansion Mem.
Richards-Gebaur A.F.B.
State Capitals
County Seats

MONTANA
CANADA
BRITISH COLUMBIA
ALBERTA
IDAHO
WYO.
State Capital
County Seats
MILES
KILOMETERS

D
108°
E
106°
F
104°
Shaunavon
Assiniboia
Radville
PROV. PARKS
Frenchman
SASKATCHEWAN
BLAINE
DANIELS
Plentywood
Opheim
Scobey
SHERIDAN
VALLEY
Poplar
Medicine L.
Chinook
Harlem
ROOSEVELT
Froid
Saco
Hinsdale
Milk
Dodson
Malta
River
Glasgow
Wolf Point
Poplar
Culbertson
NORTH
FT. BELKNAP IND. RES.
ROCKY BOY'S IND. RES.
Nashua
Frazer
River
Hays
PHILLIPS
Ft. Peck
Missouri
FT. UNION TRADING POST NAT'L HIST. SITE
48°
FT. PECK DAM
Fort Peck Lake
Fairview
RICHLAND
Sidney
MC CONE
Lambert
Richey
Winifred
PINEY BUTTES
Savage
FERGUS
Circle
DAWSON
DAKOTA
Roy
Jordan
PETROLEUM
Musselshell
GARFIELD
Glendive
Wibaux
Lewistown
Dry
Little
Winnett
PRAIRIE
River
Beach
Terry
WIBAUX
MUSSEL SHELL
Melstone
ROSEBUD
Roundup
Yellowstone
Musselshell
Plevna
GOLDEN
Miles City
Baker
Ryegate
VALLEY
BULL MTS.
Hysham
Forsyth
FALLON
YELLOWSTONE
CUSTER
MEDICINE ROCKS
STILL WATER
TREASURE
Pompeys Pillar
46°
Ekalaka
Ballantine
TONGUE R. IND. RES.
Columbus
Park City
Billings
Hardin
Laurel
Crow Agency
Lame Deer
POWDER RIVER
S. DAK.
CARTER
Joliet
CUSTER BATTLEFIELD NAT'L MON.
Broadus
CROW
CARBON
Fromberg
BIG
HORN
Bighorn
Tongue
Bridger
IND.
Lodge Grass
Red Lodge
RES.
Hammond
Missouri R.
BIGHORN CANYON NAT'L REC. AREA
Bighorn Lake
Little Bighorn
Tongue R. Res.
Alzada
River
WYOMING
Little
Powell
Sheridan
BIGHORN MTS.
DEVILS TOWER NAT'L MON.
Buffalo Bill Res.
Cody
Greybull
Shoshone
Cloud Pk. 13,165
Buffalo
Powder
Gillette
Belle Fourche
Worland
44°
© Copyright HAMMOND INCORPORATED, Maplewood, N. J.
Longitude
West
of 108°
Greenwich
106°
1
2
3

NEBRASKA
104°
102°
100°
A
B
1
2
3
42°
40°
JEWEL CAVE N.M.
WIND CAVE NAT'L PK.
Hot Sprs.
Edgemont
Cheyenne R.
Angostura Res.
BADLANDS
BADLANDS NAT'L MON.
PINE RIDGE INDIAN RES.
River
White River
South Fork
Winner
Martin
ROSEBUD IND. RES.
Keya Paha
SOUTH DAKOTA
WYOMING
Chadron
Harrison
Crawford
White
DAWES
PINE RIDGE
Gordon
Rushville
Hay Sprs.
Box Butte Res.
Niobrara
River
Valentine
KEYA
Springview
Niobrara
Merritt Res.
Snake R.
Ainsworth
BROWN
AGATE FOSSIL BEDS NAT'L MON.
SIOUX
Hemingford
BOX BUTTE
Alliance
SHERIDAN
CHERRY
North
Ashby
Mullen
BLAINE
Morrill
Mitchell
SCOTTS BLUFF
SCOTTS BLUFF NAT'L MON.
Scottsbluff
Minatare
Gering
BLUFF
Bayard
CHIMNEY ROCK NAT'L HIST. SITE
Bridgeport
Hyannis
GRANT
HOOKER
Thedford
THOMAS
Dismal
R.
Middle
Brewster
Swan L.
GARDEN
SAND HILLS
Harrisburg
BANNER
MORRILL
North Platte R.
Arthur
ARTHUR
Tryon
MC PHERSON
LOGAN
Stapleton
Arnold
Dalton
Oshkosh
CHEYENNE
L. McConaughy
KINGSLEY DAM
Potter
Kimball
Sidney
Lodgepole Cr.
KIMBALL
5,426
DEUEL
Chappell
Big Sprs.
Ogallala
Paxton
Sutherland
North Platte
KEITH
R.
COLORADO
Gothenburg
Cozad
Lexington
LINCOLN
Grant
PERKINS
Sterling Res.
South Platte
Sterling
Frenchman
CHASE
Imperial
HAYES
Hayes Ctr.
Curtis
FRONTIER
Stockville
Harry Strunk L.
Elwood
Wauneta
Cr.
Arapahoe
Cambridge
Culbertson
HITCHCOCK
Trenton
DUNDY
McCook
RED WILLOW
River
Beaver City
FURNAS
Swanson L.
Benkelman
R.
Ft. Morgan
Akron
Wray
Arikaree
Republican
Beaver
St. Francis
Oberlin
Beaver Cr.
MILES
0 10 20 40 60 80
KILOMETERS
0 20 40 60 80
State Capitals
County Seats
© Copyright HAMMOND INCORPORATED, Maplewood, N. J.
104°
A
102° Longitude West B of Greenwich 100°
South
Saline
Hill City

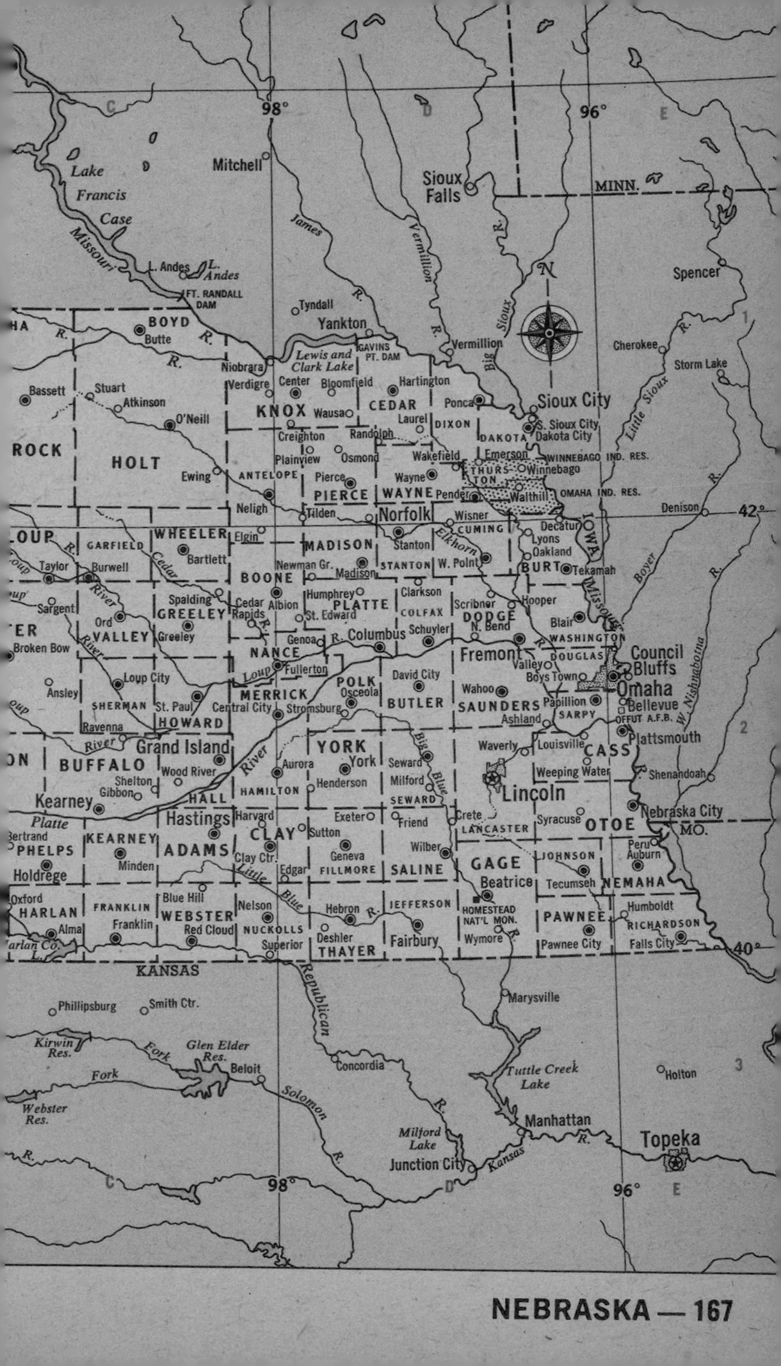

C
98°
D
96°
E
Lake
Francis
Case
Missouri
Mitchell
Sioux
Falls
MINN.
James
R.
Vermillion
R.
Sioux
Big
N
Spencer
L. Andes
L.
Andes
FT. RANDALL
DAM
Tyndall
Yankton
GAVINS
PT. DAM
Vermillion
Cherokee
Storm Lake
1
BOYD
Butte
R.
R.
Lewis and
Clark Lake
Niobrara
Verdigre
Center
Bloomfield
Hartington
Ponca
Sioux City
S. Sioux City
Dakota City
Little Sioux
Bassett
Stuart
Atkinson
O'Neill
KNOX
Wausa
CEDAR
Laurel
DIXON
DAKOTA
ROCK
HOLT
Creighton
Randolph
Plainview
Osmond
Wakefield
Emerson
WINNEBAGO IND. RES.
THURS-
TON
Winnebago
Ewing
ANTELOPE
Pierce
Wayne
PIERCE
WAYNE
Pender
Walthill
OMAHA IND. RES.
Neligh
Tilden
Norfolk
Wisner
Denison
42°
LOUP
GARFIELD
WHEELER
Elgin
MADISON
Stanton
CUMING
Decatur
Lyons
Oakland
IOWA
Taylor
Burwell
Cedar
Bartlett
Newman Gr.
Madison
STANTON
W. Point
BURT
Tekamah
Boyer
R.
BOONE
Elkhorn
Sargent
River
Spalding
Cedar
Rapids
Albion
Humphrey
PLATTE
Clarkson
COLFAX
Scribner
Hooper
Missouri
Ord
GREELEY
St. Edward
DODGE
N. Bend
Blair
VALLEY
Greeley
Genoa
R.
Columbus
Schuyler
WASHINGTON
Broken Bow
River
NANCE
Fremont
DOUGLAS
Council
Bluffs
Loup
Fullerton
Valley
Boys Town
Ansley
Loup City
POLK
David City
Wahoo
Omaha
MERRICK
Osceola
Papillion
Bellevue
W. Nishnabotna
SHERMAN
St. Paul
Central City
Stromsburg
BUTLER
SAUNDERS
SARPY
Ashland
OFFUT A.F.B.
Ravenna
HOWARD
2
River
Grand Island
YORK
Waverly
Louisville
Plattsmouth
BUFFALO
River
Aurora
York
Seward
CASS
Wood River
Weeping Water
Shenandoah
Shelton
Gibbon
HAMILTON
Henderson
Milford
Big Blue
Lincoln
Kearney
HALL
SEWARD
Platte
Hastings
Harvard
Exeter
Friend
Crete
Syracuse
OTOE
Nebraska City
Bertrand
PHELPS
KEARNEY
ADAMS
CLAY
Sutton
LANCASTER
MO.
Minden
Clay Ctr.
Geneva
Wilber
JOHNSON
Peru
Auburn
Holdrege
Little
Edgar
FILLMORE
SALINE
GAGE
Oxford
Blue Hill
Blue
Beatrice
Tecumseh
NEMAHA
HARLAN
FRANKLIN
WEBSTER
Nelson
Hebron
R.
JEFFERSON
HOMESTEAD
NAT'L MON.
PAWNEE
Humboldt
Alma
Franklin
Red Cloud
NUCKOLLS
Deshler
Fairbury
Wymore
RICHARDSON
Harlan Co.
L.
Superior
THAYER
Pawnee City
Falls City
40°
KANSAS
Republican
Marysville
Phillipsburg
Smith Ctr.
Kirwin
Res.
Glen Elder
Res.
Fork
Fork
Beloit
Concordia
Tuttle Creek
Lake
Holton
3
Solomon
Webster
Res.
R.
Manhattan
R.
Topeka
Milford
Lake
R.
Junction City
Kansas
R.
C
98°
D
96°
E

120°
118°
116°
114°
42°
40°
OREGON
IDAHO
UTAH
CALIFORNIA
GREAT SALT LAKE DESERT
WASHOE
HUMBOLDT
PERSHING
ELKO
LANDER
EUREKA
WHITE PINE
CHURCHILL
STOREY
LYON
DOUGLAS
Goose Lake
Alkali Lakes
Alkali L.
Vya
Denio
SUMMIT LAKE IND. RES.
Pahute Pk.
Kings R.
Quinn R.
McDermitt
SANTA ROSA RA.
Orovada
BLACK ROCK DESERT
Humboldt R.
Little Humboldt R.
S. Fk. Owyhee R.
Owyhee
Owyhee R.
WESTERN SHOSHONE INDIAN RES.
Mountain City
Matterhorn 10,889
Contact
Tuscarora
Midas
INDEPENDENCE MTS.
Montello
Wells
Deeth
Winnemucca
Golconda
WINNEMUCCA IND. RES.
Granite Pk. 8,990
SMOKE CREEK DESERT
Gerlach
Tungsten
Imlay
Rye Patch Res.
Battle Mountain
Beowawe
Carlin
Elko
Lamoille
Wendover
Lee
Jiggs
Ruby Valley
RUBY MTS.
Franklin L.
Ruby L.
Nelson Cr.
HUMBOLDT RA.
Honey L.
PYRAMID LAKE INDIAN RES.
Pyramid Lake
Winnemucca Lake
Lovelock
Humboldt Sink
Carson Sink
Nixon
Humboldt Salt Marsh
MOUNTAINS
Reese R.
Diamond Valley Alkali Flat
Cherry Creek
GOSHUTE IND. RES.
Newark Valley Alkali Flat
Black Sprs.
Sun Valley
Wadsworth
Fernley
Reno
Sparks
Truckee R.
Job Pk. 8,799
FALLON IND. RES.
Fallon
Steamboat
Virginia City
Crystal Bay
Carson City
Carson L.
Sand Sprs. Salt Flat
L. Tahoe
Weed Hts.
Minden
Gardnerville
Yerington
WALKER RIVER IND. RES.
Schurz
E. Walker R.
Walker Lake
Wellington
Markleeville
Austin
SHOSHONE
TOIYABE RA.
VALLEY
Alkali Flat
Gabbs
Arc Dome 11,775
Round
Eureka
Mt. Hamilton 10,741
McGill
Ruth
E. Ely
Ely
SCHELL CREEK RA.
SNAKE RA.
LEHMAN CAVES NAT'L MON.
Baker
Wheeler Pk. 13,063
Duckwater
Lund

UTAH
ARIZ.
ARIZONA
CALIFORNIA
NEVADA
L I N C O L N
C L A R K
N Y E
ESMERALDA
S I E R R A
N E V A D A
Mt. Wilson 9,296
Panaca
Pioche
Caliente
Meadow Valley Wash
Alamo
White R.
Troy Pk. 11,268
RAILROAD
Bald Mtn. 9,380
Kawich Pk. 9,500
Tonopah
Goldfield
NELLIS AIR FORCE RANGE AND AEC NUCLEAR TESTING SITE
Yucca Flat
Frenchman Flat
Mercury
Beatty
Mesquite
Virgin R.
MOAPA RIVER IND. RES.
Logandale
Overton
Sheep Pk. 9,720
Indian Sprs.
Charleston Pk. 11,919
N. Las Vegas
Las Vegas
Paradise
Henderson
Boulder City
NELLIS A.F.B.
Lake Mead
HOOVER DAM
LAKE MEAD NATIONAL RECREATION AREA
Lake Mohave
Searchlight
DAVIS DAM
Kingman
FT. MOHAVE IND. RES.
PAIUTE IND. RES.
DEATH VALLEY NAT'L MON. (Devil's Hole)
AMARGOSA RA.
DEATH VALLEY
DEATH VALLEY NATIONAL MONUMENT
Lowest Pt. in N. A. 282 Ft. below Sea Level
FT. INDEPENDENCE IND. RES.
Independence
Owens Lake
Owens R.
Los Angeles Aqueduct
Mt. Whitney 14,494
WHITE MTS.
Dyer
Salt Marsh
Columbus Salt Marsh
Mina
Hawthorne
HAWTHORNE NAV. AMM. DEPOT
Boundary Pk. 13,140
Mono L.
Bridgeport
DEVIL'S POSTPILE NAT'L MON.
YOSEMITE NATIONAL PARK
Mariposa
KINGS CANYON NAT'L PARK
SEQUOIA NAT'L PARK
TULE RIVER IND. RES.
Madera
San Joaquin R.
Fresno
Kings R.
Hanford
Visalia
Tulare
Tulare Lake
MILES
0 10 20 40 60 80
KILOMETRES
0 20 40 60 80
State Capital
© C.S. HAMMOND & Co., Maplewood, N.J.
38°
36°
114°
116°
118°
120°
Longitude West of Greenwich

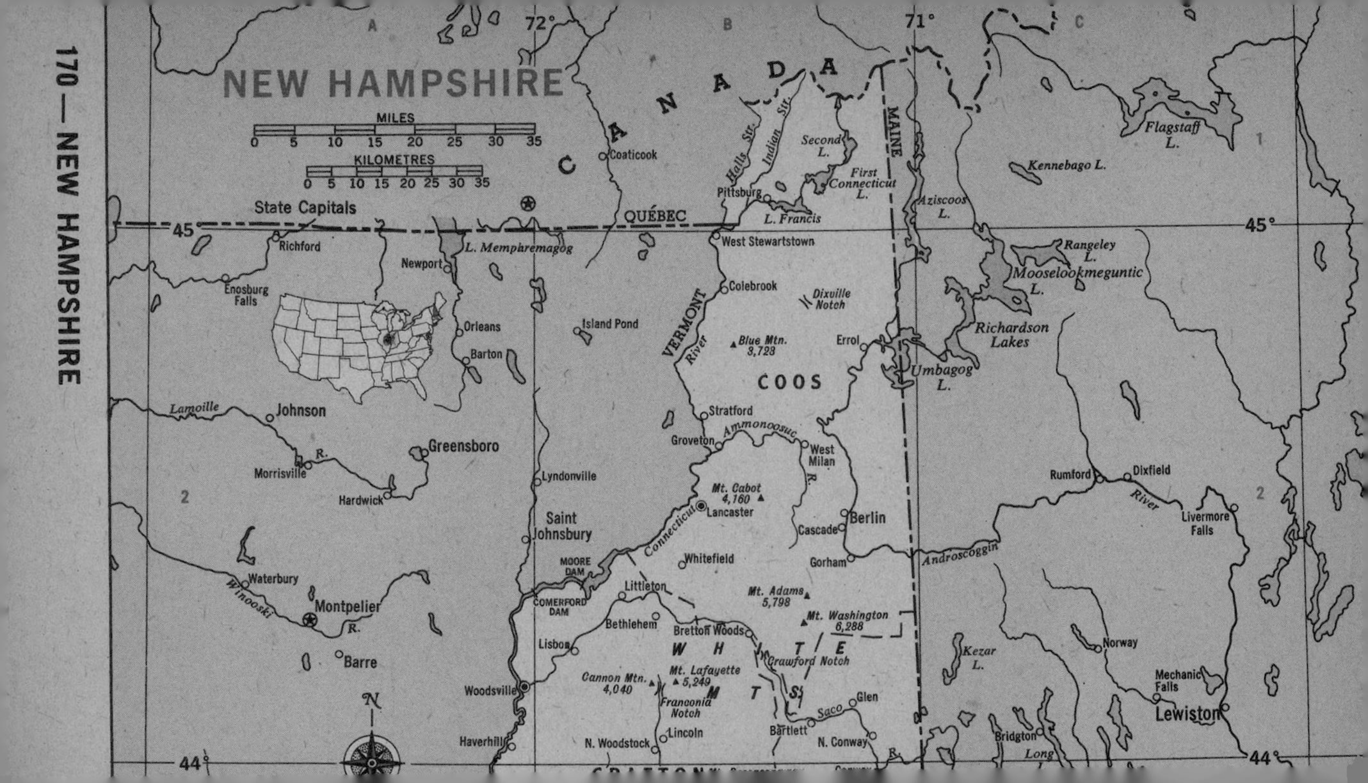
NEW HAMPSHIRE
MILES
0 5 10 15 20 25 30 35
KILOMETRES
0 5 10 15 20 25 30 35
State Capitals
A
B
C
72°
71°
45°
44°
1
2
CANADA
QUÉBEC
MAINE
VERMONT
COOS
WHITE MTS.
Coaticook
Halls Str.
Indian Str.
Second L.
First Connecticut L.
Pittsburg
L. Francis
Aziscoos L.
Flagstaff L.
Kennebago L.
Rangeley L.
Mooselookmeguntic L.
Richardson Lakes
Umbagog L.
Richford
L. Memphremagog
Newport
West Stewartstown
Enosburg Falls
Colebrook
Dixville Notch
Orleans
Island Pond
Barton
River
Blue Mtn. 3,728
Errol
Lamoille
Johnson
Stratford
Groveton
Ammonoosuc
West Milan
R.
Greensboro
Morrisville
Hardwick
Lyndonville
Mt. Cabot 4,160
Lancaster
Berlin
Cascade
Rumford
Dixfield
Livermore Falls
Saint Johnsbury
Connecticut
Whitefield
Gorham
Androscoggin
MOORE DAM
COMERFORD DAM
Littleton
Waterbury
Winooski
Montpelier
Bethlehem
Bretton Woods
Mt. Adams 5,798
Mt. Washington 6,288
Lisbon
Crawford Notch
Barre
Cannon Mtn. 4,040
Mt. Lafayette 5,249
Franconia Notch
Woodsville
Glen
Saco
Bartlett
N. Conway
Kezar L.
Norway
Mechanic Falls
Lewiston
Bridgton
Long
Haverhill
N. Woodstock
Lincoln
N

ATLANTIC OCEAN
MAINE
VERMONT
MASSACHUSETTS
CARROLL
BELKNAP
STRAFFORD
MERRIMACK
ROCKINGHAM
HILLSBORO
CHESHIRE
SULLIVAN
OSSIPEE MTS.
SANDWICH RA.
Concord
Manchester
Nashua
Portsmouth
Dover
Laconia
Keene
Lebanon
Hanover
Claremont
Franklin
Rochester
Exeter
Portland
Westbrook
Biddeford
Sanford
Lowell
Lawrence
Haverhill
Newburyport
Gloucester
Brattleboro
Merrimack River
Connecticut R.
L. Winnipesaukee
L. Sunapee
Newfound L.
Isles of Shoals
SAINT-GAUDENS NAT'L HIST. SITE
PEASE A.F.B.
Mt. Monadnock 3,166
Mt. Kearsarge 2,937
Mt. Cardigan 3,121
Smarts Mtn. 3,240
© C.S. HAMMOND & Co., Maplewood, N.J.
Longitude West of Greenwich
43°
71°
72°

NEW YORK
PENNSYLVANIA
KITTATINNY MTS.
High Pt. 1,803
DELAWARE WATER GAP NAT'L REC. AREA
SUSSEX
PASSAIC
BERGEN
WARREN
MORRIS
ESSEX
HUDSON
UNION
HUNTERDON
SOMERSET
MIDDLESEX
MERCER
MONMOUTH
Hudson River
Delaware R.
Raritan R.
Millstone R.
Musconetcong R.
Pequest R.
Paulins Kill
Wallkill R.
L. Hopatcong
Greenwood L.
Wanaque Res.
Round Valley Res.
Spruce Run Res.
PICATINNY ARSENAL
MORRISTOWN NAT'L HIST. PARK
Port Jervis
Matamoras
Montague
Newton
Sparta
Paterson
Hackensack
Passaic
Clifton
Montclair
Bloomfield
Newark
Jersey City
Elizabeth
Irvington
E. Orange
Plainfield
Linden
Rahway
Woodbridge
Perth Amboy
Edison
New Brunswick
Somerville
Flemington
Phillipsburg
Easton
Belvidere
Trenton
Princeton
Freehold
Long Branch
Asbury Park
Sandy Hook
Staten I.
Long I.
Brooklyn
Upper Bay
Lower Bay
Jamaica B.
GATEWAY NAT'L REC. AREA
Yonkers
Peekskill
Ossining
Haverstraw
Suffern
Chester
Warwick
Stroudsburg
Delaware Water Gap
Bangor
Doylestown
Hatboro
Morrisville
L. Tappan
Oradell Res.
Newark Bay
The Narrows
STATUE OF LIBERTY N.M.
EDISON NAT'L HIST. SITE
NEWARK AIRPORT
FT. WADSWORTH
FT. HANCOCK
FORT MONMOUTH
Raritan Bay
Arthur Kill
N.Y.
Manhattan
Union City
Hoboken
Bayonne
Kearny
N. Bergen
Teaneck
Fair Lawn

NEW JERSEY
ATLANTIC OCEAN
MILES
0 5 10 20 30
KILOMETERS
0 5 10 20 30
State Capitals
County Seats
PINE BARRENS
BURLINGTON
OCEAN
CAMDEN
ATLANTIC
GLOUCESTER
SALEM
CUMBERLAND
CAPE MAY
PHILADELPHIA
Camden
Wilmington
DELAWARE
MARYLAND
PA.
Delaware Bay
Delaware R.
Point Pleasant
Manasquan
Lakewood
Lakehurst
LAKEHURST NAVAL AIR STA.
New Egypt
McGUIRE A.F.B.
FORT DIX
Wrightstown
Bordentown
Florence
Burlington
Willingboro
Mt. Holly
Bristol
Riverside
Palmyra
Pennsauken
Moorestown
Medford
Medford Lakes
Cherry Hill
Haddonfield
Collingswood
Audubon
Gloucester City
Lindenwold
Berlin
Pine Hill
Somerdale
Woodbury
Paulsboro
Swedesboro
Pitman
Glassboro
Clayton
Williamstown
Hammonton
Egg Harbor City
Batsto
Mays Landing
Buena
Vineland
Millville
Elmer
Woodstown
Penns Grove
Pennsville
Salem
Seabrook
Bridgeton
Cedarville
Port Norris
Greenwich
Dorothy
Woodbine
Cape May Court House
Villas
Cape May
Wildwood
Stone Harbor
Avalon
Sea Isle City
Ocean City
Somers Point
Linwood
Northfield
Pleasantville
Absecon
Atlantic City
Ventnor City
Margate City
Brigantine
Great Egg Harbor Inlet
Little Egg Inlet
Little Egg Harbor
Great Bay
Beach Haven
Ship Bottom
Long Beach
Barnegat Light
Barnegat
Manahawkin
Tuckerton
New Gretna
Forked River
Beachwood
Toms River
Island Hts.
Lavallette
Seaside Hts.
Seaside Park
Island Beach
Barnegat Bay
Mullica R.
Wading R.
Rancocas Cr.
Great Egg Harbor R.
Maurice R.
Union L.
Cohansey R.
Oldmans Cr.
Brandywine Cr.
Schuylkill River
Hereford Inlet
Egg Island Pt.
Ben Davis Pt.
Deep Water Pt.
Chester
West Chester
Phoenixville
Norristown
Newark
Elkton
Oxford
Middletown
Smyrna
Dover
Greensboro
C.&D. Canal
Elk R.
Sassafras R.
Choptank
Tuckahoe Cr.
40°
39°
30′
76°
75°
74°
Longitude West of Greenwich
A B C D E
4 5 6
© Copyright HAMMOND INCORPORATED, Maplewood, N.J.

COLORADO
OKLA.
TEXAS
UTAH
UNION
COLFAX
HARDING
MORA
TAOS
RIO ARRIBA
SAN JUAN
SANDOVAL
LOS ALAMOS
SANTA FE
SAN MIGUEL
QUAY
CURRY
GUADALUPE
TORRANCE
BERNALILLO
VALENCIA
MC KINLEY
NAVAJO INDIAN RESERVATION
CONTINENTAL DIVIDE
SANGRE DE CRISTO MTS.
SAN JUAN MTS.
CHUSKA MTS.
MANZANO MTS.
Raton
Clayton
Maxwell
Springer
Cimarron
Eagle Nest L.
Roy
Mosquero
Tucumcari
Logan
Ute Res.
CONCHAS DAM
Conchas Lake
Wagon Mound
FT. UNION N.M.
Watrous
Las Vegas
Mora
Taos
Questa
Wheeler Pk. 13,161
Ranches of Taos
Penasco
Truchas
Chimayo
Dixon
Velarde
Espanola
San Juan Pueblo
Arroyo Hondo
Tierra Amarilla
Chama
Canjilon
El Rito
Dulce
JICARILLA IND. RES.
Navajo Res.
Los Alamos
BANDELIER NAT'L MON.
Santa Fe
Tesuque
Glorieta
Pecos
PECOS NAT'L MON.
Rowe
Santo Domingo Pueblo
San Felipe Pueblo
Jemez Pueblo
Bernalillo
Corrales
Albuquerque
KIRTLAND A.F.B.
Tijeras
Moriarty
Isleta
Los Lunas
Belen
Jarales
Estancia
Mountainair
Laguna del Perro
Encino
Vaughn
Santa Rosa
Anton Chico
Lake Sumner
Ft. Sumner
Clovis
Melrose
CANNON
Cuba
La Jara
San Mateo
Mt. Taylor 11,301
Grants
Milan
Bluewater
San Rafael
EL MORRO N.M.
Acoma
Acomita
Paguate
Thoreau
Crownpoint
CHACO CANYON NAT'L MON.
WINGATE ARMY DEPOT
Gallup
Gamerco
Manuelito
Zuni
ZUNI IND. RES.
Quemado
Farmington
Aztec
AZTEC RUINS N.M.
Bloomfield
Blanco
Shiprock
Toadlena
CANYON DE CHELLY NAT'L MON.
Durango
Pagosa Sprs.
Silverton
Cortez
MESA VERDE NAT'L PARK
SO. UTE IND. RES.
UTE MTN. IND. RES.
YUCCA HOUSE N.M.
HOVENWEEP NAT'L MON.
Only point in U.S. common to 4 states
Monticello
Mt. Wilson 14,246
Uncompahgre Pk. 14,304
Alamosa
Monte Vista
Del Norte
Center
Saguache
Crestone Pk. 14,294
GREAT SAND DUNES NAT'L MON.
Blanca Pk. 14,317
Walsenburg
Trinidad
La Junta
Las Animas
CAPULIN MTN. NAT'L MON.
Rio Grande
Pecos R.
Canadian R.
Dry Cimarron R.
Carrizo Cr.
Ute Cr.
Rio Puerco
Puerco R.
Chaco R.
San Juan R.
Animas R.
Dolores R.
Purgatoire R.
Arkansas R.
Huerfano R.
Zuni R.
Blanco Cr.
Lava Beds
38°
36°
106°
108°

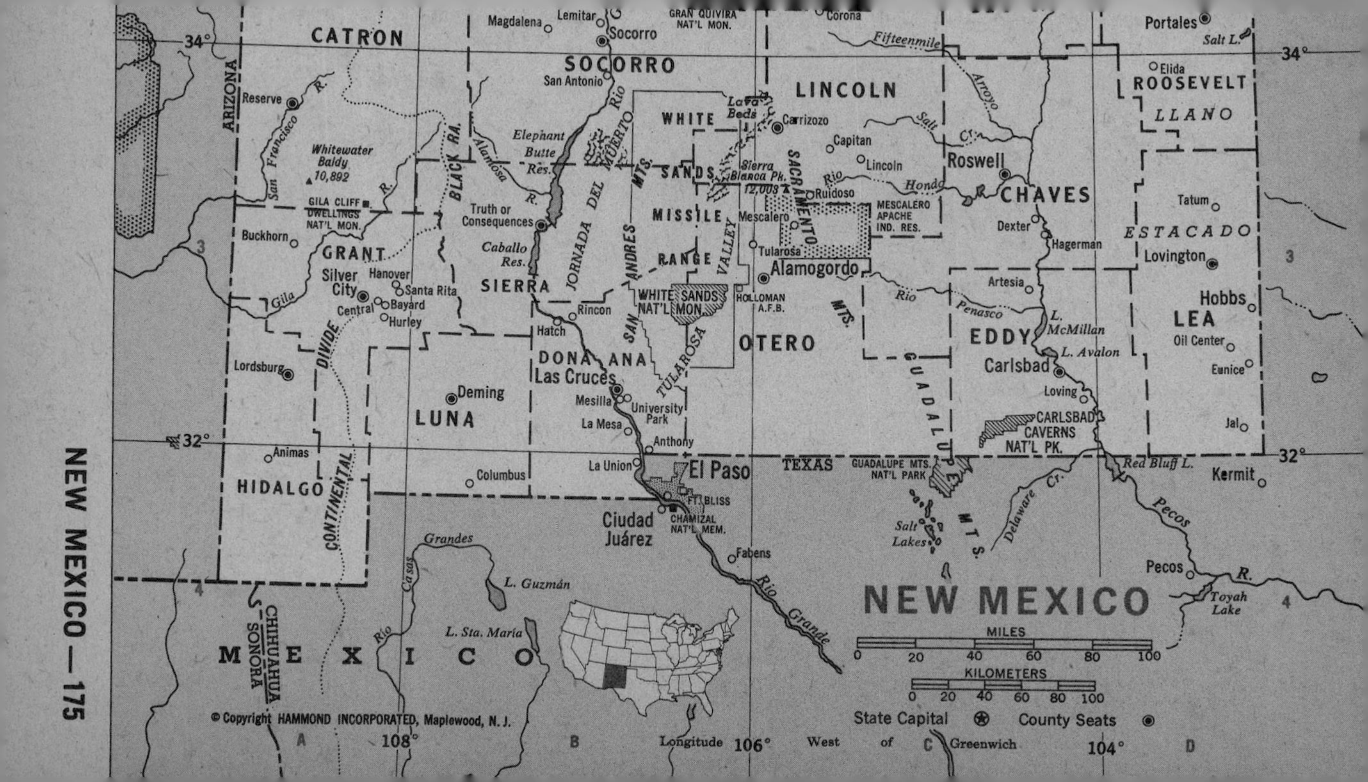
NEW MEXICO
MILES
0 20 40 60 80 100
KILOMETERS
0 20 40 60 80 100
State Capital
County Seats
Longitude 106° West of Greenwich
108°
104°
34°
32°
A
B
C
D
3
4
© Copyright HAMMOND INCORPORATED, Maplewood, N.J.
CATRON
SOCORRO
LINCOLN
ROOSEVELT
CHAVES
LEA
EDDY
OTERO
DONA ANA
SIERRA
LUNA
GRANT
HIDALGO
LLANO
ESTACADO
WHITE SANDS MISSILE RANGE
SAN ANDRES MTS.
SACRAMENTO
MTS.
GUADALUPE MTS.
JORNADA DEL MUERTO
TULAROSA VALLEY
CONTINENTAL DIVIDE
BLACK RA.
TEXAS
ARIZONA
MEXICO
CHIHUAHUA
SONORA
Portales
Salt L.
Elida
Tatum
Lovington
Hobbs
Oil Center
Eunice
Jal
Kermit
Pecos
Hagerman
Dexter
Artesia
Roswell
Carlsbad
Loving
CARLSBAD CAVERNS NAT'L PK.
GUADALUPE MTS. NAT'L PARK
Salt Lakes
Red Bluff L.
Toyah Lake
Corona
Lincoln
Capitan
Carrizozo
Ruidoso
MESCALERO APACHE IND. RES.
Mescalero
Tularosa
Alamogordo
HOLLOMAN A.F.B.
Sierra Blanca Pk. 12,003
Lava Beds
WHITE SANDS NAT'L MON.
GRAN QUIVIRA NAT'L MON.
Magdalena
Lemitar
Socorro
San Antonio
Elephant Butte Res.
Truth or Consequences
Caballo Res.
Rincon
Hatch
Las Cruces
Mesilla
University Park
La Mesa
Anthony
La Union
El Paso
FT. BLISS
CHAMIZAL NAT'L MEM.
Ciudad Juárez
Fabens
Deming
Columbus
Santa Rita
Hanover
Bayard
Hurley
Central
Silver City
Whitewater Baldy 10,892
GILA CLIFF DWELLINGS NAT'L MON.
Buckhorn
Reserve
Lordsburg
Animas
L. McMillan
L. Avalon
L. Guzmán
L. Sta. María
Rio Grande
Pecos R.
Delaware Cr.
Rio Penasco
Rio Hondo
Salt Cr.
Arroyo
Fifteenmile
Alamosa R.
San Francisco R.
Gila
Rio Casas Grandes

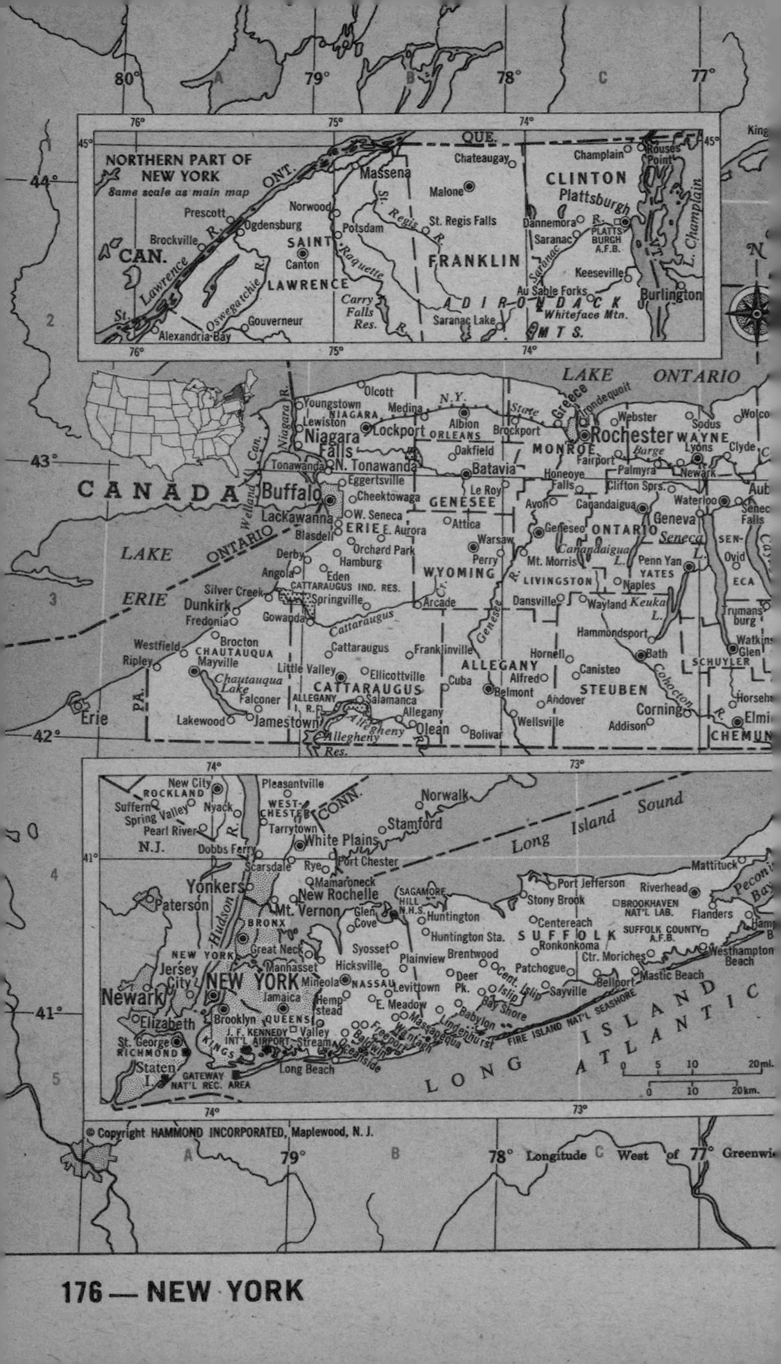

NORTHERN PART OF NEW YORK
Same scale as main map
QUE.
ONT.
CAN.
Prescott
Ogdensburg
Brockville
St. Lawrence R.
Oswegatchie R.
Alexandria Bay
Gouverneur
Norwood
Massena
Potsdam
SAINT LAWRENCE
Canton
St. Regis R.
Raquette R.
Carry Falls Res.
Chateaugay
Malone
St. Regis Falls
FRANKLIN
ADIRONDACK MTS.
Saranac Lake
Whiteface Mtn.
Champlain
Rouses Point
CLINTON
Plattsburgh
Dannemora
Saranac
Saranac R.
PLATTSBURGH A.F.B.
Keeseville
Au Sable Forks
Burlington
L. Champlain
LAKE ONTARIO
CANADA
Olcott
Youngstown
NIAGARA
Lewiston
Niagara Falls
Niagara R.
Welland Can.
Lockport
Medina
N.Y. State Barge Can.
Albion
ORLEANS
Brockport
Greece
Irondequoit
Rochester
Webster
Sodus
Wolcott
WAYNE
Lyons
Clyde
MONROE
Fairport
Palmyra
Newark
Oakfield
Batavia
Tonawanda
N. Tonawanda
Eggertsville
Buffalo
Cheektowaga
GENESEE
Le Roy
Honeoye Falls
Clifton Sprs.
Lackawanna
W. Seneca
ERIE
E. Aurora
Attica
Avon
Canandaigua
Waterloo
Seneca Falls
Geneseo
ONTARIO
Geneva
Blasdell
Orchard Park
Warsaw
Canandaigua L.
Seneca L.
LAKE ERIE
Derby
Hamburg
Eden
Perry
Mt. Morris
Penn Yan
YATES
Ovid
Angola
WYOMING
LIVINGSTON
Naples
Silver Creek
CATTARAUGUS IND. RES.
Springville
Arcade
Dansville
Wayland
Keuka L.
Dunkirk
Fredonia
Gowanda
Cattaraugus Cr.
Genesee R.
Trumansburg
Hammondsport
Westfield
Brocton
CHAUTAUQUA
Mayville
Ripley
Cattaraugus
Franklinville
Hornell
Bath
Watkins Glen
Little Valley
Ellicottville
ALLEGANY
Alfred
Canisteo
SCHUYLER
Chautauqua Lake
CATTARAUGUS
Cuba
Belmont
STEUBEN
Cohocton R.
PA.
Falconer
ALLEGANY I. R.
Salamanca
Andover
Corning
Horseheads
Erie
Lakewood
Jamestown
Allegany
Allegheny R.
Allegheny Res.
Olean
Bolivar
Wellsville
Addison
Elmira
CHEMUNG
New City
ROCKLAND
Pleasantville
Suffern
Spring Valley
Nyack
WESTCHESTER
CONN.
Norwalk
Pearl River
Tarrytown
Stamford
N.J.
Dobbs Ferry
White Plains
Long Island Sound
Scarsdale
Rye
Port Chester
Mattituck
Yonkers
Mamaroneck
Port Jefferson
Riverhead
Peconic Bay
Paterson
New Rochelle
SAGAMORE HILL N.H.S.
Stony Brook
BROOKHAVEN NAT'L LAB.
Flanders
Hudson R.
Mt. Vernon
Glen Cove
Huntington
BRONX
Centereach
SUFFOLK COUNTY A.F.B.
Huntington Sta.
SUFFOLK
Syosset
Ronkonkoma
NEW YORK
Great Neck
Brentwood
Ctr. Moriches
Westhampton Beach
Jersey City
Manhasset
Hicksville
Plainview
Patchogue
Mastic Beach
Mineola
NASSAU
Deer Pk.
Cent. Islip
Bellport
Newark
NEW YORK
Levittown
Sayville
Jamaica
Hempstead
E. Meadow
Islip
Bay Shore
Elizabeth
Brooklyn
QUEENS
Massapequa
Babylon
Lindenhurst
FIRE ISLAND NAT'L SEASHORE
J. F. KENNEDY INT'L AIRPORT
Valley Stream
Freeport
Wantagh
St. George
RICHMOND
KINGS
Baldwin
Oceanside
LONG ISLAND
ATLANTIC
Staten I.
GATEWAY NAT'L REC. AREA
Long Beach
0 5 10 20 mi.
0 10 20 km.
© Copyright HAMMOND INCORPORATED, Maplewood, N. J.
Longitude West of Greenwich

NEW YORK
76°
75°
74°
73°
E
F
G
1
2
3
4
5
44°
43°
42°
41°
Thousand Is.
Alexandria Bay
Clayton
Gouverneur
SAINT LAWRENCE
FRANKLIN
Saranac Lake
Saranac Lakes
Tupper Lake
Cranberry L.
Tupper L.
Lake Placid
Elizabethtown
Mt. Marcy 5,344
Port Henry
Middlebury
JEFFERSON
Black River
CAMP DRUM
ADIRONDACK
Champlain
Sackets Har.
Watertown
Carthage
Long L.
Crown Point
LEWIS
ESSEX
Ticonderoga
Adams
Lowville
Raquette L.
HERKIMER
HAMILTON
Schroon L.
Lake George
Lake
Otter Cr.
MOUNTAINS
Black R.
MTS.
Indian L.
Hudson
Pulaski
Salmon Res.
Rutland
Whitehall
Oswego
Fernwood
OSWEGO
Boonville
Lake Pleasant
Warrensburg
WARREN
Granville
Piseco L.
Lake George
Glens Falls
VERMONT
GREEN
Fulton
Camden
ONEIDA
Hinckley Res.
Hudson Falls
Phoenix
Baldwinsville
FT. STANWIX N. M.
GRIFFISS A.F.B.
Corinth
SARATOGA
WASHINGTON
Oneida
Rome
Great Sacandaga Lake
Saratoga Sprs.
Greenwich
N. Syracuse
Herkimer
Solvay
Canastota
Oneida
Little Falls
FULTON
Gloversville
SARATOGA NAT'L HIST. PARK
Syracuse
Manlius
Wampsville
Utica
Ilion
Johnstown
Ballston Spa
Skaneateles
Cazenovia
Waterville
Mohawk
Ft. Plain
Fonda
R.
Amsterdam
Bennington
ONONDAGA
Skaneateles L.
MADISON
Canajoharie
MONTGOMERY
Mechanicville
Hoosick Falls
Hoosic
Hamilton
Richfield Sprs.
Schenectady
Scotia
RENSSELAER
Tioughnioga R.
SCHENECTADY
Watervliet
Cohoes
R.
N. Adams
Moravia
Otsego L.
Cobleskill
Troy
Homer
Sherburne
Cooperstown
SCHOHARIE
Schoharie
Albany
Rensselaer
Adams
Cortland
Worcester
Middleburg
Castleton-on-Hudson
Groton
CORTLAND
Norwich
OTSEGO
Schoharie R.
ALBANY
Cayuga Hts.
CHENANGO
Oneonta
Ravena
MASS.
Pittsfield
Ithaca
Oxford
Stamford
MARTIN VAN BUREN N.H.S.
Greene
Greenville
Coxsackie
Chatham
Whitney Pt.
Susquehanna
W. Br.
COLUMBIA
Johnson City
Sidney
Delhi
Athens
R.
Endicott
Chenango Br.
DELAWARE
E. Br.
GREENE
Hudson
TIOGA
Walton
Pepacton Res.
Catskill
Great Barrington
Owego
Binghamton
Vestal
Endwell
BROOME
Deposit
Saugerties
W. Hurley
Waverly
CATSKILL
Hudson
Housatonic
Hancock
Ashokan Res.
Red Hook
Winsted
Livingston Manor
MTS.
Kingston
Rhinebeck
CONN.
PA.
Delaware
Liberty
ULSTER
DUTCHESS
Hyde Park
Torrington
New Paltz
F. D. ROOSEVELT N.H.S.
Ellenville
SULLIVAN
Highland
Poughkeepsie
Monticello
Marlboro
Wappingers Falls
Waterbury
Walden
Beacon
R.
Newburgh
PUTNAM
Middletown
Goshen
West Point
Carmel
Danbury
Port Jervis
Highland Falls
Peekskill
New Haven
N.J.
ORANGE
WESTCHESTER
Croton-on-Hudson
Haverstraw
ROCKLAND
New City
Ossining
Bridgeport
White Plains
New Rochelle
Yonkers
LONG ISLAND
(See inset map of Long Island for detail.)
NEW YORK
MILES
0
20
40
60
KILOMETERS
0
20
40
60
State Capital
County Seats
72°
Orient Pt.
Gardiners Bay
Gardiners I.
Greenport
Shelter I.
Montauk
41°
Sag Harbor
Amagansett
E. Hampton
Southampton
OCEAN
72°
Same scale as Long Island map
Fishers I.
R.I.
CONN.
SUFFOLK
Fishers Island
Plum I.
Block Island Sd.
Orient Pt.
72°

82°
81°
80°
79°
A
B
C
36°
35°
34°
33°
VIRGINIA
SOUTH CAROLINA
TENN.
Bristol
S. Holston L.
Galax
Philpott Res.
Martinsville
S. Boston
Danville
Dan R.
Lansing
ASHE
Jefferson
ALLEGHANY
Sparta
Toast
Mt. Airy
SURRY
Dobson
Pilot Mtn.
STOKES
Danbury
Mayodan
Eden
Madison
CASWELL
Yanceyville
Longhurst
Roxboro
PERSON
Reidsville
ROCKINGHAM
Walnut Cove
Boone
WATAUGA
Hays
WILKES
N. Wilkesboro
Wilkesboro
Elkin
Yadkin
YADKIN
Yadkinville
Rural Hall
FORSYTH
Kernersville
GUILFORD C. H. NAT'L MIL. PK.
Altamahaw
ORANGE
Mebane
DURHAM
Hillsborough
AVERY
Newland
Bakersville
Winston-Salem
Greensboro
Burlington
Graham
Haw R.
Lenoir
Taylorsville
CALDWELL
ALEXANDER
Granite Falls
DAVIE
Mocksville
High Point
Thomasville
GUILFORD
ALAMANCE
Durham
Chapel Hill
Carrboro
Archdale
Morganton
L. James
Marion
MC DOWELL
BURKE
Valdese
Hickory
Longview
Conover
Newton
CATAWBA
Maiden
IREDELL
Statesville
Mooresville
Lexington
DAVIDSON
Spencer
Salisbury
ROWAN
Landis
High Rock L.
Denton
Randleman
Liberty
N. Asheboro
Asheboro
RANDOLPH
B. Everett Jordan L.
Siler City
Pittsboro
CHATHAM
Cary
Deep R.
RUTHERFORD
Forest City
Lincolnton
Norman
LINCOLN
Cherryville
Shelby
Cliffside
Boiling Sprs.
Kings Mtn.
CLEVELAND
Bessemer City
Gastonia
GASTON
Belmont
Davidson
Kannapolis
CABARRUS
Concord
Mt. Pleasant
Rocky
Albemarle
STANLY
Norwood
Badin
MONTGOMERY
Troy
Mt. Gilead
Robbins
Carthage
MOORE
Vass
Southern Pines
Aberdeen
LEE
Sanford
Angier
Lillington
HARNETT
Mt. Holly
Charlotte
MECKLENBURG
Pineville
UNION
Monroe
Waxhaw
Marshville
Wadesboro
ANSON
Pee Dee R.
Ellerbe
RICHMOND
Rockingham
Hamlet
SCOTLAND
Laurinburg
Raeford
HOKE
POPE A.F.B.
FT. BRAGG
Spring L.
Fayetteville
Hope Mills
Red Sprs.
Maxton
St. Pauls
ROBESON
Pembroke
Lumberton
Rowland
Fairmont
Gaffney
Broad R.
Catawba L.
Spartanburg
York
Rock Hill
Union
Catawba R.
Lancaster
Laurens
Whitmire
Bennettsville
Chadbourn
Fair Bluff
Tabor City
Conway
Myrtle Beach
84°
83°
82°
Norris L.
Cherokee L.
Morristown
Greenville
Erwin
TENN.
Oak Ridge
Clinch R.
Holston R.
Douglas L.
French Broad
MITCHELL
Bakersville
AVERY
Newland
Knoxville
Hot Sprs.
MADISON
Burnsville
YANCEY
Spruce Pine
Mars Hill
BURKE
Ft. Loudoun L.
Maryville
Little
GREAT SMOKY MTS.
GREAT SMOKY MTS. NAT'L PARK
Clingmans Dome 6,642
Marshall
Mt. Mitchell 6,684
L. James
Weaverville
Marion
BUNCOMBE
MC DOWELL
Asheville
Swannanoa
Tennessee R.
SWAIN
CHEROKEE I.R.
Cherokee
HAYWOOD
Canton
Waynesville
Hazelwood
Oteen
Skyland
RUTHERFORD
Rutherfordton
Spindale
HENDERSON
Fontana L.
Bryson City
Sylva
Robbinsville
GRAHAM
JACKSON
Hendersonville
CARL SANDBURG HOME N.H.S.
POLK
Columbus
Brevard
Andrews
CHEROKEE
Murphy
CLAY
Hayesville
Franklin
MACON
Glenville
TRANSYLVANIA
L. Toxaway
L. Hiwassee
Chatuge L.
GEORGIA
WESTERN PART OF NORTH CAROLINA
Same scale as main map.

NORTH CAROLINA
Portsmouth
Norfolk
Suffolk
Franklin
Emporia
Buggs Island L.
L. Gaston
Meherrin R.
Great Dismal Swamp
Back Bay
Roanoke Rapids
Littleton
Weldon
Murfreesboro
Jackson
Warrenton
WARREN
VANCE
Henderson
Halifax
HALIFAX
NORTHAMPTON
Rich Square
Enfield
Scotland Neck
Louisburg
Franklinton
FRANKLIN
Whitakers
NASH
Nashville
Spring Hope
Rocky Mt.
Zebulon
Tar R.
Tarboro
EDGECOMBE
Pinetops
Wilson
WILSON
Clayton
Kenly
Selma
Fremont
Princeton
Neuse
Goldsboro
S. JOHNSON A.F.B.
WAYNE
Mt. Olive
Faison
Warsaw
Kenansville
DUPLIN
Rose Hill
Wallace
PENDER
Burgaw
MOORES CREEK NAT'L MIL. PK.
L. Waccamaw
Wilmington
NEW HANOVER
Wrightsville Beach
Green Swamp
BRUNSWICK
Southport
SUNNY POINT MIL. OCEAN TERM.
Smith I.
C. Fear
Carolina Beach
Bay
GATES
Gatesville
Winton
HERTFORD
Ahoskie
Aulander
BERTIE
Chowan R.
Windsor
Roanoke R.
Williamston
Robersonville
MARTIN
Bethel
Greenville
PITT
Farmville
GREENE
Snow Hill
Winterville
Ayden
Grifton
La Grange
Kinston
LENOIR
Trenton
JONES
Beulaville
Richlands
ONSLOW
Jacksonville
Maysville
Holly Ridge
Midway Pk.
CAMP LEJEUNE
New River Inlet
Onslow Bay
Washington
BEAUFORT
Bath
Pamlico R.
Vanceboro
CRAVEN
New Bern
James City
Bayboro
PAMLICO
Havelock
CHERRY PT. U.S.M.C. AIR STA.
Newport
CARTERET
Beaufort
Morehead City
CAPE LOOKOUT NAT'L SEASHORE
C. Lookout
Raleigh Bay
S. Mills
Currituck
CURRITUCK
PASQUOTANK
Camden
CAMDEN
Elizabeth City
Hertford
PERQUIMANS
CHOWAN
Edenton
Albemarle Sound
Columbia
TYRRELL
Plymouth
WASHINGTON
Phelps L.
Belhaven
HYDE
Mattamuskeet L.
Swanquarter
Bodie I.
Kitty Hawk
WRIGHT BROS. NAT'L MEM.
FT. RALEIGH N. H. S.
Manteo
Roanoke I.
DARE
CAPE HATTERAS NAT'L SEASHORE REC. AREA
Hatteras I.
Buxton
Hatteras
C. Hatteras
Pamlico Sound
Ocracoke
Ocracoke I.
Portsmouth I.
ATLANTIC OCEAN
78°
77°
76°
36°
35°
34°
33°
E
F
G
1
2
3
4
MILES
0 10 20 40 60
KILOMETERS
0 10 20 40 60
State Capital
County Seats
© Copyright HAMMOND INCORPORATED, Maplewood, N. J.
Longitude 78° West of E Greenwich 77° F 76° G

CANADA
SASKATCHEWAN
TURTLE PROV.
DIVIDE
BURKE
RENVILLE
BOTTINEAU
WILLIAMS
MOUNTRAIL
WARD
MC KENZIE
MC LEAN
MERCER
DUNN
OLIVER
GOLDEN VALLEY
BILLINGS
STARK
MORTON
SLOPE
HETTINGER
BOWMAN
ADAMS
GRANT
SIOUX
MONTANA
SOUTH
Estevan
Westby
Crosby
Noonan
Columbus
Lignite
Flaxton
Bowbells
Sherwood
Westhope
Mohall
Kenmare
Powers Lake
Grenora
Wildrose
Tioga
Ray
Stanley
Lansford
Glenburn
Upham
MINOT A.F.B.
Berthold
Surrey
Granville
Minot
Williston
FT. UNION TRADING POST NAT'L HIST. SITE
Plaza
Sawyer
Velva
New Town
Parshall
Alexander
Watford City
Sidney
Max
FORT BERTHOLD INDIAN RES.
Lake Sakakawea
Garrison
THEO. ROOSEVELT NAT'L MEM. PK. (North Unit)
GARRISON DAM
Riverdale
Underwood
Turtle Lake
L. Audubon
Killdeer
Dunn Ctr.
Halliday
Goldenvalley
Beulah
Zap
Hazen
Stanton
Washburn
THEO. ROOSEVELT NAT'L MEM. PK. (Elkhorn Ranch)
Manning
Center
Wilton
Wibaux
Beach
Medora
Belfield
Dickinson
Richardton
Hebron
New Salem
Glen Ullin
Mandan
Bismarck
THEO. ROOSEVELT NAT'L MEM. PK. (South Unit)
Sentinel Butte
New England
L. Tschida
Amidon
Clark Buttes
Flasher
White Butte 3,506
Marmarth
Regent
Mott
New Leipzig
Elgin
Carson
Solen
Rhame
Bowman
Scranton
Reeder
Hettinger
Cedar Cr.
Cannonball River
STANDING ROCK INDIAN RES.
Ft. Yates
Selfridge
McIntosh
Shadehill Res.
N. Fork
S. Fork
Grand River
Buffalo
Bison
Mobridge
Timber Lake
CHEYENNE RIVER INDIAN RES.
Missouri River
Yellowstone R.
Little Missouri R.
Beaver Cr.
Heart R.
Souris R.
Des Lacs R.
Deep R.
L. Darling
Medicine L.
COTEAU DU MISSOURI
BAD LANDS
104°
103°
102°
101°
49°
48°
47°
46°
A
B
C
2
3
4
© Copyright HAMMOND INCORPORATED, Maplewood, N. J.

00°
E
99°
F
98°
G
97°
H
A
Altona
INTERNAT'L PEACE GARDEN
MANITOBA
Pembina R.
49°
St. John
Hannah
Neche
Pembina
Walhalla
Rolla
CAVALIER
PEMBINA
Hallock
Belcourt
Cavalier
Two Rivers
Rocklake
Langdon
Red
OLETTE
Osnabrock
Glasston
St. Thomas
Bisbee
Crystal
Drayton
TOWNER
Hoople
Park
Edinburg
Cando
Adams
R.
Rugby
Esmond
Grafton
Middle
R.
WALSH
Park River
RAMSEY
River
Leeds
Lankin
Minto
Warren
Sweetwater L.
Fordville
BENSON
Devils Lake
Thief River Falls
Manvel
Minnewaukan
Devils L.
Lakota
Michigan
Gilby
Esmond
of
48°
Petersburg
GRAND FORKS A.F.B.
Grand Forks
Maddock
Ft. Totten
NELSON
Oberon
FT. TOTTEN IND. RES.
Larimore
Emerado
Lake R.
GRAND FORKS
Red
Goose
the
Tolna
McVille
Crookston
Sheyenne
Harvey
Aneta
Northwood
Sheyenne
New Rockford
Hatton
Fessenden
EDDY
Binford
Buxton
MINNESOTA
WELLS
R.
Finley
Mayville
Carrington
Cooperstown
Portland
2
FOSTER
STEELE
GRIGGS
Hillsboro
Pipestem
R.
Hope
TRAILL
Wild Rice R.
Hannaford
Kensal
Hunter
L. Ashtabula
North
Tuttle
Wimbledon
Page
Horsehead L.
Woodworth
Arthur
R.
CASS
Jamestown
47°
Res.
Buffalo
KIDDER
Valley City
Tower City
R.
Medina
Casselton
Jamestown
Fargo
Moorhead
Steele
Tappen
STUTSMAN
BARNES
Maple
R.
R.
Ypsilanti
Streeter
Litchville
Kindred
Gackle
Alfred
Marion
Enderlin
James
Walcott
Sheyenne
Napoleon
Abercrombie
3
LOGAN
LA MOURE
RICHLAND
La Moure
Lisbon
Cr.
Kulm
Edgeley
R.
RANSOM
Wyndmere
Wahpeton
Beaver
Lehr
Wild
Wishek
Gwinner
Milnor
Rice
R.
SARGENT
MCINTOSH
DICKEY
Oakes
Hankinson
Forman
Fairmount
Cogswell
Lidgerwood
S
Ashley
Ellendale
Rutland
46°
Zeeland
OTA
Eureka
NORTH DAKOTA
L. Traverse
MILES
0 10 20 40 60
Maple
4
R.
KILOMETERS
Ipswich
0 10 20 40 60
Big Stone L.
State Capital
County Seats
00°
E
Longitude 99° West of F Greenwich 98°
G
97°
H

OHIO
MILES
0 10 20 30 40 50
KILOMETERS
0 10 20 30 40 50
State Capital
County Seats
CANADA
MICHIGAN
PA.
INDIANA
LAKE ERIE
ONTARIO
L. St. Clair
DETROIT
Dearborn
Ann Arbor
Ypsilanti
Windsor
Amherstburg
Leamington
Chatham
Monroe
POINT PELEE NAT'L PARK
Pelee Pt.
RONDEAU PROV. PK.
Pt. Aux Pins
Pelee I.
Kelleys I.
Bass Islands
PERRYS VICTORY AND INTERNAT'L PEACE MEM.
Toledo
CLEVELAND
Lakewood
Lorain
Sandusky
Port Clinton
Akron
Canton
Youngstown
Warren
Columbus Grove
Lima
Findlay
Mansfield
Ashtabula
Conneaut
Edgewood
Geneva
Jefferson
Painesville
Mentor
Euclid
Cleveland Hts.
Shaker Hts.
Parma
Elyria
Medina
Wooster
Massillon
Alliance
Barberton
Cuyahoga Falls
Kent
Ravenna
Tiffin
Fremont
Norwalk
Bucyrus
Marion
Delaware
Mt. Vernon
Coshocton
New Philadelphia
Dover
Steubenville
E. Liverpool
Bowling Green
Napoleon
Defiance
Bryan
Wauseon
Van Wert
Celina
Wapakoneta
Sidney
Bellefontaine
Kenton
Upper Sandusky
ASHTABULA
TRUMBULL
GEAUGA
CUYAHOGA
LAKE
PORTAGE
SUMMIT
MAHONING
COLUMBIANA
STARK
CARROLL
JEFFERSON
HARRISON
TUSCARAWAS
HOLMES
WAYNE
MEDINA
LORAIN
ERIE
HURON
ASHLAND
RICHLAND
CRAWFORD
SENECA
SANDUSKY
OTTAWA
WOOD
HANCOCK
WYANDOT
MARION
MORROW
KNOX
COSHOCTON
DELAWARE
UNION
HARDIN
LOGAN
ALLEN
AUGLAIZE
SHELBY
MERCER
VAN WERT
PAULDING
PUTNAM
HENRY
DEFIANCE
WILLIAMS
FULTON
LUCAS
CHAMPAIGN
Campbell Hill 1,550
Maumee R.
Sandusky R.
Cuyahoga R.
Scioto R.
Mohican R.
Tuscarawas R.
42°
41°
81°
82°
83°
84°

BELMONT
Powhatan Point
Woodsfield
MONROE
NOBLE
Caldwell
WASHINGTON
Marietta
Parkersburg
Ohio River
Zanesville
MUSKINGUM
Muskingum R.
McConnelsville
MORGAN
New Straitsville
PERRY
Crooksville
New Lexington
Nelsonville
The Plains
Athens
ATHENS
Belpre
MEIGS
Pomeroy
Middleport
Gallipolis
GALLIA
Raccoon Cr.
WEST VIRGINIA
Columbus
FAIRFIELD
Lancaster
Baltimore
Groveport
Logan
HOCKING
Hocking
VINTON
McArthur
Wellston
Jackson
JACKSON
Oak Hill
LAWRENCE
Ironton
South Point
Chesapeake
Huntington
Coal Grove
New Boston
Wheelersburg
Circleville
PICKAWAY
Kingston
Scioto R.
Chillicothe
ROSS
MOUND CITY GROUP NAT'L MON.
Bainbridge
Waverly
PIKE
Piketon
Lucasville
SCIOTO
Portsmouth
W. Portsmouth
FRANKLIN
Grove City
Darby Cr.
London
MADISON
FAYETTE
Washington C.H.
Paint Cr.
Greenfield
Hillsboro
Rocky Fork L.
HIGHLAND
Peebles
ADAMS
W. Union
Manchester
KENTUCKY River
Springfield
CLARK
Fairborn
Yellow Sprs.
GREENE
Xenia
Sabina
Wilmington
CLINTON
CLINTON CO. A.F.B.
Blanchester
Lynchburg
Williamsburg
Mt. Orab
BROWN
Georgetown
Ripley
Dayton
MONTGOMERY
Vandalia
Oakwood
Kettering
Centerville
Miamisburg
Franklin
Middletown
WARREN
Lebanon
Mason
Germantown
Brookville
PREBLE
Eaton
Oxford
BUTLER
Hamilton
Trenton
Great Miami
Fairfield
HAMILTON
Reading
Cincinnati
Cheviot
Covington
Batavia
CLERMONT
Bethel
New Richmond
LAKE ERIE
CLEVELAND
Lakewood
Rocky River
Bay Village
CLEVELAND-HOPKINS AIRPORT
Fairview Pk.
Brooklyn
Parma Hts.
Parma
CUYAHOGA
Middleburg Hts.
Berea
Olmsted
Strongsville
N. Royalton
Broadview Hts.
Brecksville
Independence
Garfield Hts.
Maple Hts.
Bedford
Shaker Hts.
Cleveland Hts.
E. Cleveland
S. Euclid
Euclid
Lyndhurst
Mayfield Hts.
Univ. Hts.
Warrensville Hts.
Solon
Chagrin Falls
Twinsburg
Macedonia
SUMMIT
Boston Hts.
Cuyahoga R.
W. Br. Rocky R.
East Br.
Eastlake
Willowick
Wickliffe
Willoughby
LAKE
Chagrin R.
CLERMONT
Loveland
Milford
Owensville
Batavia
East Fork
Amelia
Withamsville
Forestville
Newtown
Sharonville
Blue Ash
Deer Park
Silverton
Reading
Wyoming
Lincoln Hts.
Greenhills
Mt. Healthy
N. College Hill
St. Bernard
Norwood
HAMILTON
Cincinnati
WILLIAM H. TAFT N.H.S.
Cheviot
Covedale
Bridgetown
Cleves
Harrison
Gt. Miami R.
Ohio R.
Ft. Thomas
Newport
Covington
KENTUCKY
Erlanger
Silver Grove
Little Miami R.
© Copyright HAMMOND INCORPORATED, Maplewood, N.J.
84° 83° Longitude West of 82° Greenwich 81°

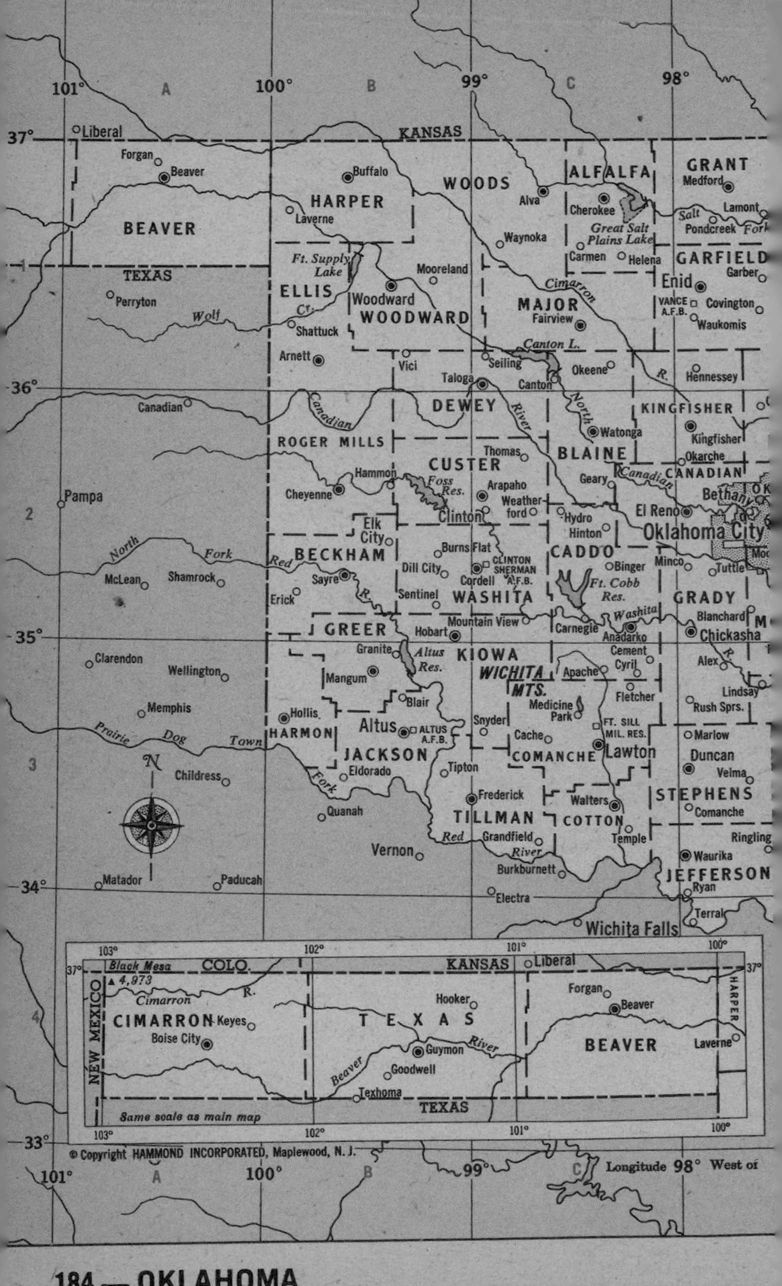

101°
A
100°
B
99°
C
98°
37°
36°
35°
34°
33°
1
2
3
4
KANSAS
TEXAS
Liberal
Forgan
Beaver
BEAVER
Buffalo
HARPER
Laverne
WOODS
Alva
ALFALFA
Cherokee
GRANT
Medford
Lamont
Salt
Pondcreek
Fork
Great Salt Plains Lake
Waynoka
Carmen
Helena
GARFIELD
Garber
Enid
VANCE A.F.B.
Covington
Waukomis
Perryton
Wolf
Cr.
Ft. Supply Lake
ELLIS
Woodward
WOODWARD
Mooreland
Shattuck
MAJOR
Fairview
Cimarron
Canton L.
Arnett
Vici
Seiling
Taloga
Canton
Okeene
R.
Hennessey
Canadian
DEWEY
River
North
KINGFISHER
Watonga
Kingfisher
ROGER MILLS
Thomas
BLAINE
Okarche
CUSTER
CANADIAN
Hammon
Foss Res.
Arapaho
Geary
Cheyenne
Weatherford
Clinton
El Reno
Bethany
Pampa
Hydro
Hinton
Oklahoma City
Elk City
North
Fork
Red
BECKHAM
Burns Flat
CLINTON SHERMAN A.F.B.
CADDO
Dill City
Cordell
Binger
Minco
Tuttle
McLean
Shamrock
Sayre
Ft. Cobb Res.
Erick
Sentinel
WASHITA
GRADY
Washita
Mountain View
Carnegie
Blanchard
Anadarko
GREER
Hobart
Chickasha
35°
Granite
Altus Res.
KIOWA
Cement
Clarendon
Cyril
Alex
Wellington
Mangum
WICHITA MTS.
Apache
Fletcher
Lindsay
Blair
Medicine Park
Rush Sprs.
Memphis
Hollis
Snyder
FT. SILL MIL. RES.
Altus
ALTUS A.F.B.
Cache
Marlow
Prairie
Dog
Town
HARMON
JACKSON
COMANCHE
Lawton
Duncan
N
Eldorado
Tipton
Velma
Childress
Frederick
Walters
STEPHENS
Quanah
TILLMAN
COTTON
Comanche
Red
Grandfield
Temple
Ringling
Vernon
River
Waurika
Burkburnett
JEFFERSON
Matador
Paducah
Electra
Ryan
Terral
Wichita Falls
103°
102°
101°
100°
Black Mesa
COLO.
KANSAS
Liberal
▲ 4,973
Cimarron
R.
Forgan
Hooker
Beaver
NEW MEXICO
CIMARRON
Keyes
T E X A S
HARPER
Boise City
River
BEAVER
Guymon
Laverne
Beaver
Goodwell
Texhoma
TEXAS
Same scale as main map
Longitude 98° West of

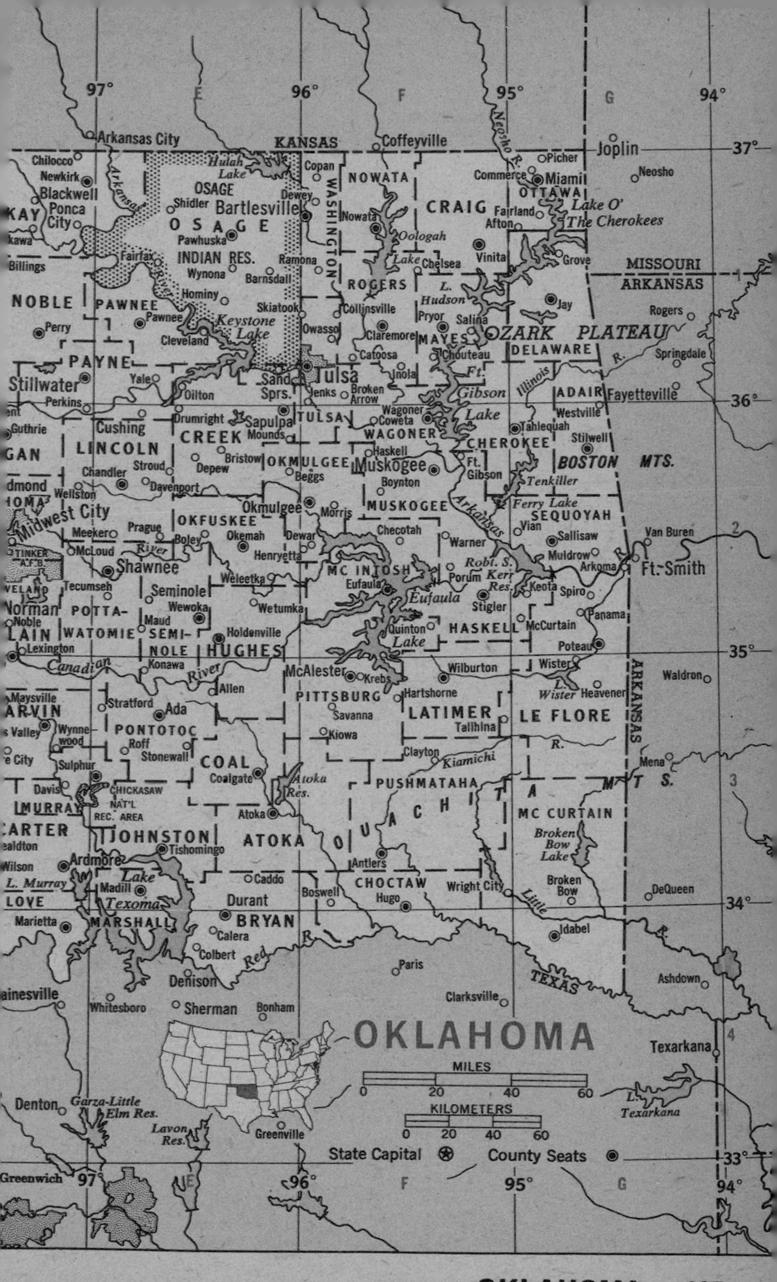
OKLAHOMA
KANSAS
MISSOURI
ARKANSAS
TEXAS
OSAGE
OSAGE INDIAN RES.
NOWATA
CRAIG
OTTAWA
WASHINGTON
ROGERS
MAYES
DELAWARE
ADAIR
CHEROKEE
KAY
NOBLE
PAWNEE
PAYNE
TULSA
WAGONER
CREEK
LINCOLN
OKMULGEE
MUSKOGEE
SEQUOYAH
OKFUSKEE
MC INTOSH
HASKELL
POTTA-WATOMIE
SEMI-NOLE
HUGHES
PITTSBURG
LATIMER
LE FLORE
PONTOTOC
COAL
PUSHMATAHA
MC CURTAIN
MURRAY
CARTER
JOHNSTON
ATOKA
CHOCTAW
BRYAN
MARSHALL
LOVE
OZARK PLATEAU
BOSTON MTS.
OUACHITA MTS.
Arkansas City
Coffeyville
Joplin
Chilocco
Newkirk
Blackwell
Ponca City
Hulah Lake
Copan
Dewey
Bartlesville
Shidler
Pawhuska
Nowata
Oologah Lake
Picher
Commerce
Miami
Neosho
Fairland
Afton
Lake O' The Cherokees
Vinita
Grove
Chelsea
L. Hudson
Jay
Rogers
Billings
Fairfax
Wynona
Ramona
Barnsdall
Hominy
Skiatook
Collinsville
Pryor
Salina
Perry
Pawnee
Keystone Lake
Cleveland
Owasso
Claremore
Catoosa
Chouteau
Springdale
Stillwater
Yale
Tulsa
Sand Sprs.
Jenks
Broken Arrow
Inola
Ft. Gibson Lake
Illinois
Fayetteville
Perkins
Oilton
Drumright
Sapulpa
Mounds
Wagoner
Coweta
Westville
Tahlequah
Guthrie
Cushing
Stilwell
Stroud
Bristow
Depew
Haskell
Muskogee
Ft. Gibson
Chandler
Beggs
Boynton
Tenkiller Ferry Lake
Edmond
Wellston
Davenport
Okmulgee
Morris
Arkansas
Midwest City
Meeker
Prague
Boley
Okemah
Dewar
Checotah
Warner
Vian
Sallisaw
Van Buren
McLoud
River
Henryetta
Muldrow
Arkoma
Ft. Smith
TINKER A.F.B.
Shawnee
Robt. S. Kerr Res.
Porum
Weleetka
Eufaula
Keota
Spiro
Tecumseh
Seminole
Eufaula
Stigler
Norman
Wewoka
Wetumka
Panama
Noble
Maud
Holdenville
Quinton
Lake
McCurtain
Poteau
Lexington
Canadian River
Konawa
Wister
McAlester
Krebs
Wilburton
Waldron
Allen
L. Wister
Heavener
Maysville
Stratford
Ada
Hartshorne
Savanna
Talihina
Wynnewood
Kiowa
Roff
Stonewall
Clayton
Kiamichi
Mena
Sulphur
Coalgate
Atoka Res.
Davis
CHICKASAW NAT'L REC. AREA
Atoka
Broken Bow Lake
Tishomingo
Ardmore
Antlers
Wilson
L. Murray
Lake Texoma
Madill
Caddo
Broken Bow
Wright City
Little R.
DeQueen
Durant
Boswell
Hugo
Marietta
Calera
Idabel
Colbert
Red R.
Paris
Denison
Ashdown
Clarksville
Gainesville
Whitesboro
Sherman
Bonham
Texarkana
L. Texarkana
Denton
Garza-Little Elm Res.
Lavon Res.
Greenville
Greenwich
MILES
0 20 40 60
KILOMETERS
0 20 40 60
State Capital
County Seats
97°
96°
95°
94°
37°
36°
35°
34°
33°
E
F
G

126°
124°
122°
46°
44°
42°
PACIFIC
OCEAN
WASH.
Vancouver
Columbia R.
PORTLAND INT'L AIRPORT
Forest Grove
Hillsboro
Beaverton
Portland
Wood Vill.
Gresham
WASHINGTON
Metzger
Tigard
Milwaukie
L. Oswego
Gladstone
W. Linn
Oregon City
YAM-HILL
Newberg
MC LOUGHLIN HOUSE NAT'L HIST. SITE
Canby
Molalla
MARION
CLACKAMAS
Woodburn
Molalla
Willamette
Pudding
Butte
Mt. Angel
Silverton
Keizer
Hayesville
POLK
Salem
10 Mi.
10 Km.
Chehalis
Cowlitz
Castle Rock
Kelso
Longview
Lewis R.
Warrenton
FT. CLATSOP NAT'L MEM.
Astoria
Clatskanie
Rainier
Kelama
Gearhart
Seaside
COLUMBIA
CLATSOP
St. Helens
Woodland
Vernonia
Nehalem
Scappoose
Vancouver
Cascade Locks
Garibaldi
Bay City
Tillamook
Hillsboro
Portland
MULTNOMAH
BONNEVILLE DAM
Kincheloe Pt.
Netarts
Newberg
Sandy
C. Lookout
TILLAMOOK
Oregon City
Pacific City
McMinnville
Carlton
Estacada
Willamina
Amity
Woodburn
Sheridan
Dallas
CLACKAMAS
Lincoln City
POLK
Salem
Falls City
Independence
Valsetz
Monmouth
Stayton
Mill City
Agate Beach
Newport
BENTON
Jefferson
Toledo
Corvallis
Albany
Philomath
Lebanon
Detroit Lake
LINCOLN
Brownsville
LINN
Waldport
Sweet Home
Harrisburg
Junction City
Fern Ridge L.
Elmira
Mapleton
Springfield
Florence
Eugene
LANE
OREGON DUNES
Creswell
Cottage Grove
Middle
Lookout Pt. Lake
Oakridge
Waldo L.
Reedsport
Umpqua
Drain
Hills Cr. Lake
NAT'L REC. AREA
North Bend
Coos Bay
Eastside
Oakland
Sutherlin
N. Umpqua
COOS
DOUGLAS
Roseburg
Coquille
Bandon
Winston
Myrtle Creek
Myrtle Point
Riddle
CRATER LAKE NAT'L PARK
Crater L.
Canyonville
C. Blanco
Powers
Glendale
Prospect
Port Orford
Rogue
CURRY
Grants Pass
JACKSON
Eagle Point
Upper Klamath L.
Gold Beach
JOSEPHINE
Central Point
Jacksonville
Medford
Kerby
OREGON CAVES N.M.
Ashland
Brookings
Merrill
CALIFORNIA
Klamath R.
REDWOOD NAT'L PARK
Crescent City
Crescent L.
Mt. Thielsen 9,182
Chiloquin
Sprague
LAVA BEDS N.M.

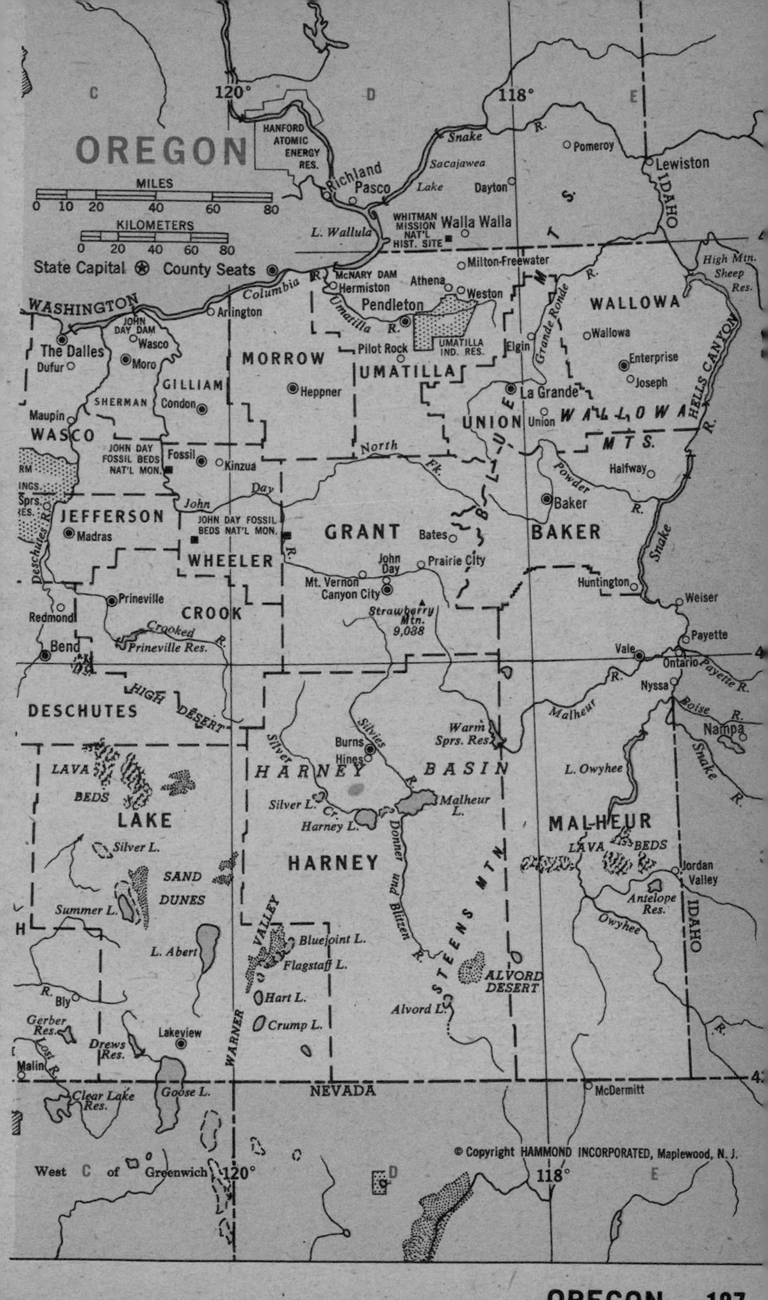
OREGON
MILES
0 10 20 40 60 80
KILOMETERS
0 20 40 60 80
State Capital
County Seats
WASHINGTON
IDAHO
NEVADA
HANFORD ATOMIC ENERGY RES.
Richland
Pasco
Sacajawea Lake
Snake R.
Pomeroy
Lewiston
Dayton
Walla Walla
WHITMAN MISSION NAT'L HIST. SITE
L. Wallula
Milton-Freewater
McNARY DAM
Hermiston
Athena
Weston
Columbia R.
Umatilla R.
Pendleton
UMATILLA IND. RES.
Arlington
JOHN DAY DAM
Wasco
The Dalles
Dufur
Moro
Pilot Rock
Elgin
Grande Ronde R.
WALLOWA
Wallowa
Enterprise
Joseph
MORROW
UMATILLA
Heppner
GILLIAM
Condon
SHERMAN
La Grande
UNION
Union
WALLOWA MTS.
BLUE MTS.
Maupin
WASCO
JOHN DAY FOSSIL BEDS NAT'L MON.
Fossil
Kinzua
North Fk.
Powder R.
Halfway
John Day R.
Baker
BAKER
JEFFERSON
Madras
Deschutes R.
JOHN DAY FOSSIL BEDS NAT'L MON.
GRANT
Bates
WHEELER
John Day
Prairie City
Mt. Vernon
Canyon City
Huntington
Snake R.
HELLS CANYON
High Mtn. Sheep Res.
Prineville
CROOK
Redmond
Strawberry Mtn. 9,038
Weiser
Crooked R.
Prineville Res.
Bend
Payette
Vale
Ontario
Payette R.
Nyssa
Malheur R.
Boise R.
Nampa
Snake R.
HIGH DESERT
DESCHUTES
Silver Cr.
Silvies R.
Warm Sprs. Res.
Burns
Hines
HARNEY BASIN
LAVA BEDS
L. Owyhee
Silver L.
Harney L.
Malheur L.
LAKE
Donner und Blitzen R.
MALHEUR
LAVA BEDS
Silver L.
SAND DUNES
HARNEY
Jordan Valley
Antelope Res.
Summer L.
STEENS MTN.
Owyhee R.
H
Bluejoint L.
Flagstaff L.
L. Abert
WARNER VALLEY
Hart L.
Crump L.
ALVORD DESERT
Alvord L.
Bly
Gerber Res.
Lost R.
Drews Res.
Lakeview
Malin
Clear Lake Res.
Goose L.
McDermitt
West C of Greenwich 120°
D
118°
E
120°
118°
C
D
E

LAKE ERIE
PITTSBURGH
ALLEGHENY
Wilkinsburg
McKeesport
Erie
ERIE
NEW YORK
Bradford
MC KEAN
WARREN
Warren
POTTER
Coudersport
CRAWFORD
Meadville
VENANGO
Oil City
FOREST
Tionesta
CAMERON
Emporium
ELK
Ridgway
MERCER
CLARION
Clarion
JEFFERSON
Brookville
CLEARFIELD
Clearfield
CENTRE
Bellefonte
State College
BUTLER
Butler
LAWRENCE
New Castle
BEAVER
ARMSTRONG
Kittanning
INDIANA
Indiana
CAMBRIA
Ebensburg
BLAIR
Altoona
Hollidaysburg
HUNTINGDON
Huntingdon
WESTMORELAND
Greensburg
Johnstown
WASHINGTON
Washington
FAYETTE
Uniontown
SOMERSET
Somerset
BEDFORD
Bedford
FULTON
McConnellsburg
Chambersburg
FRANKL
GREENE
Waynesburg
WEST VIRGINIA
OHIO
Youngstown
Cumberland
Hagerstown
Morgantown
Mt. Davis 3,213
FT. NECESSITY NAT'L BATTLEFIELD
CHESTER
W. Chester
80°
79°
78°

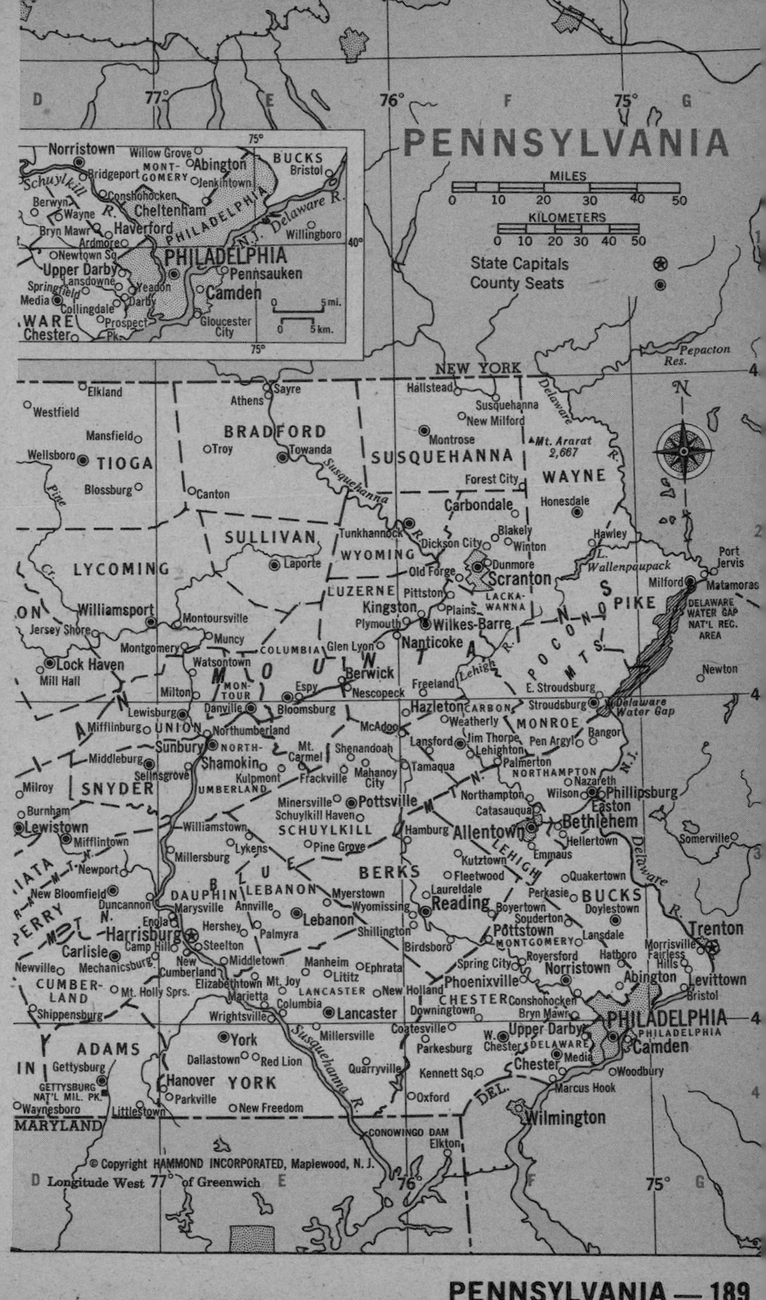

PENNSYLVANIA
MILES
0 10 20 30 40 50
KILOMETERS
0 10 20 30 40 50
State Capitals
County Seats
D
E
F
G
77°
76°
75°
Norristown
Willow Grove
Abington
BUCKS
Bristol
MONT-
GOMERY
Jenkintown
Bridgeport
Schuylkill
Conshohocken
Berwyn
Wayne
R.
Cheltenham
PHILADELPHIA
Delaware R.
Bryn Mawr
Haverford
N.J.
Willingboro
Ardmore
40°
Newtown Sq.
PHILADELPHIA
Upper Darby
Lansdowne
Pennsauken
Springfield
Yeadon
Camden
Media
Darby
Collingdale
WARE
Prospect
Chester
Pk.
Gloucester
City
0 5 mi.
0 5 km.
Pepacton
Res.
NEW YORK
N
Sayre
Elkland
Athens
Hallstead
Susquehanna
Westfield
New Milford
Delaware R.
Mansfield
BRADFORD
Montrose
Mt. Ararat
2,667
Troy
Towanda
Wellsboro
TIOGA
SUSQUEHANNA
Susquehanna
Forest City
WAYNE
Blossburg
Pine
Canton
Carbondale
Honesdale
Blakely
Hawley
SULLIVAN
Tunkhannock
R.
Dickson City
Winton
Port
Jervis
WYOMING
L.
Wallenpaupack
LYCOMING
Laporte
Old Forge
Dunmore
Milford
Matamoras
Cr.
Scranton
S
LUZERNE
Pittston
LACKA-
WANNA
PIKE
DELAWARE
WATER GAP
NAT'L REC.
AREA
ON
Williamsport
Kingston
Plains
Montoursville
Jersey Shore
Plymouth
Wilkes-Barre
Muncy
Montgomery
COLUMBIA
Glen Lyon
Nanticoke
POCONO
MTS.
Lock Haven
Watsontown
Lehigh
Newton
Mill Hall
Berwick
MON-
TOUR
Espy
Freeland
Milton
Nescopeck
E. Stroudsburg
Danville
Bloomsburg
Hazleton
CARBON
Stroudsburg
Delaware
Water Gap
Lewisburg
Weatherly
MONROE
Mifflinburg
UNION
Northumberland
McAdoo
Jim Thorpe
Pen Argyl
Bangor
Sunbury
NORTH-
Mt.
Carmel
Shenandoah
Lansford
Lehighton
Middleburg
Shamokin
Palmerton
N.J.
Selinsgrove
Kulpmont
Frackville
Mahanoy
City
Tamaqua
NORTHAMPTON
Milroy
SNYDER
UMBERLAND
Nazareth
Northampton
Wilson
Phillipsburg
Minersville
Pottsville
Burnham
Schuylkill Haven
Catasauqua
Easton
Lewistown
Williamstown
SCHUYLKILL
Hamburg
Allentown
Bethlehem
Mifflintown
Hellertown
Lykens
Pine Grove
LEHIGH
Somerville
Millersburg
Kutztown
Emmaus
Newport
BERKS
Delaware
IATA
DAUPHIN
LEBANON
Fleetwood
Quakertown
Myerstown
Laureldale
Perkasie
BUCKS
New Bloomfield
Duncannon
Anville
Wyomissing
Reading
Boyertown
Doylestown
PERRY
Marysville
Lebanon
Souderton
Enola
Hershey
Shillington
Pottstown
Lansdale
Trenton
Harrisburg
Palmyra
Carlisle
Camp Hill
Steelton
MONTGOMERY
Morrisville
Fairless
Hills
Birdsboro
Newville
Mechanicsburg
New
Cumberland
Middletown
Manheim
Ephrata
Spring City
Royersford
Hatboro
Lititz
Phoenixville
Norristown
Abington
Levittown
CUMBER-
LAND
Mt. Holly Sprs.
Elizabethtown
Mt. Joy
LANCASTER
New Holland
Bristol
Marietta
CHESTER
Conshohocken
Columbia
Shippensburg
Bryn Mawr
Wrightsville
Lancaster
Downingtown
PHILADELPHIA
Coatesville
Upper Darby
PHILADELPHIA
Millersville
W.
Chester
DELAWARE
Camden
York
Parkesburg
ADAMS
Susquehanna
Media
Dallastown
Red Lion
Chester
Gettysburg
Quarryville
Kennett Sq.
Woodbury
GETTYSBURG
NAT'L MIL. PK.
Hanover
YORK
Marcus Hook
Parkville
Oxford
Waynesboro
Littlestown
New Freedom
DEL.
Wilmington
MARYLAND
CONOWINGO DAM
Elkton
© Copyright HAMMOND INCORPORATED, Maplewood, N. J.
Longitude West 77° of Greenwich

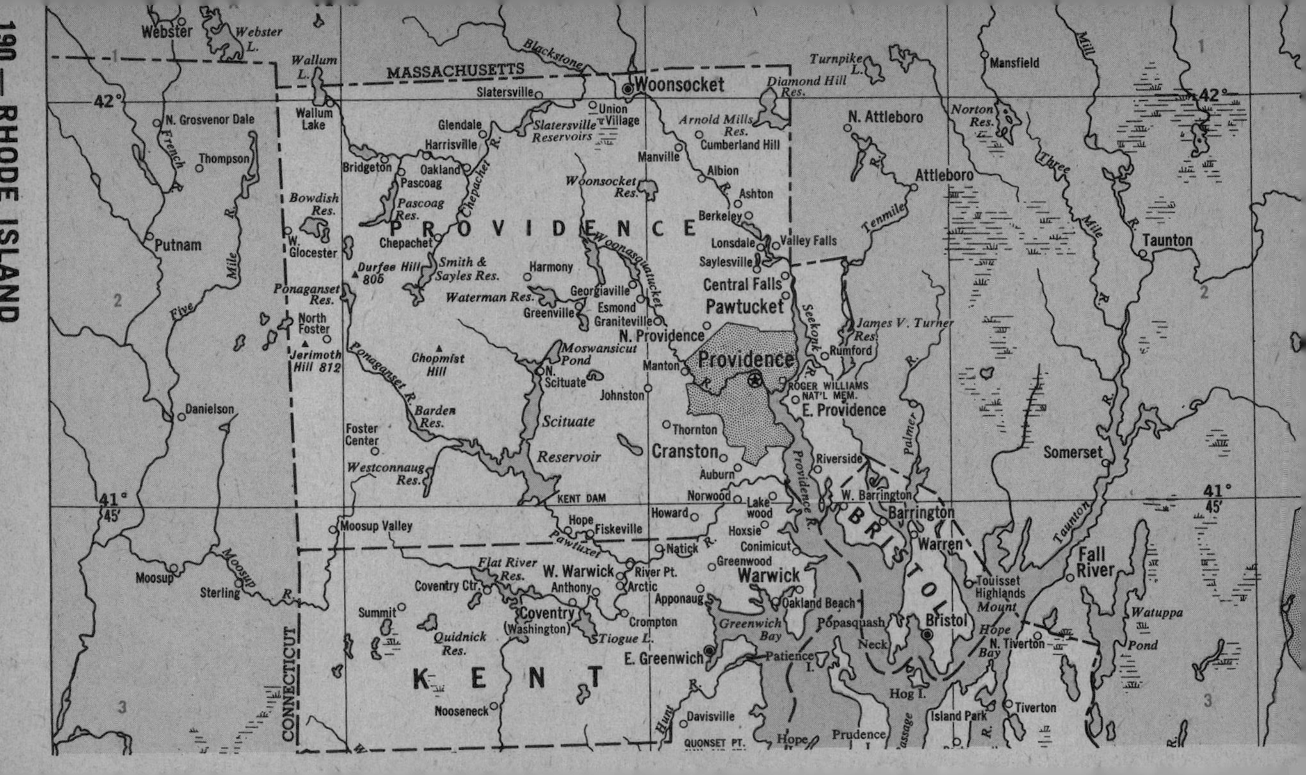
MASSACHUSETTS
CONNECTICUT
PROVIDENCE
BRISTOL
KENT
Woonsocket
Providence
Pawtucket
Central Falls
Cranston
Warwick
E. Providence
N. Providence
Fall River
Taunton
Somerset
Attleboro
N. Attleboro
Mansfield
Webster
Putnam
Danielson
Moosup
Sterling
Thompson
N. Grosvenor Dale
Bristol
Warren
Barrington
W. Barrington
Riverside
Rumford
ROGER WILLIAMS NAT'L MEM.
Valley Falls
Lonsdale
Saylesville
Ashton
Berkeley
Albion
Manville
Cumberland Hill
Union Village
Slatersville
Glendale
Harrisville
Oakland
Pascoag
Bridgeton
Chepachet
W. Glocester
Wallum Lake
North Foster
Foster Center
Moosup Valley
Summit
Coventry Ctr.
Nooseneck
Coventry (Washington)
W. Warwick
Anthony
Arctic
River Pt.
Crompton
E. Greenwich
Apponaug
Natick
Fiskeville
Hope
KENT DAM
Howard
Norwood
Auburn
Thornton
Johnston
Manton
Greenville
Georgiaville
Esmond
Graniteville
Harmony
N. Scituate
Lakewood
Hoxsie
Conimicut
Greenwood
Oakland Beach
Davisville
QUONSET PT.
Tiverton
N. Tiverton
Touisset Highlands
Island Park
Scituate Reservoir
Moswansicut Pond
Chopmist Hill
Durfee Hill 805
Jerimoth Hill 812
Ponaganset Res.
Barden Res.
Westconnaug Res.
Flat River Res.
Quidnick Res.
Tiogue L.
Waterman Res.
Smith & Sayles Res.
Pascoag Res.
Bowdish Res.
Woonsocket Res.
Slatersville Reservoirs
Arnold Mills Res.
Diamond Hill Res.
Turnpike L.
Norton Res.
James V. Turner Res.
Wallum L.
Webster L.
Watuppa Pond
Blackstone
Chepachet R.
Woonasquatucket
Seekonk R.
Providence R.
Pawtuxet
Ponaganset R.
Moosup R.
Five Mile R.
French R.
Three Mile R.
Mill
Taunton R.
Tenmile R.
Palmer R.
Hunt R.
Greenwich Bay
Popasquash Neck
Patience I.
Prudence I.
Hope I.
Hog I.
Mount Hope Bay
42°
41° 45'
1
2
3

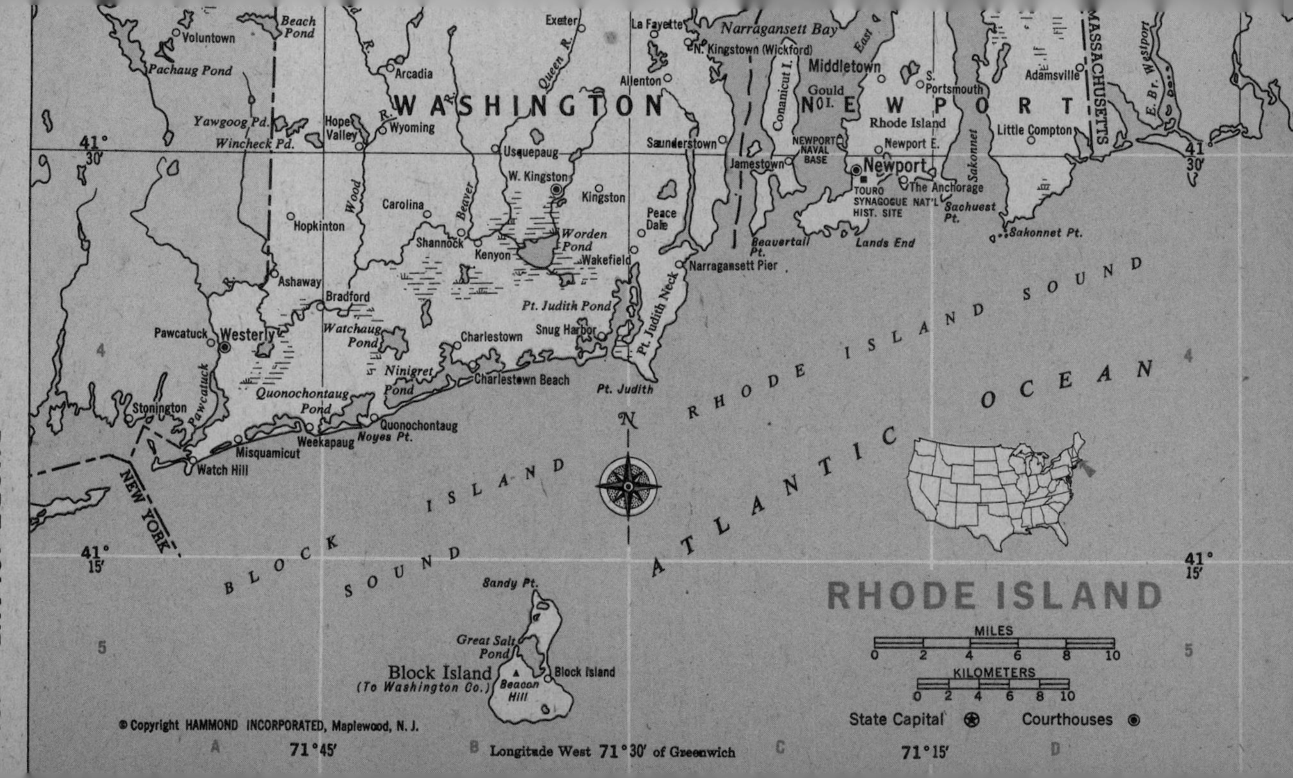
RHODE ISLAND
MASSACHUSETTS
NEW YORK
NEWPORT
WASHINGTON
RHODE ISLAND SOUND
BLOCK ISLAND SOUND
ATLANTIC OCEAN
Narragansett Bay
Newport
Middletown
S. Portsmouth
Little Compton
Adamsville
Jamestown
Conanicut I.
Westerly
Watch Hill
Block Island
(To Washington Co.)
Kingston
W. Kingston
Wakefield
Narragansett Pier
Pt. Judith
Charlestown
Hope Valley
Wyoming
Arcadia
Exeter
Usquepaug
Ashaway
Bradford
Hopkinton
Carolina
Shannock
Kenyon
Worden Pond
Peace Dale
Saunderstown
Allenton
La Fayette
N. Kingstown (Wickford)
Sakonnet Pt.
Beavertail Pt.
Sandy Pt.
Great Salt Pond
Beacon Hill
Stonington
Pawcatuck
MILES
KILOMETERS
State Capital
Courthouses
Longitude West 71° 30′ of Greenwich
71° 45′
71° 15′
41° 30′
41° 15′
© Copyright HAMMOND INCORPORATED, Maplewood, N. J.

SOUTH CAROLINA
MILES
0
20
40
60
KILOMETERS
0
20
40
60
State Capital
County Seats
NORTH CAROLINA
GEORGIA
BLUE RIDGE
GREENVILLE
PICKENS
OCONEE
ANDERSON
ABBEVILLE
GREENWOOD
MC CORMICK
EDGEFIELD
SALUDA
LAURENS
SPARTANBURG
CHEROKEE
YORK
UNION
NEWBERRY
FAIRFIELD
LEXINGTON
AIKEN
BARNWELL
ALLENDALE
Greenville
Spartanburg
Anderson
Greenwood
Augusta
Aiken
Savannah
Gaffney
Rock Hill
Gastonia
Union
Laurens
Clinton
Newberry
Easley
Greer
Sassafras Mtn. 3,560
Hartwell Lake
Clark Hill Lake
L. Murray
Lake Keowee
Lake Sinclair
SAVANNAH RIVER ATOMIC ENERGY RESERVATION
COWPENS NAT'L BATTLEFIELD
KINGS MTN. NAT'L MIL. PK.
HARTWELL DAM
CLARK HILL DAM
Savannah River
Saluda R.
Broad River
Ogeechee R.
Oconee R.
35°
34°
33°
32°
83°
82°

Charlotte
Norwood
Fayetteville
Fort Mill
Monroe
Wadesboro
Rockingham
Clinton
CATAWBA IND. RES.
NORTH CAROLINA
Laurinburg
Lancaster Mills
Lancaster
Springdale
Pageland
Chesterfield
Cheraw
McColl
Lumberton
Elizabethtown
Cape Fear R.
LANCASTER
CHESTERFIELD
Bennettsville
Clio
MARLBORO
Great Falls
Kershaw
Wateree Lake
Society Hill
DILLON
Dillon
Lumber R.
KERSHAW
DARLINGTON
Hartsville
Latta
Lake View
Whiteville
L. Waccamaw
Camden
Bishopville
Darlington
Doneraile
Windy Hill
Mullins
Nichols
Lamar
Florence
Marion
LEE
Timmonsville
FLORENCE
MARION
Boykin
Forest Acres
FT. JACKSON
Loris
Columbia
SHAW A.F.B.
Mayesville
Sumter
Pamplico
Aynor
Eastover
Lake City
HORRY
Waccamaw
Shannontown
SUMTER
Johnsonville
Conway
North Myrtle Beach
Congaree R.
CALHOUN
CLARENDON
Hemingway
Manning
WILLIAMSBURG
Myrtle Beach
St. Matthews
Summerton
Kingstree
GEORGETOWN
MYRTLE BEACH A.F.B.
Long Bay
Cameron
Elloree
SANTEE DAM
Murrells Inlet
Orangeburg
Santee
Lake Marion
Andrews
Pawleys Island
ORANGEBURG
Eutawville
St. Stephen
Georgetown
Bowman
Holly Hill
L. Moultrie
PINOPOLIS DAM
Branchville
Harleyville
North I.
St. George
Moncks Corner
DORCHESTER
Ridgeville
BERKELEY
McClellanville
Summerville
CHARLESTON
Bulls Bay
C. Romain
Walterboro
CHARLESTON A.F.B.
Bull I.
N. Charleston
Capers I.
COLLETON
Johns I.
Isle of Palms
Charleston
Mt. Pleasant
Sullivans Island
FT. SUMTER NAT'L MON.
Yemassee
Folly Beach
Kiawah I.
Edisto Island
Edisto I.
Beaufort
Port Royal
St. Helena Sound
Hunting I.
PARRIS I. MARINE BASE
Port Royal Sound
Hilton Head I.
Tybee Roads
Sea Islands
ATLANTIC OCEAN
N
© Copyright HAMMOND INCORPORATED, Maplewood, N. J.
Longitude West 80° of Greenwich
79°
32°
33°
34°
35°

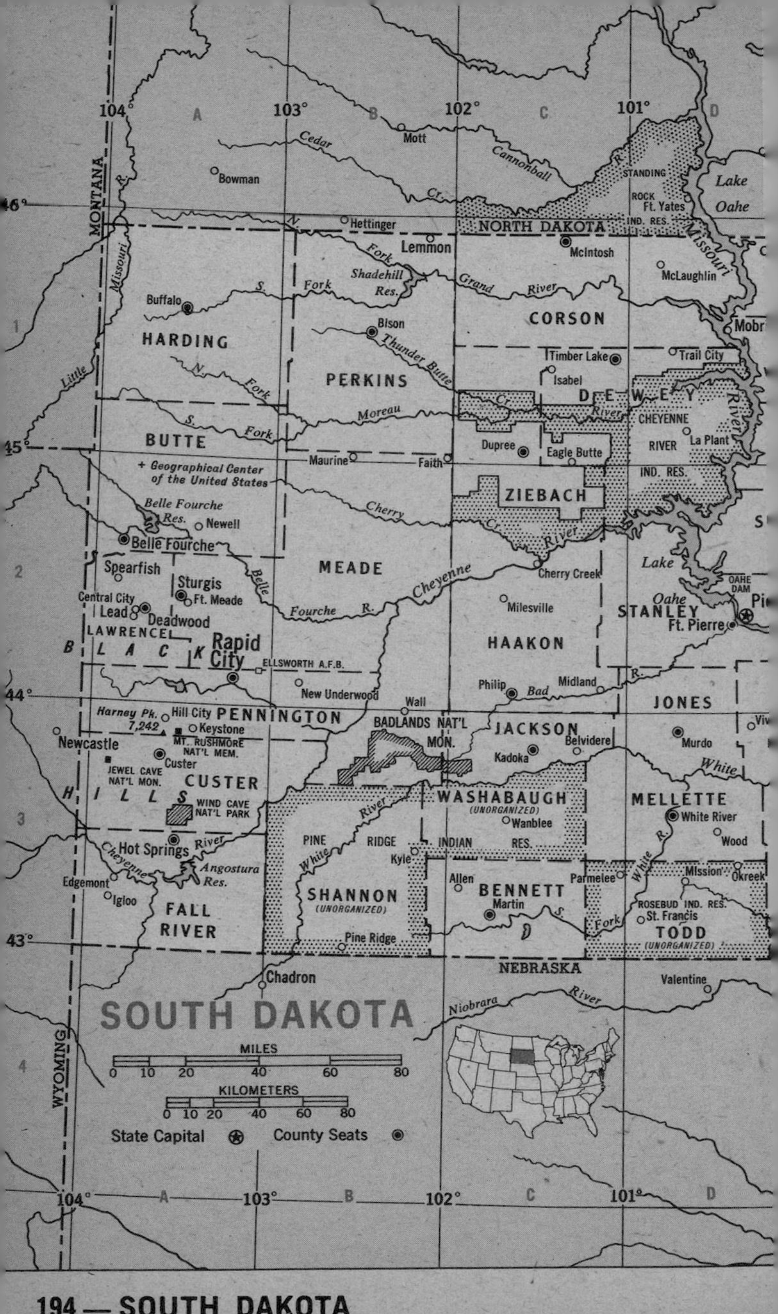
SOUTH DAKOTA
NORTH DAKOTA
NEBRASKA
MONTANA
WYOMING
HARDING
PERKINS
CORSON
DEWEY
BUTTE
ZIEBACH
MEADE
LAWRENCE
PENNINGTON
HAAKON
STANLEY
JONES
JACKSON
CUSTER
WASHABAUGH
(UNORGANIZED)
MELLETTE
SHANNON
(UNORGANIZED)
BENNETT
TODD
(UNORGANIZED)
FALL RIVER
STANDING ROCK IND. RES.
CHEYENNE RIVER IND. RES.
PINE RIDGE INDIAN RES.
ROSEBUD IND. RES.
BLACK HILLS
Bowman
Mott
Hettinger
Lemmon
McIntosh
McLaughlin
Ft. Yates
Lake Oahe
Buffalo
Bison
Timber Lake
Trail City
Isabel
La Plant
Dupree
Eagle Butte
Maurine
Faith
+ Geographical Center of the United States
Belle Fourche Res.
Newell
Belle Fourche
Spearfish
Sturgis
Ft. Meade
Central City
Lead
Deadwood
Rapid City
ELLSWORTH A.F.B.
New Underwood
Wall
Cherry Creek
Milesville
Lake Oahe
OAHE DAM
Ft. Pierre
Philip
Midland
Harney Pk. 7,242
Hill City
Keystone
MT. RUSHMORE NAT'L MEM.
Newcastle
Custer
JEWEL CAVE NAT'L MON.
WIND CAVE NAT'L PARK
BADLANDS NAT'L MON.
Kadoka
Belvidere
Murdo
Wanblee
White River
Wood
Hot Springs
Angostura Res.
Edgemont
Igloo
Kyle
Allen
Parmelee
Mission
Okreek
Martin
St. Francis
Pine Ridge
Chadron
Valentine
Cedar Cr.
Cannonball R.
Missouri R.
N. Fork Grand River
S. Fork
Shadehill Res.
Thunder Butte Cr.
N. Fork
S. Fork Moreau River
Cherry Cr.
Cheyenne River
Belle Fourche R.
Bad R.
White River
White R.
S. Fork
Niobrara River
Little Missouri R.
104°
103°
102°
101°
46°
45°
44°
43°
A
B
C
D
1
2
3
4
MILES
0 10 20 40 60 80
KILOMETERS
0 10 20 40 60 80
State Capital
County Seats

MARSHALL
BROWN
Aberdeen
MC PHERSON
EDMUNDS
FAULK
SPINK
Redfield
CLARK
HAMLIN
CODINGTON
Watertown
DEUEL
GRANT
Milbank
ROBERTS
Sisseton
DAY
Webster
HYDE
HAND
BEADLE
Huron
KINGSBURY
BROOKINGS
Brookings
LAKE
Madison
MOODY
Flandreau
SANBORN
JERAULD
AURORA
MINER
HANSON
MC COOK
MINNEHAHA
Sioux Falls
DAVISON
Mitchell
BRULE
Chamberlain
CHARLES MIX
DOUGLAS
HUTCHINSON
TURNER
LINCOLN
GREGORY
BON HOMME
YANKTON
Yankton
CLAY
Vermillion
UNION
Sioux City
MINNESOTA
IOWA
Missouri
James
Big Sioux
Lewis & Clark L.
© Copyright HAMMOND INCORPORATED, Maplewood, N.J.
Longitude 99° West of Greenwich 98°

KENTUCKY
MISSOURI
ARKANSAS
ARK.
MISSISSIPPI
ALABAMA
Nashville
Memphis
Clarksville
Springfield
Gallatin
Lebanon
Murfreesboro
Franklin
Columbia
Shelbyville
Tullahoma
Jackson
Paris
Union City
Dyersburg
Humboldt
Lawrenceburg
Bowling Green
Hopkinsville
Mayfield
Sikeston
Blytheville
Caruthersville
Corinth
Florence
Huntsville
Lake Barkley
Kentucky Lake
Pickwick Lake
Wheeler Lake
Old Hickory L.
J. Percy Priest L.
Tims Ford L.
Tennessee River
LAND BETWEEN THE LAKES REC. AREA
FORT CAMPBELL
FORT NELSON NAT'L MIL. PARK
STONES RIVER NAT'L BATTLEFIELD
SHILOH NAT'L MIL. PARK
MEMPHIS N.A.S.
SEWART A.F.B.
PICKWICK DAM
WILSON DAM
WHEELER DAM
CHEATHAM DAM
MONTGOMERY
ROBERTSON
SUMNER
MACON
ROUSDALE
WILSON
DAVIDSON
CHEATHAM
STEWART
HOUSTON
DICKSON
HUMPHREYS
BENTON
HENRY
WEAKLEY
OBION
LAKE
DYER
GIBSON
CARROLL
CROCKETT
LAUDERDALE
TIPTON
HAYWOOD
MADISON
HENDERSON
DECATUR
PERRY
HICKMAN
LEWIS
WILLIAMSON
RUTHERFORD
CANNON
MAURY
BEDFORD
COFFEE
MARSHALL
MOORE
GILES
LINCOLN
FRANKLIN
LAWRENCE
WAYNE
HARDIN
CHESTER
MC NAIRY
HARDEMAN
FAYETTE
SHELBY

TENNESSEE
MILES
KILOMETERS
State Capital
County Seats
Longitude West 84° of Greenwich
© Copyright HAMMOND INCORPORATED, Maplewood, N.J.
KENTUCKY
VIRGINIA
NORTH CAROLINA
SOUTH CAROLINA
GEORGIA
ALA.
Knoxville
Chattanooga
Oak Ridge
Kingsport
Bristol
Johnson City
Morristown
Greeneville
Cookeville
Crossville
McMinnville
Cleveland
Athens
Maryville
GREAT SMOKY MTS.
GREAT SMOKY MTS. NAT'L PARK
Clingmans Dome 6,643
Mt. Mitchell 6,684
CUMBERLAND GAP NAT'L HIST. PARK
CHICKAMAUGA & CHATTANOOGA NAT'L MIL. PARK
ANDREW JOHNSON N.H.S.
Lookout Mtn.
Dale Hollow L.
Lake Cumberland
Norris L.
Cherokee L.
Douglas L.
Watts Bar L.
Chickamauga L.
Guntersville Lake
Fontana L.
Hiwassee L.
Tennessee R.
Clinch R.
N. Fk. Holston R.
Cumberland River
French Broad R.
Pigeon R.
Nolichucky

OKLAHOMA
NEW MEXICO
OKLA.
ARK.
LA.
Shreveport
Texarkana
Amarillo
Lubbock
Wichita Falls
Abilene
Odessa
Midland
San Angelo
Fort Worth
DALLAS
Tyler
Waco
Lawton
Frederick
Ardmore
Denison
Sherman
Paris
Longview
Marshall
Big Spring
Brownwood
Plainview
Pampa
Borger
Clovis
Hobbs
LLANO ESTACADO
Red R.
Brazos
Colorado
Pecos
Canadian
L. Texoma
Canton L.
Altus Res.
Grapevine Res.
Eagle Mtn. L.
Benbrook Res.
Garland
Mesquite
Irving
Arlington
Grand Prairie
Richardson
TARRANT
DALLAS
JOHNSON
ELLIS
PARKER

TEXAS
GULF OF MEXICO
MILES
0 40 80 120
KILOMETERS
0 40 80 120
State Capital
County Seats
KEY TO NUMBERS ON MAP
County / County Seat
1 DELTA / Cooper
2 FRANKLIN / Mt. Vernon
3 CAMP / Pittsburg
4 MORRIS / Daingerfield
5 ROCKWALL / Rockwall
6 RAINS / Emory
7 GREGG / Longview
8 SOMERVELL / Glen Rose
9 SAN JACINTO / Coldspring
WESTERN PART OF TEXAS
Same scale as main map
MEXICO
CHIHUAHUA
COAHUILA
NUEVO LEON
TAMAULIPAS
NEW MEXICO
El Paso
Ciudad Juárez
San Antonio
Austin
HOUSTON
Galveston
Corpus Christi
Laredo
Nuevo Laredo
Brownsville
Matamoros
Beaumont
Port Arthur
PADRE I. NAT'L SEASHORE
BIG BEND NAT'L PARK
GUADALUPE MTS. NAT'L PARK
CARLSBAD CAVERNS NAT'L PK.
BALCONES ESCARPMENT
HARRIS
HOUSTON
Pasadena
Baytown
Galveston Bay
GULF OF MEXICO
Longitude West of Greenwich
© Copyright HAMMOND INCORPORATED, Maplewood, N.J.

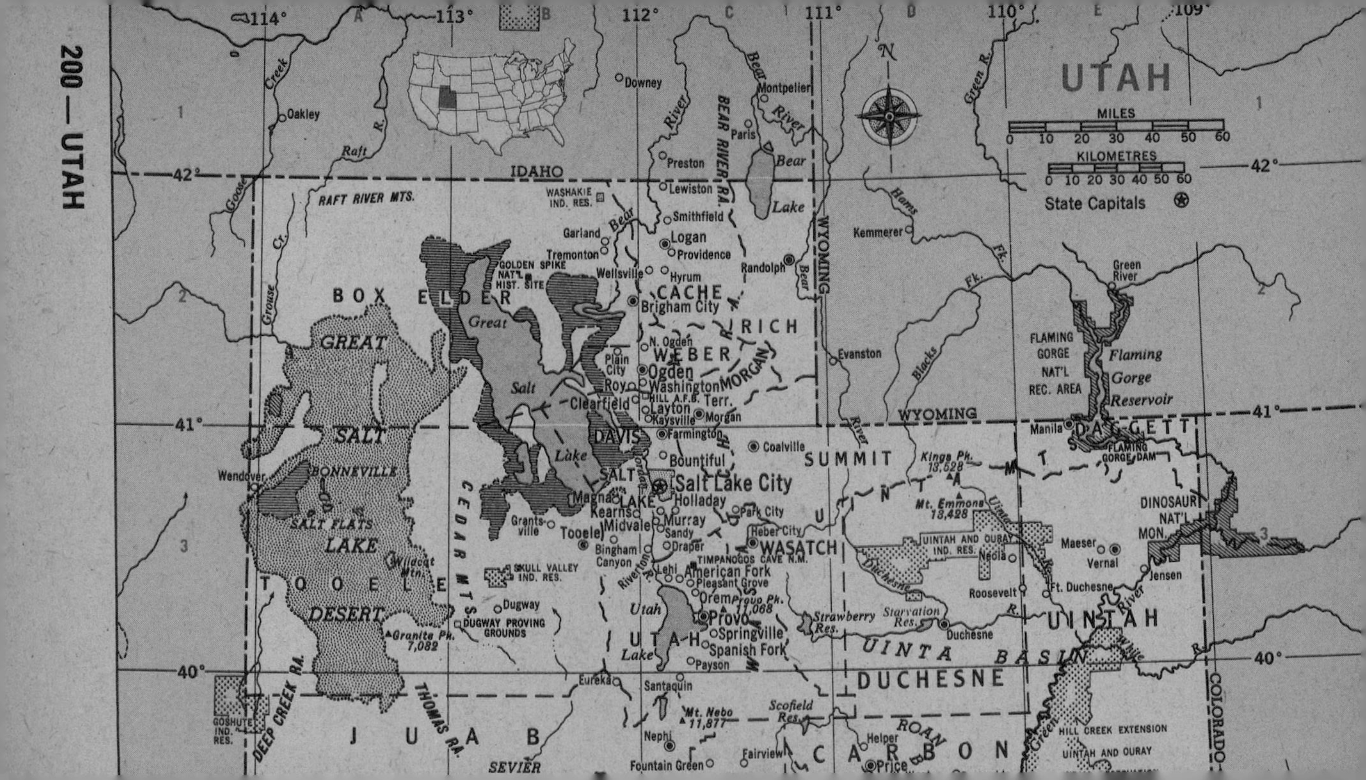
UTAH
MILES
0 10 20 30 40 50 60
KILOMETRES
0 10 20 30 40 50 60
State Capitals
N
IDAHO
WYOMING
COLORADO
114°
113°
112°
111°
110°
109°
42°
41°
40°
A
B
C
D
E
1
2
3
Oakley
Downey
Preston
Montpelier
Paris
Bear Lake
Bear River
BEAR RIVER RA.
RAFT RIVER MTS.
Raft R.
Goose Creek
Grouse Cr.
Lewiston
Smithfield
Logan
Providence
Hyrum
Wellsville
Garland
Tremonton
WASHAKIE IND. RES.
GOLDEN SPIKE NAT'L HIST. SITE
Randolph
Kemmerer
Hams Fk.
Blacks Fk.
Green River
Evanston
BOX ELDER
CACHE
RICH
Brigham City
N. Ogden
WEBER
Plain City
Ogden
Roy
Washington Terr.
HILL A.F.B.
MORGAN
Clearfield
Layton
Kaysville
Morgan
Farmington
DAVIS
Coalville
Bountiful
SUMMIT
Great Salt Lake
GREAT SALT LAKE DESERT
BONNEVILLE SALT FLATS
Wendover
SALT LAKE
Salt Lake City
Magna
Kearns
Holladay
Midvale
Murray
Sandy
Draper
Park City
Heber City
WASATCH
TIMPANOGOS CAVE N.M.
Grantsville
Tooele
TOOELE
Bingham Canyon
Riverton
Lehi
American Fork
Pleasant Grove
Orem
Provo Pk. 11,068
Provo
Springville
Spanish Fork
Utah Lake
UTAH
Payson
Santaquin
Mt. Nebo 11,877
Nephi
Fountain Green
Fairview
Eureka
CEDAR MTS.
SKULL VALLEY IND. RES.
Dugway
DUGWAY PROVING GROUNDS
Wildcat Mtn.
Granite Pk. 7,082
DEEP CREEK RA.
GOSHUTE IND. RES.
THOMAS RA.
JUAB
SEVIER
Kings Pk. 13,528
Mt. Emmons 13,428
UINTA MTS.
UINTAH AND OURAY IND. RES.
Neola
Duchesne R.
Strawberry Res.
Starvation Res.
Duchesne
Roosevelt
Ft. Duchesne
Maeser
Vernal
Jensen
UINTAH
UINTA BASIN
DUCHESNE
Scofield Res.
Helper
Price
ROAN
CARBON
Manila
DAGGETT
FLAMING GORGE NAT'L REC. AREA
Flaming Gorge Reservoir
FLAMING GORGE DAM
DINOSAUR NAT'L MON.
Green R.
White R.
HILL CREEK EXTENSION
UINTAH AND OURAY

MILLARD
SANPETE
EMERY
GRAND
SEVIER
BEAVER
PIUTE
WAYNE
SAN JUAN
IRON
GARFIELD
WASHINGTON
KANE
ARCHES NAT'L PARK
CANYONLANDS NATIONAL PARK
CAPITOL REEF NAT'L PARK
BRYCE CANYON NAT'L PARK
ZION NAT'L PARK
GRAND CANYON NAT'L PARK
CEDAR BREAKS NAT'L MON.
NATURAL BRIDGES NAT'L MON.
HOVENWEEP NAT'L MON.
RAINBOW BRIDGE NAT'L MON.
NAVAJO NAT'L MON.
PIPE SPRING NAT'L MON.
GLEN CANYON NAT'L REC. AREA
NAVAJO INDIAN RESERVATION
KAIBAB IND. RES.
KANOSH IND. RES.
KOOSHAREM IND. RES.
PAIUTE IND. RES.
SHIVWITS IND. RES.
MONUMENT VALLEY
LABYRINTH CANYON
BROWN CLIFFS
HENRY MTS.
TUSHAR MTS.
PAVANT MTS.
MINERAL MTS.
WAH WAH MTS.
CRICKET MTS.
HOUSE RA.
CONFUSION RA.
SAN RAFAEL SWELL
WASATCH PLATEAU
AQUARIUS PLAT.
KAIPAROWITS PLATEAU
PAUNSAUGUNT PLAT
PINK CLIFFS
SEVIER DESERT
ESCALANTE DESERT
Mt. Waas 12,331
Mt. Peale 12,721
Abajo Pk. 11,360
Mt. Ellen 11,522
Mt. Pennell 11,371
Navajo Mtn. 10,388
Thousand Lake Mtn. 11,306
Mt. Marvine 11,600
Monroe Pk. 11,226
Delano Pk. 12,173
Little Creek Pk. 10,142
Brian Head 11,307
Hinckley
Delta
Scipio
Holden
Ephraim
Manti
Gunnison
Centerfield
Redmond
Mayfield
Salina
Hiawatha
Huntington
Castle Dale
Orangeville
Ferron
Emery
Green River
Garrison
Fillmore
Kanosh
Aurora
Richfield
Elsinore
Monroe
Marysvale
Loa
Bicknell
Hanksville
Moab
La Sal
Milford
Beaver
Junction
Minersville
Circleville
Antimony
Beryl
Parowan
Panguitch
Boulder
Escalante
Monticello
Blanding
Cedar City
Tropic
Enterprise
Orderville
La Verkin
Washington
Hurricane
St. George
Kanab
Page
Mexican Hat
Sevier
Sevier Bridge Res.
Sevier Lake
Beaver
Fish L.
Fremont R.
Muddy Cr.
San Rafael R.
Green River
Colorado
Dirty Devil R.
Escalante R.
Paria R.
E. Fork
Lake Powell
San Juan River
Virgin
Kanab Cr.
Only point in U. S. common to 4 states
ARIZONA
NEVADA
39°
38°
37°
114°
113°
Longitude 112° West of Greenwich
109°
© C.S. HAMMOND & Co., Maplewood, N.J.

CANADA
QUEBEC
NEW YORK
NEW HAMPSHIRE
FRANKLIN
ORLEANS
ESSEX
LAMOILLE
CALEDONIA
CHITTENDEN
WASHINGTON
ORANGE
ADDISON
GRAND ISLE
LAKE CHAMPLAIN
WHITE MOUNTAINS
Missisquoi Bay
Lake Memphremagog
Rouses Point
Alburg
Highgate Ctr.
Swanton
Missisquoi R.
Enosburg Falls
Richford
N. Troy
Jay Pk. 3,861
Troy
Newport
Derby Line
Derby
Norton
Gore Mtn. 3,330
Canaan
Beecher Falls
Colebrook
Seymour L.
Island Pond
Isle La Motte
N. Hero
St. Albans
Bakersfield
Irasburg
Orleans
Barton
Willoughby L.
Bloomfield
Connecticut R.
Plattsburgh
Grand Isle
Eden
Black R.
W. Burke
Fairfax
Lamoille River
Johnson
S. Hero
Milton
Cambridge
Jeffersonville
Hyde Park
Caspian L.
Greensboro
Burke Mtn. 3,267
Guildhall
Groveton
Smugglers Notch
Morrisville
Colchester
Essex
Mt. Mansfield 4,393
Keeseville
Lyndon
Lyndonville
Winooski
Hardwick
Lunenburg
Lancaster
Essex Jct.
Stowe
Au Sable Forks
Burlington
S. Burlington
Richmond
Danville
St. Johnsbury
Concord
Gilman
Waterbury Res.
Cabot
Shelburne
Winooski
Marshfield
Moore Res.
MOORE DAM
Waterbury
E. Barnet
Hinesburg
COMERFORD DAM
Camels Hump 4,083
E. Montpelier
Plainfield
Barnet
Littleton
Ammonoosuc
River
Montpelier
Barre
Groton
Vergennes
Waitsfield
S. Barre
S. Ryegate
Northfield
Mad R.
Graniteville
Websterville
Wells River
Woodsville
Bristol
Mt. Ellen 4,135
Williamstown
Addison
Otter Cr.
Roxbury
Newbury
Port Henry
Middlebury
Bread Loaf Mtn. 3,823
Bradford
Lincoln
E. Middlebury
Granville
Chelsea
Randolph
45°
44°
73°
72°
30'

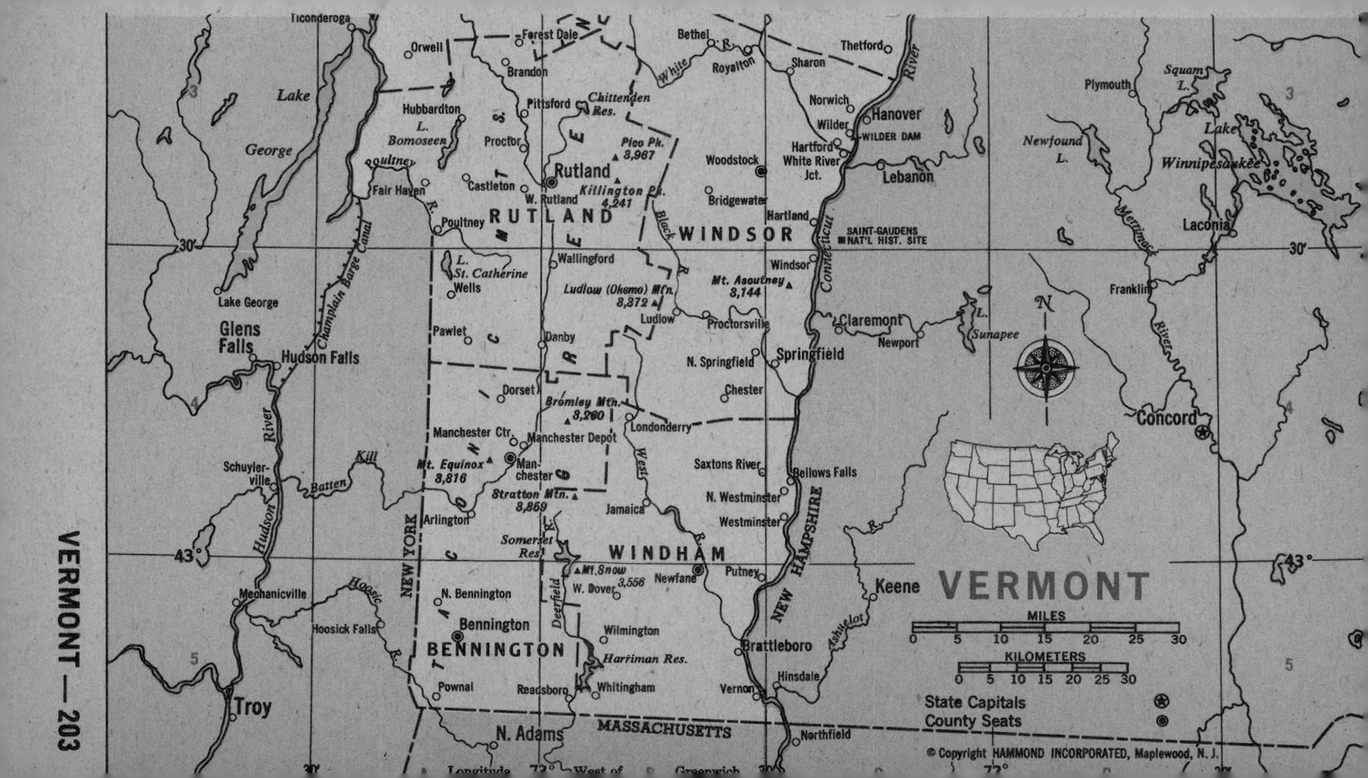
VERMONT
MILES
0 5 10 15 20 25 30
KILOMETERS
0 5 10 15 20 25 30
State Capitals
County Seats
© Copyright HAMMOND INCORPORATED, Maplewood, N.J.
RUTLAND
WINDSOR
WINDHAM
BENNINGTON
NEW HAMPSHIRE
NEW YORK
MASSACHUSETTS
Lake George
Lake Winnipesaukee
Squam L.
Newfound L.
L. Sunapee
L. Bomoseen
L. St. Catherine
Champlain Barge Canal
Hudson River
Connecticut River
Merrimack River
White R.
Black R.
West R.
Poultney R.
Hoosic R.
Batten Kill
Ashuelot R.
Deerfield R.
Chittenden Res.
Harriman Res.
Somerset Res.
Pico Pk. 3,967
Killington Pk. 4,241
Mt. Ascutney 3,144
Ludlow (Okemo) Mtn. 3,372
Bromley Mtn. 3,260
Mt. Equinox 3,816
Stratton Mtn. 3,859
Mt. Snow 3,556
SAINT-GAUDENS NAT'L HIST. SITE
WILDER DAM
Ticonderoga
Orwell
Forest Dale
Brandon
Bethel
Royalton
Sharon
Thetford
Norwich
Hanover
Wilder
Hartford
White River Jct.
Lebanon
Hubbardton
Pittsford
Proctor
Rutland
Castleton
W. Rutland
Fair Haven
Poultney
Woodstock
Bridgewater
Hartland
Windsor
Wallingford
Wells
Ludlow
Proctorsville
Claremont
Newport
Lake George
Glens Falls
Hudson Falls
Pawlet
Danby
N. Springfield
Springfield
Dorset
Chester
Londonderry
Manchester Ctr.
Manchester Depot
Manchester
Schuylerville
Saxtons River
Bellows Falls
N. Westminster
Westminster
Arlington
Jamaica
Newfane
Putney
Keene
Mechanicville
N. Bennington
W. Dover
Hoosick Falls
Bennington
Wilmington
Brattleboro
Hinsdale
Pownal
Readsboro
Whitingham
Vernon
Troy
N. Adams
Northfield
Plymouth
Laconia
Franklin
Concord
30'
43°
72°
73°
Longitude 72° West of Greenwich

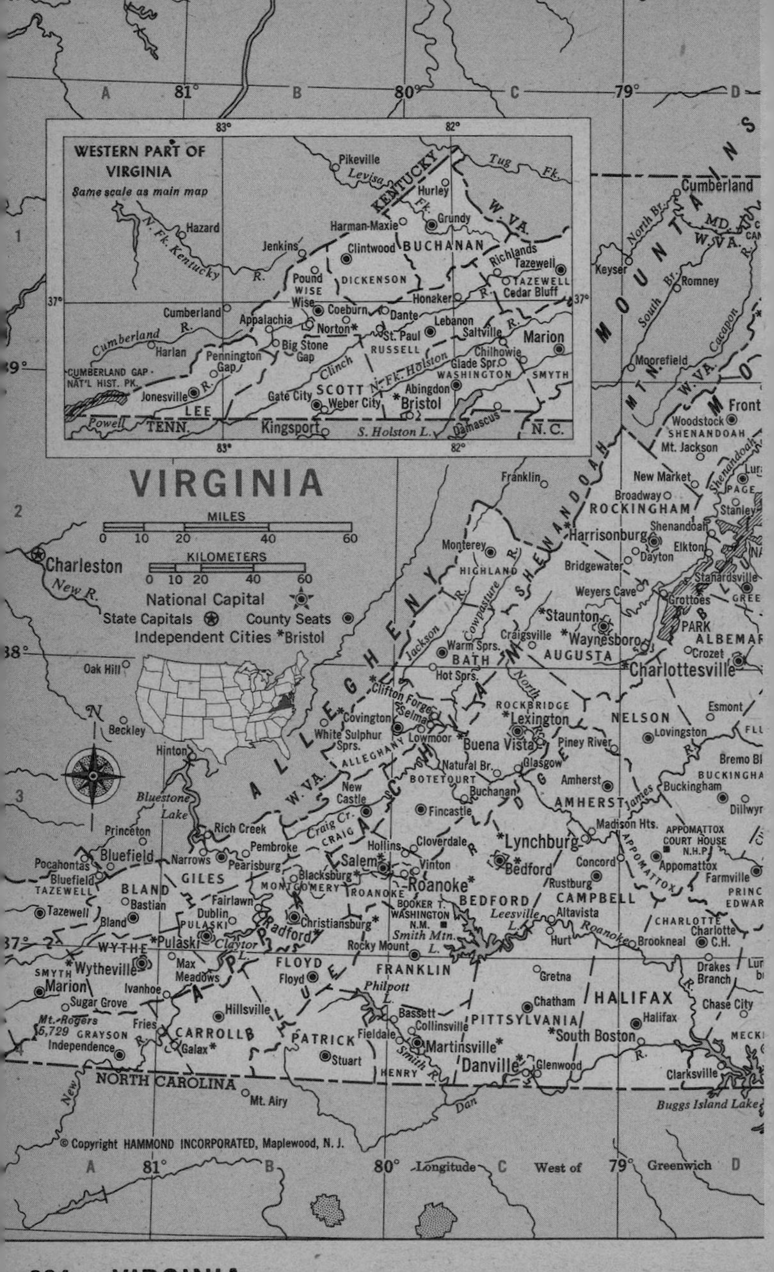
VIRGINIA
WESTERN PART OF VIRGINIA
Same scale as main map
MILES
0 10 20 40 60
KILOMETERS
0 10 20 40 60
National Capital
State Capitals
County Seats
Independent Cities *Bristol
Pikeville
Levisa
KENTUCKY
Tug
Fk.
Hurley
W. VA.
Hazard
N. Fk. Kentucky
R.
Jenkins
Harman-Maxie
Grundy
Clintwood
BUCHANAN
Richlands
Tazewell
Pound
WISE
DICKENSON
TAZEWELL
Cedar Bluff
Honaker
Wise
Coeburn
Dante
Cumberland
Appalachia
Norton
Lebanon
Saltville
Marion
Cumberland
Harlan
Big Stone Gap
Pennington Gap
St. Paul
RUSSELL
Chilhowie
Clinch
N. Fk. Holston
Glade Spr.
CUMBERLAND GAP NAT'L HIST. PK.
WASHINGTON
SMYTH
Jonesville
Gate City
SCOTT
Abingdon
Weber City
*Bristol
LEE
Powell
TENN.
Kingsport
S. Holston L.
Damascus
N.C.
37°
83°
82°
A
81°
B
80°
C
79°
D
Cumberland
North Br.
MD.
W.VA.
Keyser
Romney
South Br.
Cacapon
MOUNTAINS
Moorefield
W. VA.
Front
Woodstock
SHENANDOAH
Mt. Jackson
New Market
Shenandoah
PAGE
Stanley
Franklin
Broadway
ROCKINGHAM
SHENANDOAH
Harrisonburg
Shenandoah
Elkton
Monterey
Dayton
Bridgewater
HIGHLAND
Weyers Cave
Stanardsville
Grottoes
*Staunton
*Waynesboro
PARK
ALBEMARLE
Craigsville
AUGUSTA
Crozet
Warm Sprs.
BATH
*Charlottesville
Jackson
Cowpasture
Hot Sprs.
ALLEGHENY
Clifton Forge
Selma
ROCKBRIDGE
*Lexington
NELSON
Esmont
Covington
Lovingston
White Sulphur Sprs.
Lowmoor
*Buena Vista
Piney River
ALLEGHANY
Natural Br.
Glasgow
BOTETOURT
Amherst
Buchanan
BLUE RIDGE
W. VA.
New Castle
AMHERST
James
Buckingham
Fincastle
Dillwyn
Madison Hts.
Craig Cr.
CRAIG
APPOMATTOX COURT HOUSE N.H.P.
Hollins
Cloverdale
*Lynchburg
APPOMATTOX
Salem*
Vinton
Concord
Appomattox
Bedford
Farmville
Blacksburg*
Roanoke*
Rustburg
MONTGOMERY
ROANOKE
CAMPBELL
BEDFORD
BOOKER T. WASHINGTON N.M.
Leesville L.
Altavista
CHARLOTTE
Christiansburg*
Charlotte C.H.
Radford*
Smith Mtn. L.
Hurt
Roanoke
Brookneal
Rocky Mount
FLOYD
FRANKLIN
Drakes Branch
Floyd
Gretna
Philpott L.
Chatham
HALIFAX
Chase City
Bassett
Collinsville
PITTSYLVANIA
Halifax
Fieldale
PATRICK
Martinsville
*South Boston
Stuart
Smith R.
Danville*
Glenwood
HENRY
Clarksville
Dan
Buggs Island Lake
Charleston
New R.
Oak Hill
Beckley
Hinton
Bluestone Lake
Princeton
Rich Creek
Pembroke
Bluefield
Narrows
Pearisburg
Pocahontas
Bluefield
GILES
TAZEWELL
BLAND
Bastian
Fairlawn
Tazewell
Bland
Dublin
PULASKI
*Pulaski
Claytor L.
WYTHE
*Wytheville
Max Meadows
SMYTH
Marion
Ivanhoe
Sugar Grove
Hillsville
Mt. Rogers 5,729
GRAYSON
Fries
CARROLL
Independence
Galax*
NORTH CAROLINA
Mt. Airy
New
© Copyright HAMMOND INCORPORATED, Maplewood, N. J.
Longitude
West of
Greenwich

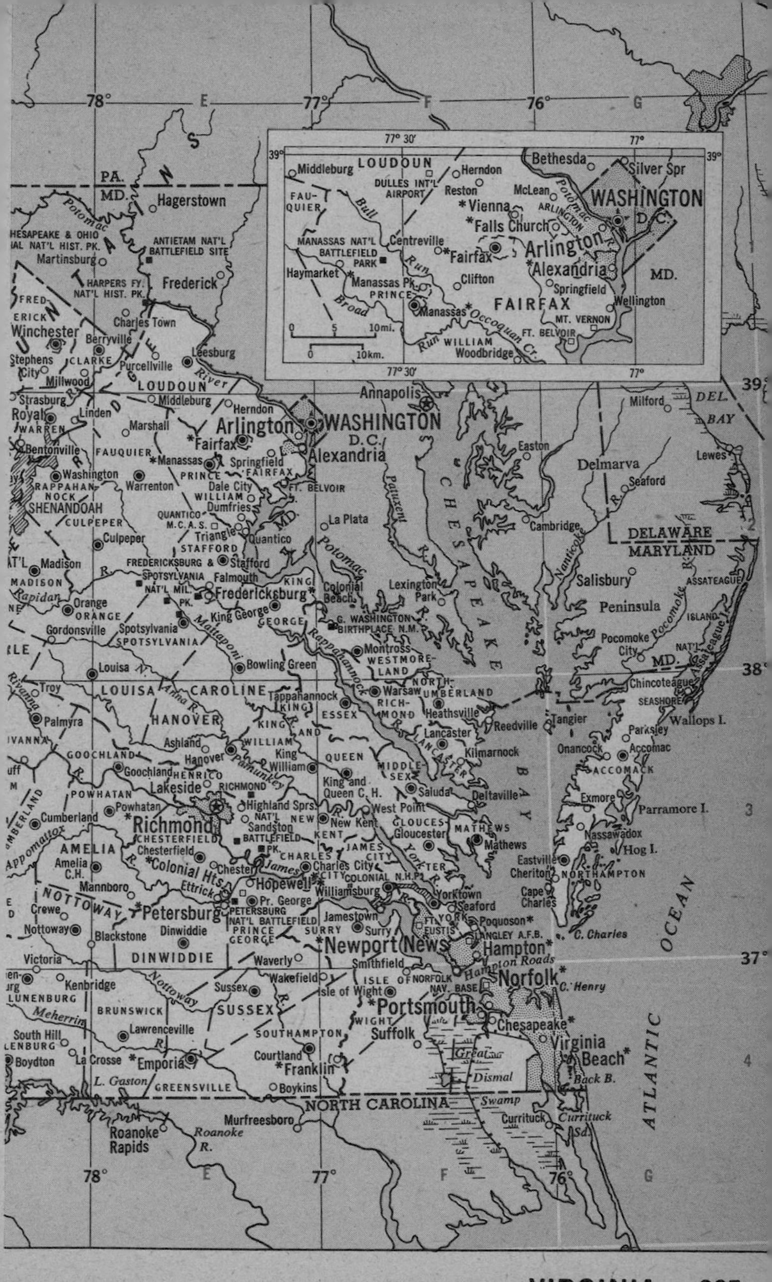
WASHINGTON D.C.
Arlington
Alexandria
Fairfax
Richmond
Petersburg
Newport News
Hampton
Norfolk
Portsmouth
Chesapeake
Virginia Beach
Suffolk
Fredericksburg
Winchester
Hagerstown
Frederick
Annapolis
Salisbury
Delmarva Peninsula
CHESAPEAKE BAY
ATLANTIC OCEAN
NORTH CAROLINA
DELAWARE
MARYLAND
PA.
MD.
LOUDOUN
FAIRFAX
SHENANDOAH
Leesburg
Herndon
Manassas
Springfield
Culpeper
Hopewell
Williamsburg
Yorktown
Emporia
Franklin
Roanoke Rapids

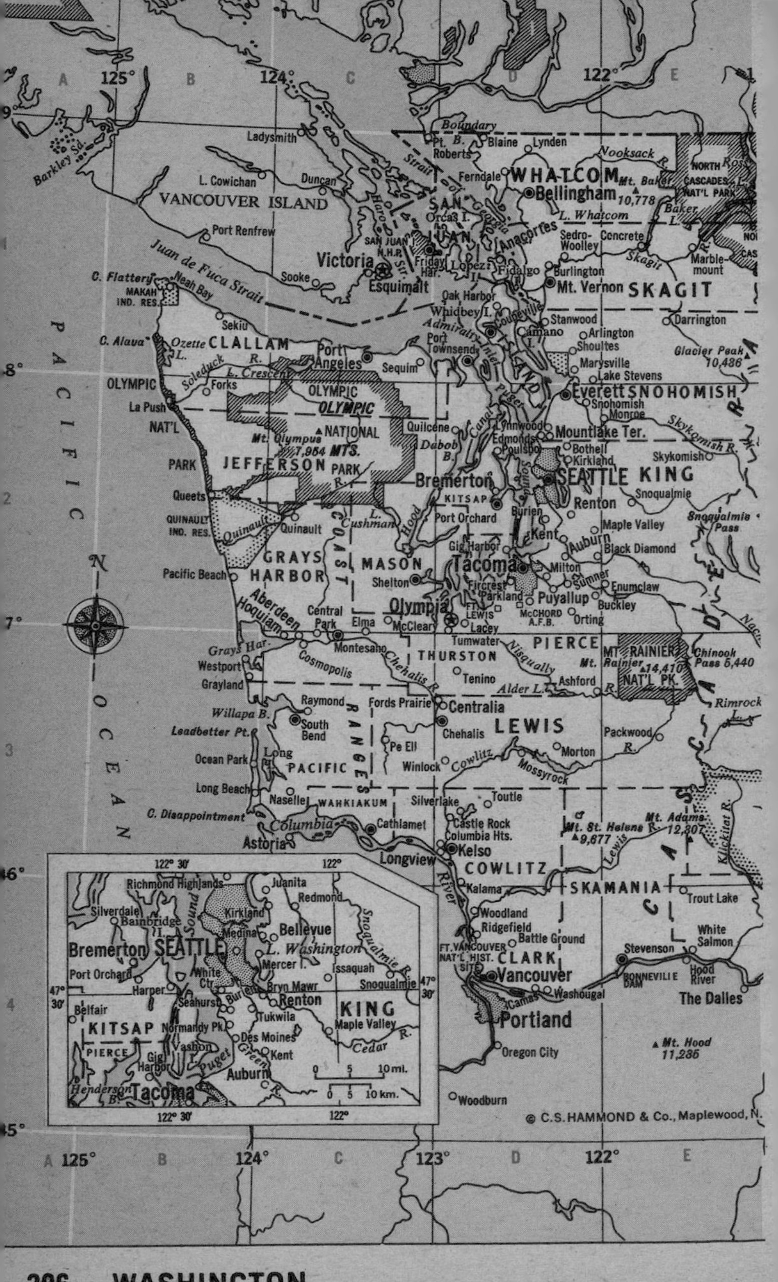
125°
124°
122°
123°
A
B
C
D
E
Ladysmith
Barkley Sd.
L. Cowichan
Duncan
VANCOUVER ISLAND
Port Renfrew
Juan de Fuca Strait
Victoria
Esquimalt
Sooke
SAN JUAN I. N.H.P.
Haro Str.
Boundary B.
Pt. Roberts
Strait of Georgia
Blaine
Lynden
Nooksack R.
WHATCOM
Ferndale
Bellingham
Mt. Baker 10,778
L. Whatcom
NORTH CASCADES NAT'L PARK
Baker L.
SAN JUAN
Orcas I.
Friday Har.
Lopez I.
Anacortes
Fidalgo
Sedro-Woolley
Concrete
Skagit
Marblemount
Burlington
Mt. Vernon
SKAGIT
C. Flattery
Neah Bay
MAKAH IND. RES.
Oak Harbor
Whidbey I.
Coupeville
Stanwood
Camano I.
Darrington
Sekiu
Admiralty Inlet
Port Townsend
Arlington
C. Alava
Ozette L.
CLALLAM
Port Angeles
Sequim
Shoultes
Glacier Peak 10,486
Marysville
Lake Stevens
Soleduck R.
L. Crescent
OLYMPIC
Forks
Everett
SNOHOMISH
Snohomish
Monroe
La Push
OLYMPIC
NAT'L
PARK
OLYMPIC MTS.
NATIONAL
Mt. Olympus 7,954
PARK
Quilcene
Dabob B.
Hood Canal
Lynnwood
Edmonds
Poulsbo
Mountlake Ter.
Bothell
Kirkland
Skykomish R.
Skykomish
JEFFERSON
SEATTLE
KING
Bremerton
KITSAP
Renton
Snoqualmie
Queets
QUINAULT IND. RES.
Quinault
L. Cushman
Port Orchard
Burien
Maple Valley
Snoqualmie Pass
Kent
Auburn
Black Diamond
COAST
GRAYS HARBOR
MASON
Gig Harbor
Tacoma
Milton
Sumner
Enumclaw
Pacific Beach
Shelton
Fircrest
Parkland
Puyallup
Buckley
Aberdeen
Hoquiam
Central Park
Elma
Olympia
McCleary
FT. LEWIS
Lacey
McCHORD A.F.B.
Orting
Montesano
Tumwater
PIERCE
MT. RAINIER NAT'L PK.
Mt. Rainier 14,410
Chinook Pass 5,440
Grays Har.
Westport
Cosmopolis
Chehalis R.
THURSTON
Nisqually
Tenino
Ashford
Grayland
Alder L.
Raymond
Fords Prairie
Centralia
Rimrock L.
Willapa B.
South Bend
RANGES
Chehalis
LEWIS
Leadbetter Pt.
Packwood
Pe Ell
Morton
Ocean Park
Long
PACIFIC
Winlock
Cowlitz
Mossyrock
Long Beach
Naselle
WAHKIAKUM
Silverlake
Toutle
C. Disappointment
Columbia
Cathlamet
Castle Rock
Columbia Hts.
Mt. St. Helens 9,677
Lewis R.
Mt. Adams 12,307
Klickitat R.
Astoria
Longview
Kelso
COWLITZ
River
Kalama
SKAMANIA
Trout Lake
Woodland
Ridgefield
Battle Ground
White Salmon
FT. VANCOUVER NAT'L HIST. SITE
CLARK
Stevenson
BONNEVILLE DAM
Hood River
Vancouver
Camas
Washougal
The Dalles
Portland
Mt. Hood 11,235
Oregon City
Woodburn
P A C I F I C
O C E A N
CASCADE
© C.S.HAMMOND & Co., Maplewood, N.
122° 30'
Richmond Highlands
Juanita
Redmond
Silverdale
Bainbridge I.
Sound
Kirkland
Medina
Bellevue
Bremerton
SEATTLE
L. Washington
Snoqualmie R.
Mercer I.
Issaquah
Port Orchard
White Ctr.
Bryn Mawr
Snoqualmie
47° 30'
Harper
Seahurst
Burien
Renton
KING
Belfair
Tukwila
KITSAP
Normandy Pk.
Maple Valley
Vashon I.
Des Moines
Cedar R.
PIERCE
Gig Harbor
Puget
Green
Kent
Auburn
0 5 10 mi.
0 5 10 km.
Henderson B.
Tacoma

CANADA
BRITISH COLUMBIA
OKANOGAN
FERRY
STEVENS
PEND
OREILLE
CHELAN
DOUGLAS
LINCOLN
SPOKANE
GRANT
KITTITAS
ADAMS
WHITMAN
FRANKLIN
GARFIELD
COLUMBIA
ASOTIN
YAKIMA
BENTON
WALLA WALLA
KLICKITAT
OREGON
IDAHO
Spokane
Yakima
Wenatchee
Ellensburg
Ephrata
Moses Lake
Pasco
Kennewick
Richland
Walla Walla
Pullman
Colfax
Clarkston
Lewiston
Coeur d'Alene
Okanogan
Omak
Oroville
Tonasket
Republic
Colville
Newport
Davenport
Ritzville
Othello
Dayton
Pomeroy
Prosser
Goldendale
HANFORD ATOMIC ENERGY RESERVATION
COLVILLE INDIAN RES.
COULEE DAM NATIONAL RECREATION AREA
GRAND COULEE DAM
Columbia River
Snake R.
WASHINGTON
MILES
0 20 40 60 80
KILOMETERS
0 20 40 60 80
State and Provincial Capitals
County Seats
Longitude 120° West of Greenwich
119°
118°
117°
49°
48°
47°
46°
45°

A
82°
B
81°
C
Senecaville Res.
Glen Dale
Moundsville
MARSHALL
Cameron
Woodsfield
N
Muskingum
OHIO
River
New Martinsville
Paden City
Sistersville
WETZEL
Pine Grove
Grant To
TYLER
Middlebourne
Mannington
Enterprise
Shinnston
Lumberport
1
Nelsonville
Hocking
Marietta
Williamstown
St. Marys
Ohio
Middle I.
Cr.
Athens
R.
Raccoon
Vienna
PLEASANTS
Belpre
West Union
Clarksburg
Pennsboro
Salem
Parkersburg
Harrisville
New Milton
Nutter Fort
WOOD
RITCHIE
DODDRIDGE
HARRISON
Wellston
Pomeroy
Elizabeth
Little
LEWIS
Weston
Jackson
Middleport
Mason
39°
West Fk.
New Haven
Ravenswood
WIRT
Kanawha
Glenville
Grantsville
CALHOUN
GILMER
Cottageville
Burnsville
Point Pleasant
Ripley
Gallipolis
River
Spencer
JACKSON
Kanawha
ROANE
BRAXTON
MASON
Gassaway
Sutton L.
Walton
Sutton
PUTNAM
Ironton
Ashland
2
Winfield
Sissonville
Ohio
CABELL
Clendenin
Procious
Huntington
Milton
Hurricane
Clay
Nitro
Institute
Elk
Pinch
WEBSTER
Kenova
Ceredo
Barboursville
St. Albans
Dunbar
Charleston
CLAY
Cowen
S. Charleston
NICHOLAS
KANAWHA
Summersville
W. Hamlin
Big Sandy
Hamlin
Griffithsville
Marmet
Belle
Cedar Gr.
Summersville Lake
Richwood
Wayne
Chesapeake
Smithers
Gauley
Hubball
East Bank
Boomer
E. Lynn
Guyandotte
Coal
Montgomery
Gauley Bridge
LINCOLN
Ansted
R.
Fort Gay
Powellton
Madison
Fayetteville
WAYNE
BOONE
Page
38°
Whitesville
FAYETTE
Charmco
KY.
Jeffrey
Oak Hill
Rainelle
Rupert
Chapmanville
Gordon
Glen Jean
Minden
Levisa Fk.
Crum
Henlawson
Wharton
Dorothy
Mt. Hope
GREENBRI
Tug
Verdunville
LOGAN
Barrett
Lawton
Logan
Kermit
Holden
Mt. Gay
RALEIGH
Lewisburg
Dewey L.
MINGO
Switzer
Amherstdale
New
Nolan
Man
Beckley
Ronceverte
Delbarton
Kopperston
Mabscott
Beaver
Alderson
Chattaroy
Mallory
Sophia
Williamson
Oceana
Hinton
Talcott
MONROE
Russell
Gilbert
WYOMING
Matewan
Pineville
Mullens
SUMMERS
Union
Fk.
Forest Hill
3
Wharncliffe
Corinne
Bluestone Lake
Pikeville
Iaeger
Davy
Welch
Kimball
Keystone
Bluestone R.
Athens
Lake
Roderfield
Coalwood
McDowell
MERCER
KENTUCKY
VIRGINIA
Caretta
Gary
Thorpe
Maybeury
Princeton
River
War
Bramwell
MC DOWELL
Bluefield
Grundy
Blacksburg
Bluefield
Jenkins
Clintwood
A
82° Longitude West
B
of Greenwich
81°
C

WEST VIRGINIA
NORTHERN PART OF WEST VIRGINIA
Same scale as main map
PENNSYLVANIA
MARYLAND
VIRGINIA
State Capital
County Seats
© Copyright HAMMOND INCORPORATED, Maplewood, N. J.

LAKE SUPERIOR
APOSTLE ISLANDS
APOSTLE ISLANDS NAT'L LAKESHORE
MICHIGAN
MINNESOTA
DOUGLAS
BAYFIELD
ASHLAND
IRON
VILAS
ONEIDA
FOREST
FLORENCE
MARINETTE
LANGLADE
LINCOLN
MARATHON
TAYLOR
PRICE
SAWYER
RUSK
WASHBURN
BURNETT
POLK
BARRON
CHIPPEWA
DUNN
ST. CROIX
MENOMINEE
OCONTO
WASHINGTON
OZAUKEE
MILWAUKEE
WAUKESHA
DODGE
JEFFERSON
WALWORTH
RACINE
KENOSHA
Superior
Duluth
Ashland
Washburn
Bayfield
Hurley
Ironwood
Hayward
Spooner
Rhinelander
Antigo
Merrill
Wausau
Medford
Ladysmith
Phillips
Park Falls
Eagle River
Marinette
Menominee
Escanaba
Iron Mountain
Chippewa Falls
Menomonie
Hudson
Milwaukee
Waukesha
Racine
Kenosha
Wauwatosa
Brookfield
Mequon
Lake Michigan
Chequamegon Bay
Timms Hill 1,952
Rib Mtn.

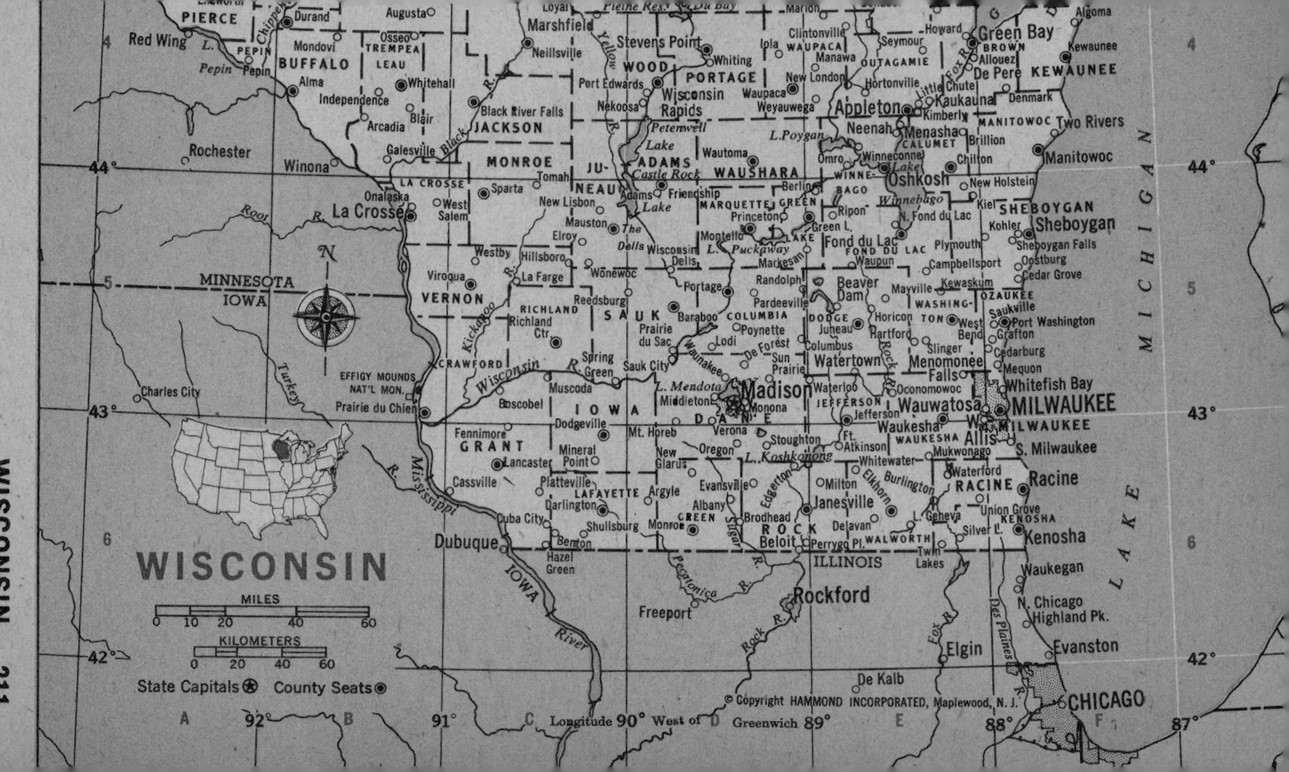
WISCONSIN
MILES
0 10 20 40 60
KILOMETERS
0 20 40 60
State Capitals
County Seats
LAKE MICHIGAN
© Copyright HAMMOND INCORPORATED, Maplewood, N.J.
Longitude 90° West of Greenwich 89°
MILWAUKEE
Madison
CHICAGO
Green Bay
La Crosse
Dubuque
Rockford
MINNESOTA
IOWA
ILLINOIS
EFFIGY MOUNDS NAT'L MON.

110°
108°
BIGHORN CANYON
NAT'L REC. AREA
CROW INDIAN RES.
MONTANA
Yellowstone
Nat'l Park
Mt. Washburn
10,243
Mt. Holmes
10,336
YELLOWSTONE
NATIONAL
PARK
Hebgen L.
W. Yellowstone
Old Faithful
Shoshone L.
Yellowstone Lake
ABSAROKA
Bighorn Lake
Cowley
Byron
Powell
Lovell
Dayton
Shoshone R.
BIG HORN
Bighorn
Greybull
Buffalo Bill Res.
Cody
PARK
BIGHORN
Otto
Basin
Fortress Mtn.
12,073
Cloud Pk
13,165
Manderson
Nowood
Meeteetse
Needle Mtn.
12,130
Ten Sleep
Worland
44°
GRAND TETON
NAT'L PK.
Jackson Lake
Moran
WASHAKIE
Mt. Crosby
12,435
Hamilton Dome
Grand Teton
13,770
IDAHO
TETON
CONTINENTAL
HOT SPRINGS
OWL CREEK MTS.
Thermopolis
WIND RIVER CANYON
Jackson
Dubois
DIVIDE
WIND RIVER RANGE
Wind River
Palisades Res.
Doubletop Pk.
11,715
Boysen Res.
Snake
WYOMING RANGE
WIND RIVER
Pavillion
Shoshoni
Gannett Pk.
13,804
FREMONT
Etna
SALT RIVER RANGE
Fremont Pk.
13,730
Fort Washakie
INDIAN RES.
Riverton
Thayne
Green
Hudson
Grover
Daniel
Pinedale
Mt. Bonneville
12,530
Lander
Afton
SUBLETTE
Atlantic Pk.
12,734
RATTLESNAKE
Wyoming Pk.
11,418
Big Piney
Sweetwater
Jeffrey City
GREEN MTN.
La Barge
Big Sandy Res.
Bairoil
LINCOLN
Fontenelle Res.
Farson
Cokeville
Sandy
Eden
GREAT DIVIDE
BASIN
42°
Big
Green
SWEETWATER
FOSSIL BUTTE
NAT'L MONUMENT
Frontier
Kemmerer
Diamondville
Superior
S. Superior
Rawlins
River
Reliance
Wamsutter
Bitter Cr.
Granger
Green River
Rock Springs
CA
Bear
Fort Bridger
Lyman
Evanston
Mountain View
FLAMING GORGE
NAT'L REC. AREA
UINTA
Flaming Gorge Res.
McKinnon
Baggs
Dixon
UTAH
FLAMING GORGE DAM
110°
108°

WYOMING
MILES
0 10 20 40 60
KILOMETRES
0 20 40 60
State Capital
106°
104°
D
E
Tongue River Res.
Sheridan
Clearmont
Ucross
Buffalo
JOHNSON
Powder R.
CAMPBELL
Gillette
Keyhole Res.
Hulett
DEVILS TOWER NAT'L MON.
CROOK
BEAR LODGE MTS.
Sundance
Moorcroft
Upton
BLACK HILLS
Rapid City
44°
MT. RUSHMORE NAT'L MEM.
Custer
JEWEL CAVE NAT'L MON.
WIND CAVE NAT'L PK.
Belle Fourche R.
WESTON
Osage
Newcastle
Kaycee
Linch
Midwest
Edgerton
Teapot Dome
Antelope Cr.
Cheyenne R.
Hot Sprs.
Angostura Res.
PINE RIDGE IND. RES.
SOUTH DAKOTA
NEBRASKA
NIOBRARA
Lance Creek
CONVERSE
NATRONA
Mountain View
Glenrock
Mills
Evansville
Casper
Paradise Valley
Douglas
Orin
Manville
Lusk
Chadron
Glendo Res.
Glendo
Niobrara R.
N
Alcova Res.
Pathfinder Res.
LARAMIE
Laramie Pk.
North Platte
Guernsey Res.
Sunrise
Guernsey
Ft. Laramie
FT. LARAMIE NAT'L HIST. SITE
Lingle
AGATE FOSSIL BEDS NAT'L MON.
Torrington
Wheatland
MTS.
Seminoe Res.
42°
PLATTE
GOSHEN
Scottsbluff
SCOTTS BLUFF NAT'L MON.
River
CHIMNEY ROCKS NAT'L HIST. SITE
Hanna
Medicine Bow
Wheatland Res.
Chugwater
Hawk Sprs.
Elk Mtn.
Rock River
McFadden
MEDICINE BOW RANGE
MOUNTAINS
ALBANY
Saratoga
LARAMIE
Lodgepole Creek
Laramie
Laramie R.
N. Platte R.
FRANCIS E. WARREN A.F.B.
Cheyenne
Pine Bluffs
Encampment
Foxpark
Orchard Valley
Fox Farm
COLORADO
© Copyright HAMMOND INCORPORATED, Maplewood, N.J.
Longitude West 106° of Greenwich
104°
1
2
3

Hamilton Inlet
George I.
Labrador
Rigolet
Lake Melville
MEALY MTS.
3,700
North R.
Huntingdon I.
Sandwich B.
Cartwright
Table Bay
Sea
Separation Point
Sand Hill R.
I. of Ponds
LABRADOR
Eagle
Paradise R.
HIGH MTS.
2,080
Hawke I.
Stony I.
Gilbert R.
Square Islands
Alexis R.
1,920
Port Hope Simpson
Fox Har.
St. Lewis R.
Battle Har.
Mary's Har.
R. St. Paul
QUEBEC
Henley Har.
Belle Isle
R. St-Augustin
Red Bay
W. St. Modeste
L'Anse-au-Loup
Forteau
Blanc-Sablon
Strait of Belle Isle
Pistolet B.
Griquet
Cook's Har.
St. Anthony
R. du Petit Mécatina
St-Augustin
Flowers Cove
Hare Bay
Ten Mile L.
Shoal Cove
Groais I.
St. John I.
Port au Choix
Roddickton
Grey Is.
Englee
Baie-des-Moutons
Bell I.
Ingornachoix B.
Canada B.
Harrington Harbour
Gulf
Great
LABRADOR
(PART OF NEWFOUNDLAND)
0 50 100 150 ml.
0 50 100 150 km.
Button Is.
Killinek I.
C. Chidley
Akpatok I.
N. Aulatsivik I.
Ungava Bay
TORNGAT MTS.
Cirque Mtn. 5,160
Saglek B.
Labrador Sea
Ft-Chimo
R. George
R. à la Baleine
QUEBEC
S. Aulatsivik I.
Nain
Tunungayualok I.
Davis Inlet
Hopedale
Makkovik
C. Harrison
Schefferville
Canairiktok
Hamilton Inlet
Menihek Lakes
Smallwood Res.
Rigolet
L. Melville
Cartwright
Churchill Falls
North West River
Labrador City
Lac Joseph
Churchill R.
Goose Bay
Atikonak L.
Eagle R.
Port Hope Simpson
Ashuanipi L.
R. Moisie
SAGUENAY PROVINCIAL PARK
Havre-St-Pierre
R. du Petit Mécatina
St-Augustin
Str. of Belle Isle
St. Anthony
Port au Choix
Harrington Har.
Sept-Îles
(Seven Islands)
NEWFOUNDLAND

ATLANTIC
OCEAN
St. Lawrence
NEWFOUNDLAND
ST. PIERRE & MIQUELON
(France)
LONG RANGE MTS.
Cabot Strait
St. John's
Gander
Corner Brook
Grand Falls
Bishop's Falls
Botwood
Lewisporte
Windsor
Stephenville
Stephenville Crossing
Deer Lake
Bonavista
Trinity
Clarenville
Carbonear
Harbour Grace
Placentia
Marystown
Burin
St. Lawrence
Grand Bank
Fortune
Burgeo
Channel-Port aux Basques
Avalon Pen.
Conception Bay
Trinity Bay
Bonavista Bay
Placentia Bay
Fortune B.
Notre Dame Bay
Hamilton Sound
Fogo I.
Fogo
Twillingate
New World I.
Grand Lake
Red Indian L.
Gander L.
Victoria L.
Meelpaeg L.
Great Burnt L.
GROS MORNE NAT'L PK.
TERRA NOVA NAT'L PARK
Mt. Sylvester 1,250
Main Topsail 1,822
St-Pierre
Miquelon
C. Race
C. Ray
C. Anguille
C. St. John
MILES
0 10 20 40 60 80 100
KILOMETRES
0 10 20 40 60 80 100
Provincial Capital
Longitude West of Greenwich
© C.S. HAMMOND & Co., Maplewood, N.J.

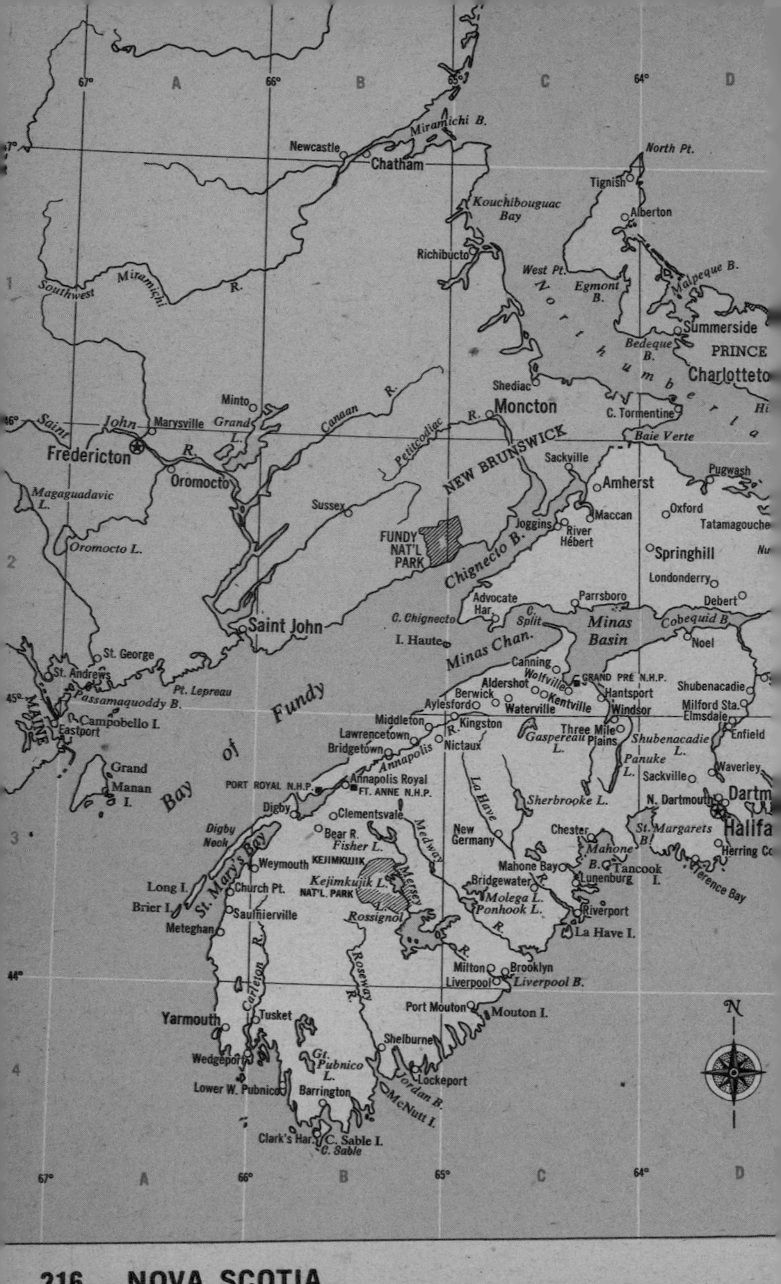
NEW BRUNSWICK
MAINE
PRINCE
Fredericton
Saint John
Moncton
Chatham
Newcastle
Oromocto
Marysville
Minto
Sussex
Shediac
Richibucto
Sackville
Amherst
Maccan
Joggins
River Hébert
Springhill
Oxford
Pugwash
Tatamagouche
Londonderry
Debert
Parrsboro
Advocate Har.
Noel
Summerside
Charlottetown
Tignish
Alberton
C. Tormentine
St. George
St. Andrews
Eastport
Campobello I.
Grand Manan I.
FUNDY NAT'L PARK
Bay of Fundy
Chignecto B.
Minas Chan.
Minas Basin
Cobequid B.
Canning
Wolfville
GRAND PRÉ N.H.P.
Hantsport
Windsor
Kentville
Aldershot
Berwick
Waterville
Aylesford
Kingston
Middleton
Lawrencetown
Bridgetown
Nictaux
Annapolis Royal
PORT ROYAL N.H.P.
FT. ANNE N.H.P.
Digby
Clementsvale
Bear R.
Weymouth
Church Pt.
Saulnierville
Meteghan
Long I.
Brier I.
KEJIMKUJIK NAT'L PARK
Kejimkujik L.
Shubenacadie
Milford Sta.
Elmsdale
Enfield
Waverley
Sackville
N. Dartmouth
Halifax
Herring Cove
Chester
Mahone Bay
Tancook I.
Lunenburg
Bridgewater
Riverport
La Have I.
New Germany
Milton
Brooklyn
Liverpool
Liverpool B.
Port Mouton
Mouton I.
Shelburne
Lockeport
Jordan B.
McNutt I.
Yarmouth
Tusket
Wedgeport
Lower W. Pubnico
Barrington
Clark's Har.
C. Sable I.
C. Sable
N

Magdalen Is. (Que.)
St. Paul I.
Gulf of St. Lawrence
C. North
Cape North
Aspy Bay
CAPE BRETON HIGHLANDS NAT'L PK.
1,747
Ingonish
Petit-Étang
Chéticamp
Grand-Étang
CAPE BRETON ISLAND
St. Ann's B.
Boularderie I.
Florence
Sydney Mines
New Waterford
Glace Bay
Margaree
Margaree Forks
Goose Cove
N. Sydney
Dominion
Reserve Mines
Donkin
Inverness
L. Ainslie
Baddeck
Sydney
Mabou
Port Hood
1,075
Mira B.
Mira R.
Scatarie I.
C. Breton
Louisbourg
LOUISBOURG N.H.P.
Bras d'Or Lake
Gabarus
Gabarus B.
East Pt.
St. Peters Bay
Souris
Georgetown
Montague
Cardigan B.
C. Bear
Strait
Pictou I.
C. George
Georges Bay
Str. of Canso
Port Hawkesbury
St. Peters
Louisdale
Arichat
I. Madame
Pictou
Trenton
Antigonish
New Glasgow
Westville
Thorburn
Stellarton
Mulgrave
Hopewell
Boylston
Chedabucto B.
Guysborough
Canso
Andrew I.
Upper Stewiacke
Goldboro
Tor B.
Sherbrooke
Country Har.
Middle Musquodoboit
Ecum Secum
Sheet Har.
Moser R.
Ship Har.
ATLANTIC OCEAN
West Pt.
Sable I.
NOVA SCOTIA
MILES
0 10 20 40 60
KILOMETRES
0 20 40 60
Provincial Capitals
© C.S. HAMMOND & Co., Maplewood, N.J.
Longitude 62° West of Greenwich
63° 62° 61° 60°
E F G H
47° 46° 45° 44°
1 2 3 4

NEW BRUNSWICK
MILES
0 10 20 40 60
KILOMETRES
0 10 20 40 60
Provincial Capitals
© C.S. HAMMOND & Co., Maplewood, N.J.
UNITED STATES
QUEBEC
MAINE
Campbellton
Rivière-du-Loup
Edmundston
Fredericton
Presque Isle
Caribou
Houlton
Millinocket
Woodstock
Grand Falls
St. Stephen
Calais
St. Andrews
Campobello
Eastport
Machias
Mt. Katahdin 5,268

Dalhousie
Heron I.
Chaleur Bay
QUE.
Balmoral
Jacquet R.
Lorne
Pointe-Verte
Blue Mtn. 1,733
Petit Rocher
Nepisiguit B.
Beresford
Robertville
Bathurst
Miscou Pt.
Miscou Ctre.
Miscou I.
Shippegan I.
Blue Cove
Maisonnette
Lamèque
Grande-Anse
Ste-Marie-sur-Mer
Caraquet
Centre-St-Simon
Shippegan
Inkerman
St-Isidore
St. Sauveur
Allardville
Sheila
Tracadie
Légère
Nepisiguit R.
Brantville
Northwest Miramichi R.
Lavillette
Neguac
Portage I.
Miramichi B.
Fox I.
Pt. Escuminac
Loggieville
Newcastle
Chatham
Chatham Head
Nelson-Miramichi
Manuels
Southwest Miramichi R.
Pt. Sapin
Kouchibouguac Bay
GULF OF ST. LAWRENCE
North Pt.
Alberton
PRINCE EDWARD ISLAND
Blackville
Rogersville
St-Louis-de-Kent
Richibucto
Rexton
Egmont B.
Northumberland Strait
PR. EDWARD ISLAND NAT'L PARK
Blissfield
Miramichi R.
Doaktown
Buctouche
Harcourt
Summerside
Astle
St-Antoine
Salmon R.
Notre-Dame
Charlottetown
Chipman
Shediac
Cap-Pelé
Cape Tormentine
Hillsborough Bay
Canaan R.
Minto
Moncton
Lewisville
Dieppe
St-Anselme
Memramcook
Port Elgin
Baie Verte
Riverview Hts.
St-Joseph
College Br.
Grand L.
Petitcodiac
Hillsborough
Sackville
C. Cliff
Maugerville
Dorchester
FORT BEAUSÉJOUR NAT'L HIST. PARK
Burton
Hopewell Cape
Amherst
Oromocto
Gagetown
Penobsquis
Geary
Riverside-Albert
Oxford
Sussex
Norton
FUNDY NAT'L PARK
Alma
Springhill
Chignecto Bay
NOVA SCOTIA
Hampton
Fairvale
Parrsboro
Grand Bay
Rothesay
C. Chignecto
C. Split
Minas Basin
Cobequid B.
Truro
St. Martins
Saint John
Isle Haute
Minas Chan.
C. Spencer
Stewiacke
Kentville
Pt. Lepreau
Bay of Fundy
Windsor
Annapolis R.
Middleton
Annapolis Basin
Digby
Dartmouth
Halifax
Longitude 65° West of Greenwich
66°
65°
64°
63°
48°
47°
46°
45°
E
F
G

220 — PRINCE EDWARD ISLAND

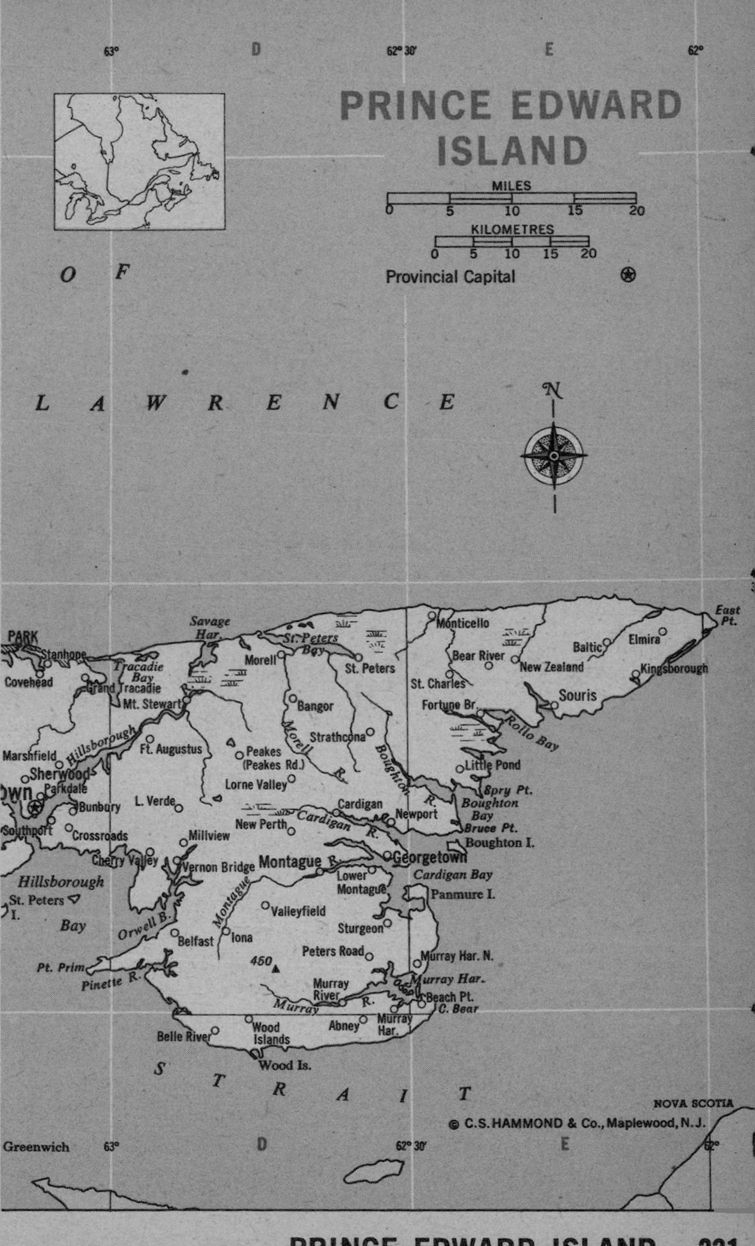
PRINCE EDWARD ISLAND
MILES
0
5
10
15
20
KILOMETRES
0
5
10
15
20
Provincial Capital
63°
D
62° 30'
E
62°
O F
L A W R E N C E
N
Savage Har.
St. Peters Bay
Monticello
East Pt.
PARK
Stanhope
Tracadie Bay
Morell
St. Peters
Bear River
Baltic
Elmira
Covehead
Grand Tracadie
St. Charles
New Zealand
Kingsborough
Mt. Stewart
Bangor
Fortune Br.
Souris
Rollo Bay
Hillsborough
Ft. Augustus
Strathcona
Marshfield
Peakes (Peakes Rd.)
Morell R.
Boughton R.
Little Pond
Sherwood
Parkdale
Lorne Valley
Spry Pt.
Boughton Bay
Bunbury
L. Verde
Cardigan
Newport
Cardigan R.
New Perth
Bruce Pt.
Southport
Crossroads
Millview
Boughton I.
Cherry Valley
Montague R.
Georgetown
Vernon Bridge
Lower Montague
Cardigan Bay
Hillsborough Bay
Panmure I.
St. Peters I.
Montague
Valleyfield
Orwell B.
Sturgeon
Belfast
Iona
Peters Road
Pt. Prim
450
Murray Har. N.
Pinette R.
Murray River
Murray R.
Murray Har.
Beach Pt.
C. Bear
Belle River
Wood Islands
Abney
Murray Har.
Wood Is.
S T R A I T
NOVA SCOTIA
© C.S.HAMMOND & Co., Maplewood, N.J.
Greenwich
63°
D
62° 30'
E
62°

QUÉBEC
Mistassini
Normandin
Dolbeau
St-Félicien
Roberval
L. des Commissaires
R. Windigo
R. Trenche
L. Blanc
R. Croche
Rapide-Blanc
L. Edouard
R. Vermilion
La Tuque
R. St. Maurice
L. Wayagamak
R. Batiscan
Res. Matawin
Res. Baskatong
Ferme-Neuve
Mont-Laurier
Maniwaki
MONT-TREMBLANT PROV. PARK
Mt. Tremblant 3,150
St-Michel-des-Saints
Ste-Thècle
St-Tite
St-Casimir
Grand'Mère
Shawinigan
La Pérade
St-Alexis-des-Monts
Cap-de-la-Madeleine
Trois-Rivières
Becancour
L'Annonciation
Labelle
St-Jovite
St-Donat-de-Montcalm
R. L'Assomption
St-Gabriel
Louiseville
L. Gagnon
L. St-Pierre
Nicolet
St-Félix-de-Valois
Berthierville
Pierreville
L. du Poisson-Blanc
Val-David
Ste-Agathe
Joliette
Victoriaville
L. Simon
Tracy
Sorel
St-Jacques
St-Sauveur
Ste-Adèle
R. Rouge
Drummondville
R. St-François
L'Épiphanie
Contrecoeur
St-Jérôme
L'Assomption
Verchères
GATINEAU PARK
Brownsburg
Lachute
St-Hyacinthe
Richmond
Buckingham
Montebello
Thurso
Ste-Thérèse
Pte-aux-Trembles
Acton Vale
Gatineau
Hawkesbury
Laval
Longueuil
La Providence
Hull
Ottawa R.
ONTARIO
St-Laurent
MONTRÉAL
Bromptonville
Aylmer
Ottawa
Rigaud
Dorion
Verdun
Marieville
Granby
Sherbrooke
Alexandria
Lachine
Pte-Claire
Beauharnois
St-Jean
Iberville
Waterloo
MT-ORFORD PROV. PK.
St-Rémi
Valleyfield
Farnham
Carleton Place
Ormstown
Napierville
Cowansville
Magog
Kemptville
Cornwall
St. Lawrence R.
Lac-Brome
Huntingdon
R. Richelieu
Bedford
Sutton
L. Memphremagog
Smiths Falls
Massena
NEW YORK
Rouses Pt.
L. Champlain
Newport
76°
75°
74°
73° Longitude
West of
A
B
C
D
80°
70°
60°
50°
QUÉBEC
0 100 200 mi.
0 100 200 km.
N.W.T.
Saglouc
Baffin I.
Hudson Str.
Mansel I.
Nouveau-Québec Crater
Maricourt
Povungnituk
Ungava Pen.
Ungava B.
Hudson Bay
L. Payne
Port-Nouveau-Québec
ATLANTIC OCEAN
R. aux Feuilles
Ft-Chimo
LABRADOR
Nain
Belcher Is.
Lac à l'Eau Claire
Hopedale
Poste-de-la-Baleine
Schefferville
L. Bienville
Smallwood Res.
James B.
La Grande-Rivière
Labrador City
Ft-George
Akimiski I.
R. Eastmain
NEWF.
Rupert House
SAGUENAY PROV. PK.
L. Mistassini
Havre-St-Pierre
Sept-Îles
ONT.
Chibougamau
Île d'Anticosti
La Sarre
Chicoutimi
Noranda
Gaspé
Gulf of St. Lawrence
Rouyn
Val-d'Or
Rimouski
LAURENTIDES PROV. PK.
LA VERENDRYE PROV. PK.
Québec
N. B.
P. E. I.
Magdalen Is.
Charlottetown
Mt-Laurier
Sydney
St. Lawrence
Fredericton
Ottawa
MONTREAL
N. S.
ME.
N. Y.
Halifax

E
F
G
H
71°
70°
69°
68°
R. Péribonca
Canton Bégin
St-Coeur-de-Marie
Alma
Desbiens
Hébertville
Jonquière
Kénogami
Arvida
Chicoutimi
Bagotville
Port-Alfred
L. Kénogami
R. Saguenay
R. Portneuf
Forestville
Sault-au-Mouton
ST. LAWRENCE RIVER
Mont-Joli
Price
Luceville
Rimouski
Bic
St-Fabien
Trois-Pistoles
Tadoussac
Île Verte
L'Isle-Verte
48°
NEW BRUNSWICK
LAURENTIDES
PROVINCIAL
PARK
3,925
I. aux Lièvres
St-Siméon
Rivière-du-Loup
L. Temiscouata
Cabano
Clermont
La Malbaie
Pte-au-Pic
Notre-Dame-du-Lac
Dégelis
St-Pascal
St-Éleuthère
St-Joseph-de la-Rivière-Bleue
Edmundston
St-Pacôme
Baie-St-Paul
I. aux Coudres
La Pocatière
St-Jean-Port-Joli
N.B.
QUEBEC
R. Jacques-Cartier
I. aux Oies
L'Islet
Ste-Anne-de-Beaupré
Cap-St-Ignace
MILES
0
10
20
40
60
47°
Montmorency
Charlesbourg
Ile d'Orléans
Montmagny
St-Pamphile
MAINE
R.
KILOMETRES
0
20
40
60
Québec
Lévis
St-Pierre-Montmagny
St-Raphaël
Ste-Foy
St-Romuald
d'Etchemin
St-Gervais
Lac-Frontière
Saint John
National Capital
Provincial Capitals
Donnacona
St-Agapitville
St-Anselme
Ste-Claire-de-Joliette
N
Ste-Marie
R. Chaudière
Lac-Etchemin
St-Joseph-de-Beauce
Bécancour
Plessisville
Tring-Jct.
Beauceville-Est
Prince-ville
Bernierville
St-Georges
St-Zacharie
Arthabaska
Thetford Mines
Linière
Black L.
Bolduc
46°
Warwick
Disraëli
L. Aylmer
L. St-François
La Guadeloupe
Asbestos
Windsor
Weedon-Centre
Lac-Mégantic
Scotstown
E. Angus
Cookshire
Lennoxville
Waterville
Coaticook
N. H.
MAINE
45°
Greenwich
66°
64°
St. Lawrence R.
Grande-Vallée
Cap-Chat
Ste-Anne-des-Monts
Mt. Jacques Cartier
4,160
Murdochville
GASPÉSIE PROV. PK.
FORILLON NAT'L PARK
C. du Gaspé
Gaspé
Les Méchins
Matane
R. Cascapédia
Price
Sayabec
Gaspé Peninsula
Percé
Mont-Joli
Amqui
Causapscal
Grande-Rivière
Chandler
Gulf of St. Lawrence
Bonaventure
Matapédia
New Carlisle
Chaleur B.
Campbelltown
NEW BRUNSWICK
0
20
40 mi.
0
20
40 km.
68°
66°
64°
1
2
3
4
© C.S.HAMMOND & CO., Maplewood, N.J.

Elliot Lake
Thessalon
Sudbury
Nickel Centre
Walden
Warren
Verner
Sturgeon Falls
North Bay
L. Nipissing
Callander
Spanish
Massey
Spanish R.
Panache L.
Espanola
Blind River
North Channel
Drummond I.
Noelville
Wanapitei R.
Powassan
French R.
Gore Bay
Little Current
Cockburn I.
Manitoulin I.
Duck Is.
Providence Bay
Manitowaning
South River
Sundridge
Magnetawan R.
Byng Inlet
Burk's Falls
McKellar
Fitzwilliam I.
Georgian Bay
Main Channel
GEORGIAN BAY IS. NAT'L PK.
Tobermory
Cabot Head
Nobel
Parry Sound
KILLBEAR PT. PROV. PK.
Parry I.
L. Joseph
L. Rosseau
Huntsville
MacTier
Bracebridge
Alpena
Thunder B.
LAKE HURON
Bruce Pen.
Lion's Head
C. Croker
GEORGIAN BAY IS. NAT'L PK.
L. Muskoka
Gravenhurst
Christian I.
MICH.
Harrisville
Wiarton
Owen Sd.
Penetanguishene
Midland
Victoria Har.
UNITED STATES
Oscoda
Southampton
Owen Sound
Meaford
Thornbury
Nottawasaga Bay
Orillia
Collingwood
Stayner
Barrie
L. Simcoe
Beaverton
Port Elgin
Chesley
Markdale
INVERHURON PROV. PK.
Angus
Kincardine
Saugeen R.
Sutton
Walkerton
Hanover
Durham
Alliston
Bradford
Port Austin
Shelburne
Newmarket
Uxbridge
Wingham
Harriston
Mt. Forest
Arthur
Orangeville
Aurora
Richmond Hill
Harbor Beach
Palmerston
Listowel
Fergus
Goderich
Grand R.
Brampton
Bramalea
Acton
Clinton
Cass City
Milverton
Elmira
Guelph
Georgetown
Mississauga
Bridgeport
Seaforth
Sandusky
Mitchell
Waterloo
Kitchener
Milton
Oakville
Stratford
Burlington
PINERY PROV. PK.
Exeter
Cambridge
Hamilton
Welland Canal
MICHIGAN
IPPERWASH PROV. PK.
Dundas
St. Mary's
Paris
Parkhill
Grimsby
Lapeer
Woodstock
Stoney Cr.
Pt. Edward
Forest
Ingersoll
Brantford
St. Catharines
Port Huron
London
Caledonia
Thorold
Sarnia
Strathroy
Niagara Falls
Corunna
Norwich
Hagersville
Haldimand
Petrolia
Lambeth
Waterford
Welland
St. Clair
St. Clair R.
St. Thomas
Tillsonburg
Delhi
Simcoe
Dunnville
Port Colborne
Wallaceburg
Glencoe
Sydenham R.
Thames R.
Dresden
Pontiac
Aylmer
Nanticoke
Mt. Clemens
W. Lorne
Port Rowan
Port Stanley
L. St. Clair
Long Pt.
DETROIT
Chatham
Ridgetown
Tecumseh
Blenheim
Dunkirk
Windsor
Belle R.
Tilbury
RONDEAU PROV. PK.
Pt. aux Pins
Westfield
Amherstburg
Essex
Wheatley
LAKE ERIE
L. Chautauqua
Harrow
Leamington
Erie
Jamestown
Kingsville
PT. PELEE NAT'L PARK
Pt. Pelee
PA.

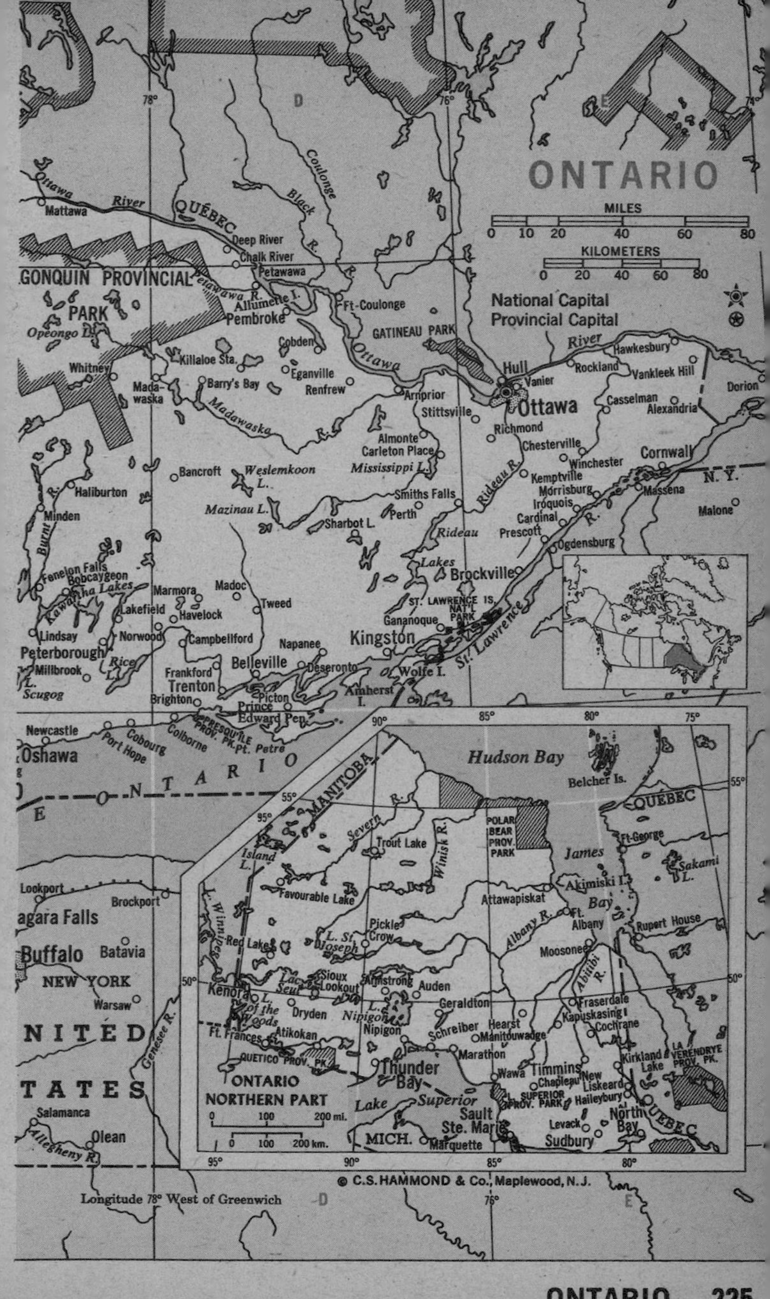
ONTARIO
MILES
0 10 20 40 60 80
KILOMETERS
0 20 40 60 80
National Capital
Provincial Capital
Ottawa
Hull
Vanier
Kingston
Belleville
Trenton
Peterborough
Oshawa
Brockville
Cornwall
Pembroke
ALGONQUIN PROVINCIAL PARK
GATINEAU PARK
QUÉBEC
N.Y.
NEW YORK
UNITED STATES
Buffalo
Niagara Falls
Hudson Bay
James Bay
ONTARIO NORTHERN PART
0 100 200 mi.
0 100 200 km.
Thunder Bay
Sudbury
Sault Ste. Marie
Timmins
North Bay
Kenora
Lake Superior
MANITOBA
MICH.
© C.S.HAMMOND & Co., Maplewood, N.J.
Longitude 78° West of Greenwich

102°
A
100°
B
98°
C
Cedar L.
Grand Rapids
Red Deer L.
Long Pt.
Dawson B.
Lake
LAKE
Poplar
Mafeking
Kawinaw L.
PORCUPINE HILLS
Swan Lake
Pelican L.
Birch I.
Chitek L.
Reindeer Island
Berens River
Birch River
R.
Swan
Bowsman
Swan River
Minitonas
Winnipegosis
Waterhen Lake
Berens I.
Kenville
Campberville
Sturgeon B.
Benito
DUCK MTN.
DUCK MTN. PROV. PK.
Pine River
Gypsumville
Boggy Creek
Winnipegosis
Fork River
Ethelbert
Portage B.
Lake St. Martin
Moose I.
Baldy Mtn. 2,727
Crane River
Steep Rock
Sifton
Rorketon
Lake
Dallas
Roblin
Dauphin L.
Moosehorn
Valley R.
Dauphin
Ashern
Hodgson
Grandview
Gilbert Plains
Ste-Rose-du-Lac
Fisher Branch
Dog L.
SASKATCHEWAN
Ochre River
Inglis
RIDING MTN.
Riverton
RIDING MTN. NAT'L PARK
Laurier
Eriksdale
Arborg
Russell
Angusville
McCreary
Assiniboine
Binscarth
Rossburn
Lundar
Kelwood
Gimli
Oakburn
Amaranth
Elphinstone
Manitoba
Oak Point
Inwood
Winnipeg Beach
St-Lazare
Shoal Lake
Sandy Lake
Erickson
Riding Mt.
Glenella
Marius
Shoal Lakes
Matlock
Birtle
Strathclair
Clanwilliam
Plumas
St. Laurent
Teulon
Newdale
Minnedosa
Arden
Langruth
R.
Miniota
Hamiota
Rapid City
Neepawa
Gladstone
LOWER FT. GARRY N.H.P.
Oak River
Rivers
Stonewall
Portage la Prairie
Poplar Pt.
Stony Mtn.
Elkhorn
Kenton
R.
Winnipeg
W. Kildonan
Austin
Mac Gregor
Douglas
St. James-Assiniboia
Virden
Carberry
Oakville
Alexander
Brandon
Transcona
Pipestone Cr.
Ft. Garry
St. Boniface
Oak Lake
Assiniboine
Starbuck
Lorette
Souris
Wawanesa
St. Claude
Elm Creek
Holland
R.
Reston
Pipestone
Treherne
Niverville
Ste. Agathe
R.
Hartney
Glenboro
Cypress River
Carman
Kleefeld
Notre Dame de Lourdes
Red
Souris
Elgin
Belmont
St. Pierre
Ninette
Somerset
Roland
Morris
Grunthal
Melita
Baldur
Miami
Whitewater L.
St. Jean Baptiste
Ste. Elizabeth
La Rivière
Manitou
Pierson
Deloraine
Boissevain
Morden
Plum Coulee
Roseau
Killarney
Pilot Mound
Pembina R.
Waskada
Crystal City
Winkler
Altona
Dominion City
TURTLE MTN. PROV. PK.
Cartwright
INTERNAT'L PEACE GARDEN
Gretna
Emerson
N. DAKOTA
© C.S.HAMMOND & Co., Maplewood, N.J.
A
100°
B
Longitude
98°
West of
C

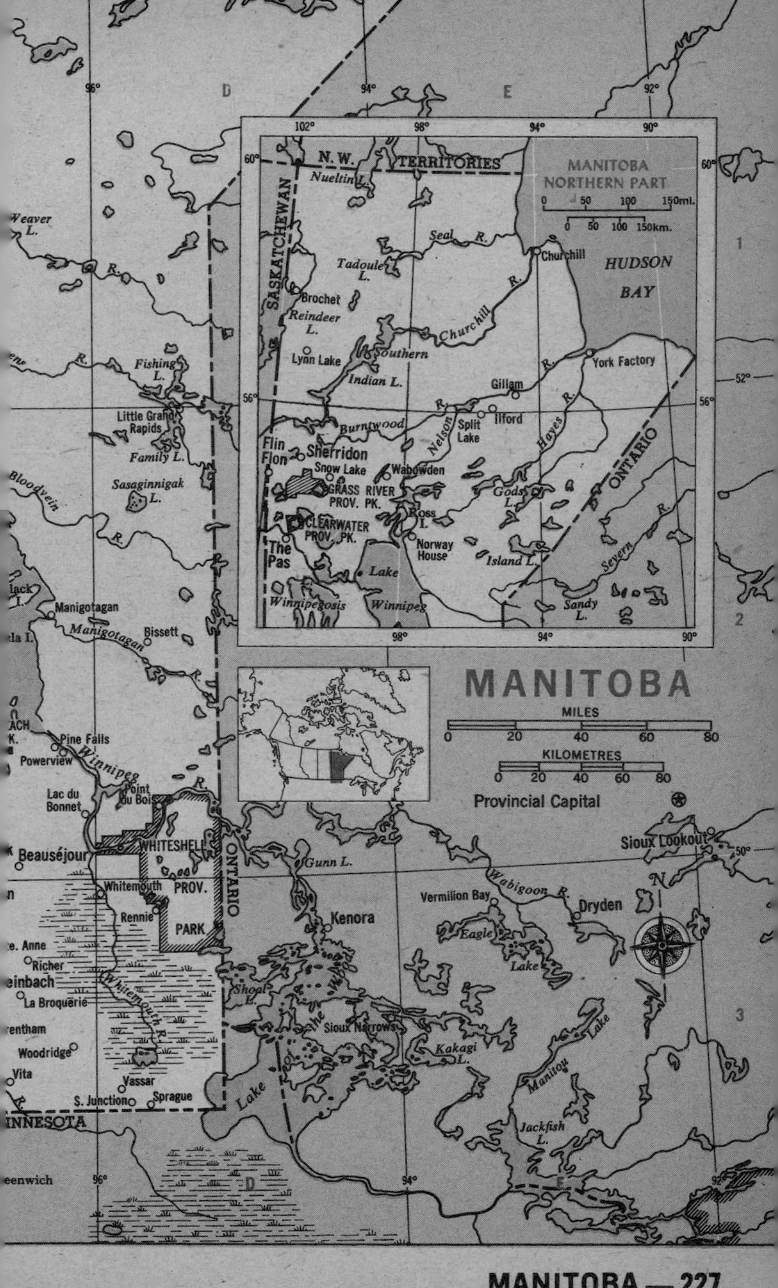
MANITOBA
MILES
0 20 40 60 80
KILOMETRES
0 20 40 60 80
Provincial Capital
MANITOBA NORTHERN PART
0 50 100 150ml.
0 50 100 150km.
N.W. TERRITORIES
SASKATCHEWAN
ONTARIO
HUDSON BAY
Nueltin L.
Seal R.
Tadoule L.
Churchill
Brochet
Reindeer L.
Churchill R.
Lynn Lake
Southern Indian L.
York Factory
Gillam
Burntwood R.
Nelson R.
Split Lake
Ilford
Hayes R.
Flin Flon
Sherridon
Snow Lake
Wabowden
GRASS RIVER PROV. PK.
Gods L.
CLEARWATER PROV. PK.
Ross I.
The Pas
Norway House
Island L.
Severn R.
Lake Winnipegosis
Lake Winnipeg
Sandy L.
Weaver L.
Fishing L.
Little Grand Rapids
Family L.
Bloodvein R.
Sasaginnigak L.
Manigotagan
Manigotagan R.
Bissett
Pine Falls
Powerview
Winnipeg R.
Lac du Bonnet
Point du Bois
Beauséjour
WHITESHELL PROV. PARK
ONTARIO
Whitemouth
Rennie
Richer
La Broquerie
Whitemouth R.
Woodridge
Vita
Vassar
S. Junction
Sprague
Gunn L.
Kenora
Vermilion Bay
Wabigoon R.
Dryden
Sioux Lookout
Eagle Lake
Shoal L.
Lake of the Woods
Sioux Narrows
Kakagi L.
Manitou Lake
Jackfish L.
N

A
B
C
D
110°
108°
106°
54°
52°
50°
1
2
3
N
Frog L.
Loon Lake
Meadow Lake
Cowan Lake
Delaronde L.
Kingsmere L.
Crean L.
Montréal Lake
Montréal Lake
PRINCE ALBERT NATIONAL PARK
Waskesiu Lake
Candle L.
Chitek L.
Big River
St. Walburg
Turtle L.
Leoville
Sturgeon
Paradise Hill
Witchekan L.
Debden
Christopher Lake
Smeaton
Paddockwood
Choiceland
North
Turtleford
Lloydminster
Glaslyn
Canwood
Spiritwood
R.
Edam
Medstead
Shell Lake
Saskatchewan
Lashburn
Jackfish L.
Shellbrook
Rabbit Lake
Prince Albert
Birch Hills
Paynton
Battle R.
Meota
Leask
Marsden
Neilburg
Blaine Lake
Duck Lake
St-Louis
Kinistino
Manito L.
Cut Knife
N. Battleford
Hafford
Melfort
Battleford
Redberry L.
Rosthern
Wakaw
Basin L.
St-Brieux
Lenore Lake
Maymont
Waldheim
Middle L.
Provost
Unity
Wilkie
Radisson
Hague
Cudworth
Naicam
Macklin
Scott
River
Dalmeny
Lake Lenore
Eyehill
Sonningdale
Langham
Warman
Aberdeen
Bruno
Denzil
Tramping Lake
Landis
Cr.
Luseland
Biggar
Perdue
Asquith
Saskatoon
Humboldt
Watson
ALBERTA
Kerrobert
Viscount
Fusilier
Delisle
Allan
Lanigan
Coleville
Dodsland
Eagle
Creek
Dundurn
Young
Harris
River
Hanley
Watrous
Kindersley
Conquest
Rosetown
Outlook
Kenaston
Nokomis
Alsask
Brock
Milden
Simpson
Semans
SOUTH SASK. RIVER DAM
Eatonia
Dinsmore
Davidson
Imperial
Plato
Govan
Eston
Elrose
Dunblane
Loreburn
Last Mountain Lake
Elbow
Strasbourg
Lucky Lake
Craik
Leader
Saskatchewan
Southey
South
Prelate
Kyle
Central Butte
Holdfast
Qu'Appelle R.
Abbey
Beechy
Regina Beach
THE COTEAU
Cabri
Thunder Cr.
Bethune
GREAT SAND HILLS
Lumsden
Fox Valley
Chaplin
Moose Jaw
Regina
Herbert
Mortlach
Bigstick L.
Swift Current
Morse
Chaplin Lake
Pense
Cr.
Crane L.
Gull Lake
Hodgeville
Moose Jaw
Rouleau
Tompkins
Old Wives Lake
Maple Creek
Current
R.
CYPRESS HILLS PROV. PK.
Swift
Vanguard
Cr.
Avonlea
4,546
Gravelbourg
Mossbank
HILLS
CYPRESS HILLS PROV. PK.
CYPRESS
Shaunavon
Notukeu
Ponteix
Lafleche
Eastend
Kincaid
Wood
Assiniboia
Ogema
Frenchman
Consul
Mankota
Battle Cr.
Willow Bunch
Bengough
Climax
Val-Marie
R.
Rockglen
Coronach
MONTANA
Scobey
© C.S. HAMMOND & Co., Maplewood, N.J.

SASKATCHEWAN NORTHERN PART
N.W. TERRITORIES
ALBERTA
MANITOBA
Uranium City
Eldorado
Gunnar
Stony Rapids
Black L.
Lake Athabasca
Fond du Lac R.
Wollaston L.
Wollaston Lake
Cree L.
Geikie R.
Wathaman R.
Clearwater R.
Frobisher L.
Reindeer L.
La Loche
Buffalo Narrows
Île-à-la-Crosse
Churchill R.
Sandy Bay
La Ronge
Beauval
Flin Flon
Creighton
MEADOW LAKE PROV. PARK
LAC LA RONGE PROV. PK.
Goodsoil
Meadow Lake
PRINCE ALBERT NAT'L PK.
SASKATCHEWAN
Provincial Capital
MILES
KILOMETRES
Cumberland L.
The Pas
Tobin Lake
Squaw Rapids
Nipawin
Carrot River
Aylsham
Arborfield
Zenon Park
PASQUIA HILLS
Red Deer L.
Hudson Bay
Red Deer R.
Tisdale
Porcupine Plain
Weekes
GREENWATER LAKE PROV. PK.
PORCUPINE HILLS
MANITOBA
Archerwill
Rose Valley
Kelvington
Assiniboine
Quill Lake
Preeceville
Wadena
Sturgis
Norquay
Pelly
Little Quill L.
Invermay
Whitesand
Buchanan
Canora
DUCK MTN. PROV. PK.
Foam Lake
Sheho
GOOD SPIRIT PROV. PK.
Kamsack
Theodore
Lestock
Kelliher
Springside
Rhein
Roblin
Ituna
Yorkton
Saltcoats
RIDING MOUNTAIN NATIONAL PARK
Goodeve
Cupar
Lipton
Melville
Bredenbury
Balcarres
Langenburg
Russell
Abernethy
Neudorf
Lemberg
Qu'Appelle River
Esterhazy
Indian Head
Balgonie
Grenfell
Broadview
Rocanville
Wolseley
Whitewood
Wapella
Montmartre
Glenavon
Sedley
Moose Mountain
Kipling
Moosomin
MOOSE MTN. PROV. PK.
Wawota
Virden
Brandon
Assiniboine River
Fillmore
Maryfield
Treherne
Oak L.
Souris
Yellow Grass
Weyburn
Stoughton
Arcola
Carlyle
Manor
Redvers
Reston
Belmont
Radville
Midale
Lampman
Whitewater L.
Boissevain
Alameda
Oxbow
Frobisher
Carievale
Killarney
Torquay
Bienfait
Carnduff
TURTLE MTN. PROV. PK.
INTERNAT'L PEACE GARDENS
Estevan
NORTH DAKOTA
Westhope
Rolla
Crosby
Bottineau
Longitude West of Greenwich

Bitumount
Ft. MacKay
River
Ft. McMurray
THICKWOOD HILLS
Methy Lake
Peter Pond Lake
N
Winefred L.
Conklin
Primrose Lake
Cold L.
Cold L.
Grand Centre
Ft. Kent
Bonnyville
Glendon
Therien
Mallaig
St. Paul
Elk Point
Vilna
Imperial Mills
Lac La Biche
Athabasca
Calling Lake
N. Wabasca L.
Wabasca
Legend L.
Wabasca
River
Trout Mt. 2,709
Peerless Lake
Utikuma Lake
Atikameg
Lesser Slave L.
Grouard Mission
Canyon Creek
Slave Lake
Kinuso
Faust
Enilda
House Mt. 3,950
Wallace Mt. 4,180
Swan Hills
Smith
Athabasca
Colinton
Boyle
Rochester
Flatbush
Fawcett
Dapp
Westlock
Barrhead
Pickardville
Ft. Assiniboine
Blue Ridge
Whitecourt
Mayerthorpe
Sangudo
Smoky L.
Thorhild
Waskatenau
Redwater
Legal
Andrew
Willingdon
Lamont
ELK I. N.P.
Ft. Saskatchewan
Sherwood Pk.
Edmonton
Tofield
Ryley
Devon
Morinville
St. Albert
Stony Plain
Entwistle
Evansburg
Wildwood
Edson
Bickerdike
Marlboro
Hinton
Athabasca
N. Saskatchewan R.
Derwent
Myrnam
Two Hills
Clandonald
Mundare
Vegreville
Innisfree
Mannville
Vermilion
Kitscoty
Marwayne
Peace River
Nampa
Grimshaw
Berwyn
Manning
North Star
River
Peace
Hines Creek
Whitelaw
Bluesky
Fairview
Girouxville
Falher
Donnelly
McLennan
Eaglesham
Wanham
High Prairie
Valleyview
Fox Creek
River
Smoky
Little Smoky
River
Grande Prairie
Sexsmith
Rycroft
Spirit River
Woking
Beaverlodge
Wembley
Hythe
Gordondale
Dawson Creek
Wapiti R.
Grande Cache
WILLMORE WILDERNESS PARK
Mt. Robson 12,972
CONTINENTAL
R
BR. COLUMBIA
110°
112°
114°
116°
118°
120°
54°
56°
A
B
C
D
E
1
2

ALBERTA
SASKATCHEWAN
MONTANA
IDAHO
WASH.
BR. COL.
Medicine Hat
Calgary
Red Deer
Lethbridge
Drumheller
Banff
Jasper
Wetaskiwin
Camrose
Wainwright
Lacombe
Ponoka
Stettler
Hanna
Brooks
Redcliff
High River
Claresholm
Ft. Macleod
Pincher Cr.
Cardston
Raymond
Taber
Coaldale
Innisfail
Olds
Rocky Mtn. House
Coleman
Blairmore
BANFF NAT'L PK.
KOOTENAY NAT'L PK.
YOHO NAT'L PK.
JASPER NATIONAL PARK
WATERTON LAKES NAT'L PARK
WATERTON GLACIER INT'L PEACE PK.
GLACIER NAT'L PK.
CYPRESS HILLS PROV. PK.
WELLS GRAY PROV PARK
HAMBER PROV. PK.
Mt. Eisenhower 9,030
Mt. Columbia 12,294
MT. ASSINIBOINE PROV. PK. 11,870
Saskatchewan River
Red Deer River
Bow R.
Oldman River
Milk River
Battle R.
Columbia River
Pakowki L.
Pigeon L.
Gull L.
Big Bend Res.
MILES
KILOMETERS
Provincial Capital
ALBERTA NORTHERN PART
N.W. TERRITORIES
WOOD BUFFALO NATIONAL PARK
CARIBOU MTS.
BIRCH MTS.
Lake Athabasca
Slave R.
Peace River
Hay River
Ft. Vermilion
Ft. Chipewyan
Bitumount
Meander River
Chateh
Habay
Keg River
© C.S. HAMMOND & Co., Maplewood, N.J.
Longitude West of Greenwich

BRITISH
COLUMBIA
MILES
0 25 50 100 150
KILOMETERS
0 25 50 100 150
Provincial and State Capitals
PACIFIC OCEAN
QUEEN CHARLOTTE SOUND
QUEEN CHARLOTTE ISLANDS
HECATE STRAIT
Dixon Entrance
COAST MTS.
SKEENA MTS.
HAZELTON MTS.
OMINECA MTS.
ROCKY MTS.
CONTINENTAL DIVIDE
COAST MOUNTAINS
ALASKA
Kupreanof I.
Petersburg
Mitkof I.
Zarembo I.
Wrangell
Etolin I.
Prince of Wales I.
Revillagigedo I.
Ketchikan
Dall I.
Iskut R.
Premier
Stewart
Portland Canal
Alice Arm
Nass R.
Aiyansh
Kincolith
Kitwanga
Hazelton
New Hazelton
Portland Inlet
Port Simpson
Usk
Terrace
Skeena R.
Dundas I.
C. Knox
Prince Rupert
Port Edward
Port Essington
Masset
Graham I.
Port Clements
Porcher I.
McCauley I.
Pitt I.
Queen Charlotte
Skidegate
Banks I.
Moresby I.
Kitimat
Kemano
Butedale
Gil I.
Princess Royal I.
Aristazabal I.
Klemtu
Ocean Falls
Bella Coola
Hagensborg
King I.
Campbell I.
Hunter I.
Kunghit I.
C. St. James
Calvert I.
Rivers Inlet
Queen Charlotte Str.
Mt. Waddington 13,260
Simoom Sound
C. Scott
Port Hardy
Quatsino
Sointula
Quatsino Sd.
Alert Bay
Port Alice
Kelsey Bay
C. Cook
VANCOUVER ISLAND
Campbell River
Refuge Cove
Powell River
Tahsis
STRATHCONA PROV. PARK
Nootka I.
Nootka Sd.
Courtenay
Comox
Cumberland
Flores I.
Port Alberni
Tofino
Ucluelet
Barkley Sd.
Bamfield
Nanaimo
Ladysmith
Duncan
Esquimalt
Victoria
C. Flattery
Juan de Fuca Str.
N. Vancouver
Vancouver
Ladner
New West.
White Rock
Bell
Anacor
Squamish
Britannia Beach
Pt. Coquitla
Haney
Mis
GARIBALDI PROV. PK.
Lillooet R.
Gold Bridge
Bralorne
Chilko L.
Alkali Lake
Williams Lake
Chilcotin R.
Fraser R.
Narcosli Creek
Quesnel
Prince George
South Ft. George
Giscome
Aleza Lake
Willow River
Vanderhoof
Fraser Lake
Fort Fraser
KENNEY DAM
Tetachuck L.
TWEEDSMUIR PROV. PARK
Dean R.
NANIKA DAM
François Lake
François L.
Burns Lake
Decker Lake
Houston
Quick
Telkwa
Smithers
Babine L.
Stuart L.
Fort St. James
Takla L.
Manson Cr.
Finlay Forks
Peace R.
Parsnip R.
Ft. Grahame
Finlay R.
Ware
Thutade L.
Sikanni Chief R.
Sikanni River
Chet
132°
128°
124°
Longitude West
A
B
C
D

BRITISH COLUMBIA
NORTHERN PART
YUKON TERR.
N. W. T.
PACIFIC OCEAN
Juneau
Sitka
Atlin L.
Tulsequah
Cassiar
Lower Post
Liard R.
Old Ft. Nelson
Ft. Nelson
MUNCHO LAKE PROV. PK.
GLACIER BAY NAT'L MON.
Stikine R.
CONTINENTAL DIVIDE
Vancouver
N. Vancouver
W. Vancouver
Port Moody
Pt. Coquitlam
New Westminster
Burnaby
Richmond
Delta
Surrey
Langley
White Rock
Mission City
Haney
Hatzic
Chilliwack
Agassiz
GARIBALDI PROV. PK.
Pt. Hammond
Lynn Cr.
Roberts Cr.
Gibson's Ldg.
Qualicum Beach
Parksville
Alberni
Port Alberni
Departure Bay
Nanaimo
Cassidy
Ladysmith
Chemainus
Youbou
Lake Cowichan
Duncan
Sidney
Centre Saanich
Saanich
San Juan I.
Victoria
Esquimalt
Sooke
Port Renfrew
Juan de Fuca Strait
VANCOUVER ISLAND
Alberni Inlet
Cowichan L.
Bellingham
Orcas I.
Anacortes
WASH.
Str. of Georgia
Fort St. John
Taylor
Rolla
Pouce-Coupé
East Pine
Tupper
ALBERTA
WILLMORE WILDERNESS PARK
McBride
MT. ROBSON PROV. PK.
Valemount
JASPER NAT'L PARK
HAMBER PROV. PK.
Blue River
WELLS GRAY PROV. PK.
Quesnel L.
Clearwater
Little Fort
70 Mile House
Louis Cr.
Chase
Kamloops
Thompson R.
Salmon Arm
Enderby
Armstrong
Vernon
Okanagan L.
Merritt
Lytton
Kelowna
Peachland
Summerland
Penticton
Princeton
Hope
Hedley
Copper Mtn.
Oliver
Osoyoos
Greenwood
Grand Forks
Rossland
Trail
Warfield
Fruitvale
Kinnaird
Castlegar
Nelson
KOKANEE GLACIER PROV. PK.
Kaslo
Kootenay Lake
Nakusp
Arrow Lakes
Revelstoke
MT. REVELSTOKE NAT'L PK.
GLACIER NAT'L PK.
Golden
YOHO NAT'L PK.
BANFF NATIONAL PARK
Banff
KOOTENAY NAT'L PK.
MT. ASSINIBOINE PROV. PARK
Edgewater
Invermere
Canal Flats
Kimberly
Marysville
Cranbrook
Fernie
Sparwood
Creston
Grasmere
Claresholm
Taber
Lethbridge
Fort Macleod
WATERTON LAKES NAT'L PK.
WATERTON GLACIER INTERNAT'L PEACE PK.
GLACIER NATIONAL PARK
Whitefish
Kalispell
Cut Bank
MONTANA
IDAHO
WASHINGTON
Libby Res.
Kootenai R.
Sandpoint
Colville
Columbia R.
Okanogan
NORTH CASCADES NAT'L PARK
E. C. MANNING PROV. PK.
ROSS L. N.R.A.
CHELAN NAT'L REC. AREA
SELKIRK MTS.
MONASHEE MTS.
PURCELL MTS.
CARIBOO MTS.
ROCKY MTS.
©C. S. HAMMOND & Co., Maplewood, N. J.

YUKON TERRITORY
MILES
0 25 50 75 100 125
KILOMETERS
0 25 50 75 100 125
Territorial and State Capitals
BEAUFORT SEA
Demarcation Pt.
Herschel I.
Herschel
Mackenzie Bay
Firth R.
BRITISH MTS.
Old Crow R.
Old Crow
Porcupine
River
Eagle R.
RICHARDSON MOUNTAINS
McDougall Pass
Aklavik
Ft. McPherson
N. W. TERRS.
Kugmallit Bay
Richards I.
Tuktoyaktuk
Eskimo (Husky) Lakes
Reindeer Sta.
Sitidgi L.
Inuvik
Kugaluk R.
Liverpool Bay
Travaillant Lake
Arctic Red River
Iroquois R.
L. Belot
L. des Bois
Mackenzie
Arctic Circle
Arctic Red River
Ft. Good Hope
Smith Arm
Great Bear Lake
Ft. Franklin
FRANKLIN
Norman Wells
Canol
Ft. Norman
Mountain R.
MACKENZIE
ROCKY
Snake R.
Bonnet Plume R.
Peel River
Hart R.
Chapman Lake
Mt. Burgess 5,580
OGILVIE R.
Ogilvie
MOUNTAINS
ALASKA
Yukon
Eagle
70°
140°
135°
130°
125°
65°
A B C D E
1 2

MOUNTAINS
NAHANNI NAT'L PARK
N.W. TERRS.
Wrigley
Redstone
Keele
Christie Pass
South Nahanni
Mt. Sir James MacBrien 9,000
Hyland R.
Frances Lake
Coal R.
Smith River
Liard River
MUNCHO LAKE PROV. PK.
STONE MTN. PROV. PK.
Kechika R.
Watson Lake
Upper Liard
Lower Post
Cassiar
McDame
Dease Lake
CASSIAR MTS.
Longitude West 130° of Greenwich
Liard
Swift River
Keele Pk. 9,750
Macmillan Pass
Mt. Selous 7,140
Ross River
Faro
Hess R.
Macmillan R.
Teslin
Johnsons Crossing
Teslin Lake
YUKON VALLEY
Big Salmon R.
Big Salmon
Lower Laberge
Upper Laberge
Teslin R.
Whitehorse
Tagish Lake
Atlin
Atlin Lake
Tulsequah
Juneau
ALASKA
Keno Hill
Wernecke
Elsa
Mayo
Stewart River
Clear Creek
Barlow
McQuesten
Pelly Crossing
Pelly
Minto
McCabe Crossing
Yukon Crossing
Carmacks
Yukon
L. Laberge
Takhini
MacRae
Robinson
Carcross
Kusawa L.
BRITISH COLUMBIA
Chilkoot Pass
Skagway
Haines
Champagne
Canyon
Aishihik L.
Aishihik
Klukshu
Chilkat Pass
GLACIER BAY NAT'L MON.
Glacier Bay
Dawson
Klondike R.
Sulphur
Granville
Bear Cr.
Stewart River
Ft. Selkirk
Coffee Creek
Nisling R.
Kluane Lake
Kluane
Kluane Jct.
SHAKWAK VALLEY
ST. ELIAS MOUNTAINS
KLUANE NAT'L PARK
Burwash Landing
Destruction Bay
Alsek
Yakutat
Yakutat Bay
GULF OF ALASKA
Donjek
Koidern
Snag
White
Beaver Creek
Glacier Creek (Sixtymile)
Mt. Lucania 17,150
Mt. Logan 19,850
Mt. St. Elias 18,008
© C.S. HAMMOND & Co., Maplewood, N.J.
60°
125°
135°
140°

NORTHWEST TERRITORIES
MILES
0 50 100 200 300 400
KILOMETRES
0 50 100 200 300 400
State and Territorial Capitals
ARCTIC OCEAN
Beaufort Sea
ELIZAB
ISLANDS
BROOKS RA.
UNITED STATES
ALASKA
Yukon River
Fairbanks
Barrow
Pt. Barrow
Herschel Bay
Mackenzie Bay
Tuktoyaktuk
Aklavik
Old Crow
Inuvik
Ft. McPherson
Mackenzie
Ft. Good Hope
Norman Wells
Ft. Franklin
Ft. Norman
Great Bear Lake
Port Radium
Dawson
Mayo
Elsa
Keele Pk. 9,720
Macmillan Pass
Ross R.
Pelly
Beaver Cr.
Destruction Bay
Carmacks
Kluane Jct.
KLUANE NAT'L PK.
Mt. Logan 19,850
Whitehorse
Carcross
Skagway
Teslin
GLACIER BAY N.M.
Juneau
Petersburg
YUKON TERR.
MACKENZIE MTS.
DISTRICT OF MACKENZIE
Wrigley
River
Mt. Sir James MacBrien 9,062
NAHANNI NAT'L PARK
Ft. Simpson
Watson L.
BRITISH COLUMBIA
ROCKY MTS.
Liard
Ft. Liard
MUNCHO L. PROV. PK.
Ft. Nelson
Lac la Martre
Rae-Edzo
Yellowknife
Ft. Providence
Great Slave Lake
Snowdrift
Hay River
Pine Pt.
Ft. Resolution
Slave R.
Nonacho L.
WOOD BUFFALO NAT'L PK.
Ft. Smith
ALBERTA
Meander R.
SASKATCHEWAN
Uranium City
Wholdaia L.
Thelon
West Longitude
Contwoyto L.
Back
Arctic Circle
Coppermine
Coronation Gulf
Bluenose L.
Paulatuk
C. Bathurst
C. Lambton
Amundsen Gulf
Holman I.
DISTRICT
Victoria Island
Wollaston Pen. 1,700
Cambridge Bay
Queen
Banks I.
Sachs Har.
C. Pr. Alfred
M'Clure Str.
Parry
Viscount Melville Sound
Hadley Bay
Stefansson I.
Prince Patrick I.
Mould Bay
Lands End
Borden I.
Mackenzie King I.
Isachsen
Hazen Str.
PARRY ISLANDS
Melville I.
North Magnetic Pole
6,500
7,400
160°
140°
120°
A
B
C
2
3
4
70°
60°

GREENLAND
(Denmark)
Baffin Bay
Davis Strait
Baffin Island
FRANKLIN
DISTRICT OF KEEWATIN
Hudson Bay
Hudson Strait
Foxe Basin
Foxe Chan.
Ungava Peninsula
Ungava Bay
QUEBEC
MANITOBA
Ellesmere Island
UNITED STATES RA.
Axel Heiberg I.
Devon I.
Lancaster Sd.
Jones Sd.
Gulf of Boothia
Boothia Pen.
Melville Pen.
Southampton I.
Coats I.
Mansel I.
Nottingham I.
Resolution I.
Akpatok I.
Bylot I.
Somerset I.
Prince of Wales I.
King William I.
Prince Charles I.
Cumberland Sd.
Frobisher B.
AUYUITTUQ NAT'L PARK
Kane Basin
Nares Str.
Kennedy Chan.
Smith Sd.
Hayes Pen.
Melville Bay
Thule
Grise Fiord
Resolute Bay
Pond Inlet
Arctic Bay
Clyde
Igloolik
Repulse Bay
Pangnirtung
Frobisher Bay
Lake Harbour
Chesterfield Inlet
Rankin Inlet
Whale Cove
Eskimo Pt.
Baker L.
Povungnituk
Saglouc
Ft-Chimo
Eureka
Alert
Gjoa Haven
Nouveau-Québec Crator
All islands in Hudson and James Bays lie within the District of Keewatin.
© C.S. HAMMOND & Co., Maplewood, N.J.

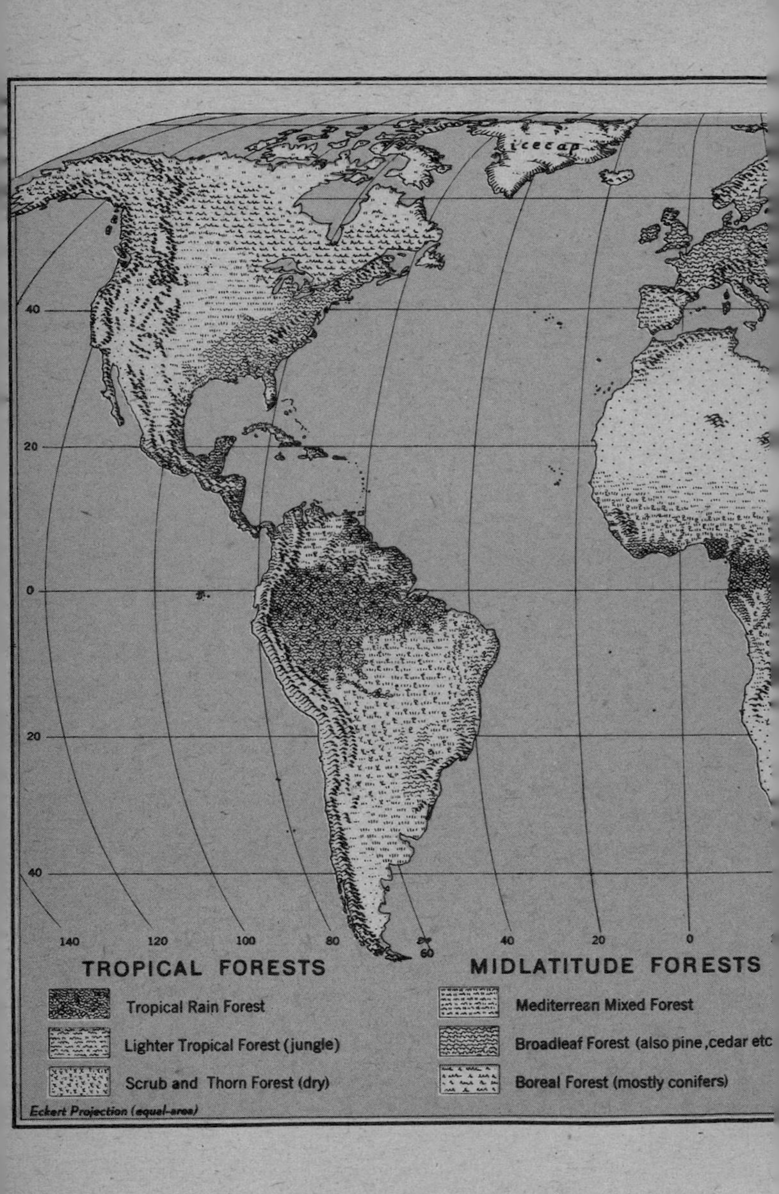
icecap
40
20
0
20
40
140
120
100
80
60
40
20
0
TROPICAL FORESTS
Tropical Rain Forest
Lighter Tropical Forest (jungle)
Scrub and Thorn Forest (dry)
MIDLATITUDE FORESTS
Mediterrean Mixed Forest
Broadleaf Forest (also pine, cedar etc
Boreal Forest (mostly conifers)
Eckert Projection (equal-area)

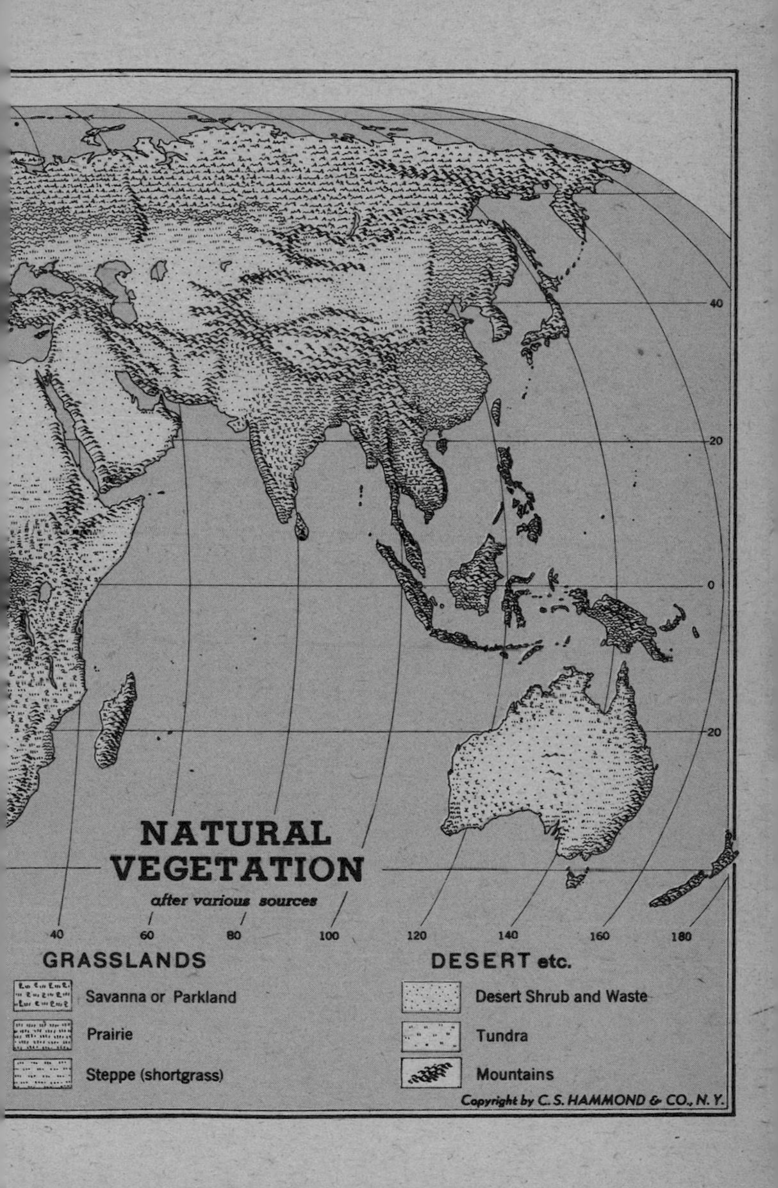
40
20
0
20
NATURAL
VEGETATION
after various sources
40
60
80
100
120
140
160
180
GRASSLANDS
Savanna or Parkland
Prairie
Steppe (shortgrass)
DESERT etc.
Desert Shrub and Waste
Tundra
Mountains
Copyright by C. S. HAMMOND & CO., N. Y.

40
20
0
20
40
140
120
100
80
60
40
20
0
very wet
humid
subhumid
This classification is based on effective rain or snowfall, taking into account faster evaporation in warmer climates.
For temperature conditions the latitude, ocean currents etc. have also to be considered.
Eckert Projection (equal-area)

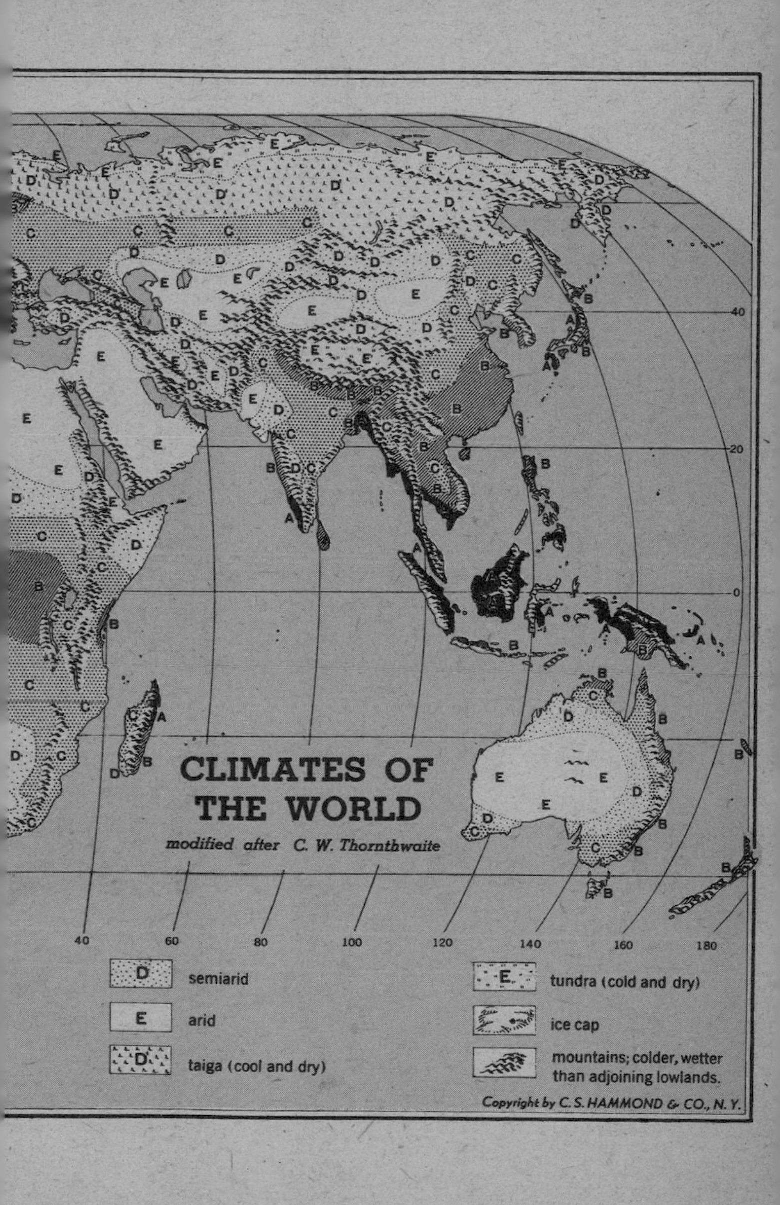
CLIMATES OF THE WORLD
modified after C. W. Thornthwaite
40
60
80
100
120
140
160
180
40
20
0
D semiarid
E arid
D taiga (cool and dry)
E tundra (cold and dry)
ice cap
mountains; colder, wetter than adjoining lowlands.
Copyright by C. S. HAMMOND & CO., N. Y.

DENSITY OF POPULATION. *One of the most outstanding facts of human geography is the extremely uneven distribution of people over the Earth. One-half of the Earth's surface has less than 3 people per square mile, while in the lowlands of India, China, Java and Japan rural density reaches*

the incredible congestion of 2000-3000 per square mile. Three-fourths of the Earth's population live in four relatively small areas; Northeastern United States, North-Central Europe, India and the Far East.

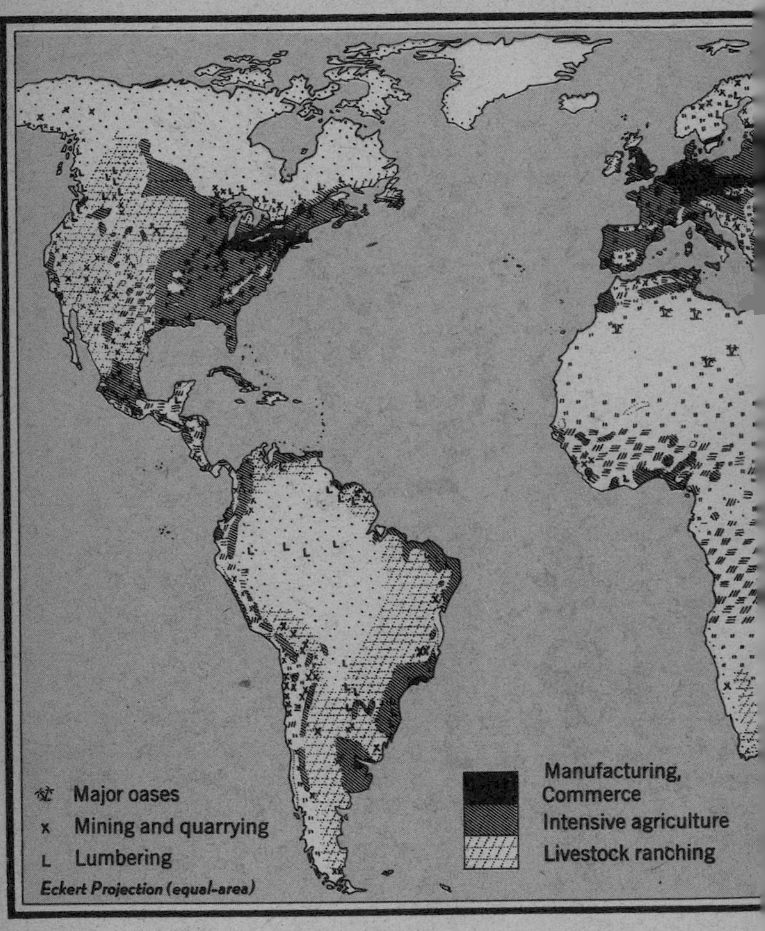

OCCUPATIONS. *Correlation with the density of population shows that the most densely populated areas fall into the regions of manufacturing and intensive farming. All other economies require considerable space. The most*

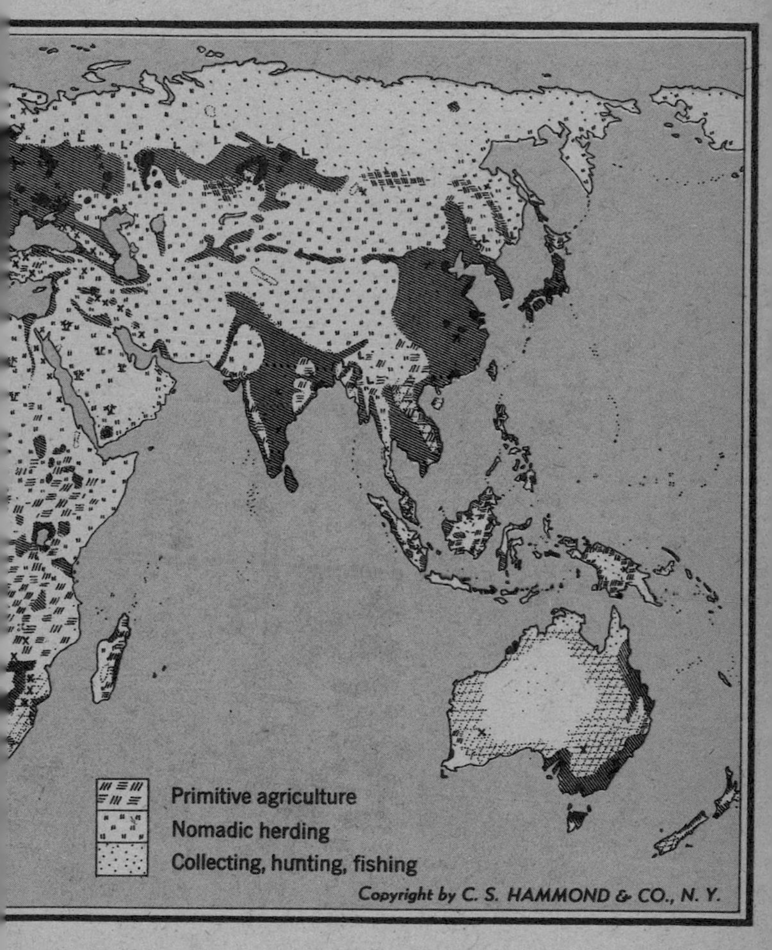

sparsely inhabited areas are those of collecting, hunting and fishing. Areas with practically no habitation are left blank.

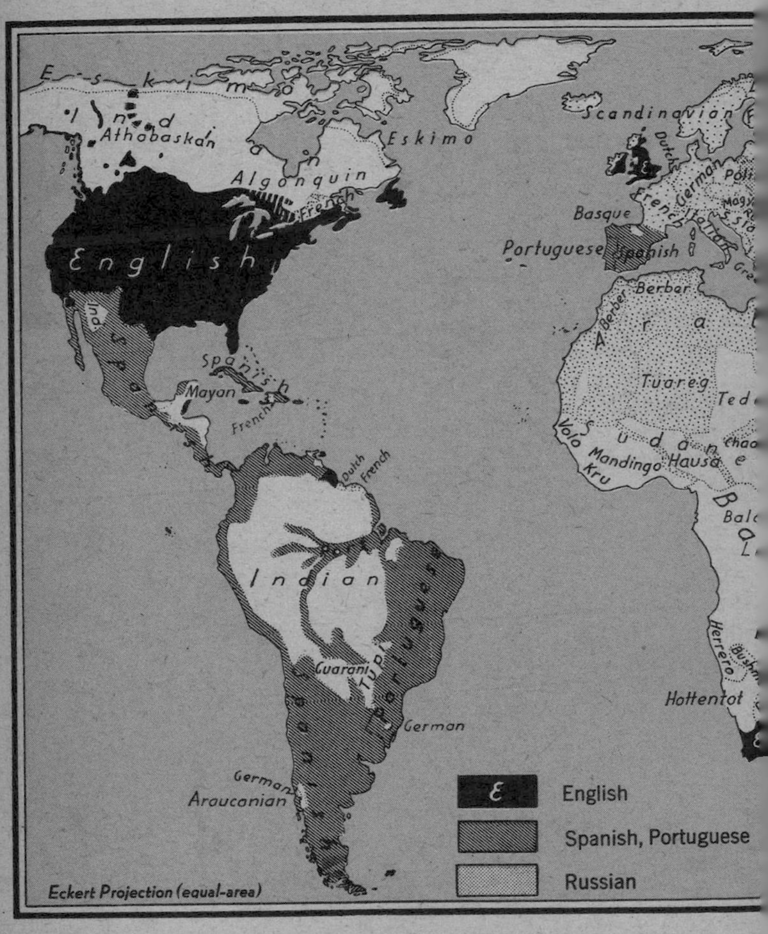

LANGUAGES. *Several hundred different languages are spoken in the World, and in many places two or more languages are spoken, sometimes by the same people. The map above shows the dominant languages in each*

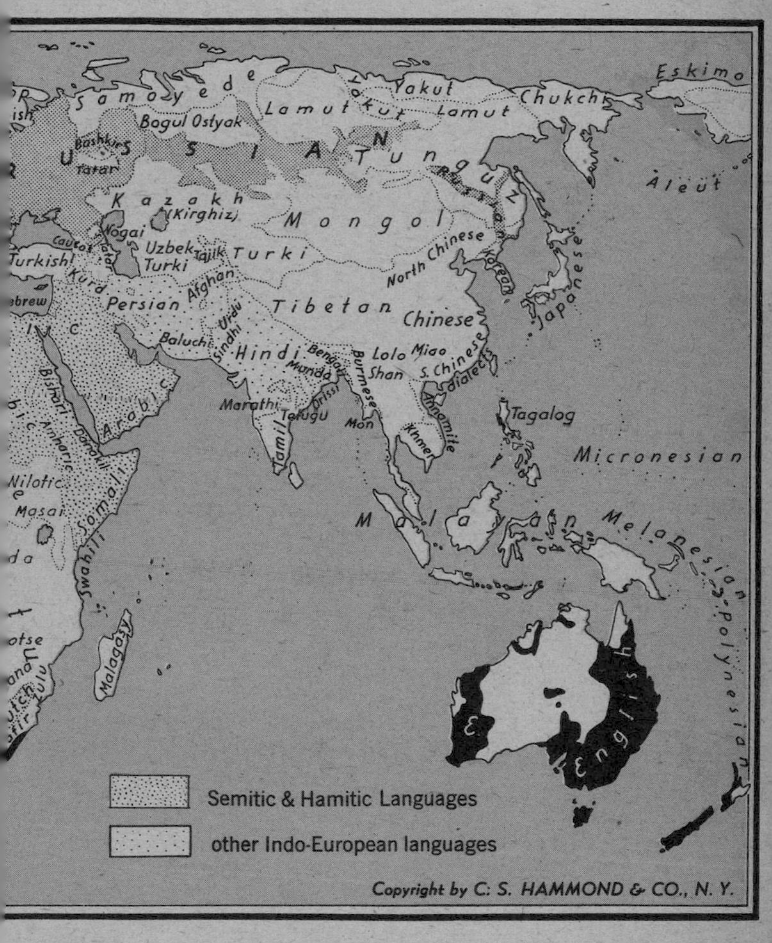

locality. English, French, Spanish, Russian, Arabic and Swahili are spoken by many people as a second language for commerce or travel.

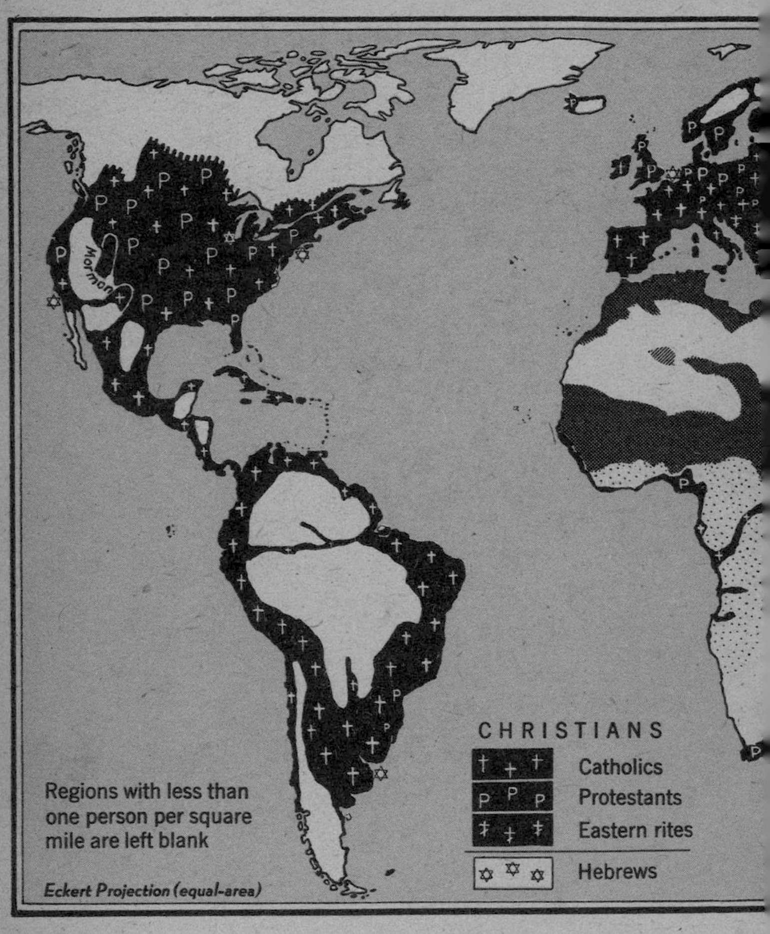

RELIGIONS. *Most people of the Earth belong to four major religions: Christians, Mohammedans, Brahmans, Buddhists and derivatives. The Eastern rites of the Christians include the Greek Orthodox, Greek Catholic, Armenian, Syrian, Coptic and more minor churches. The lamaism of Tibet and Mongolia*

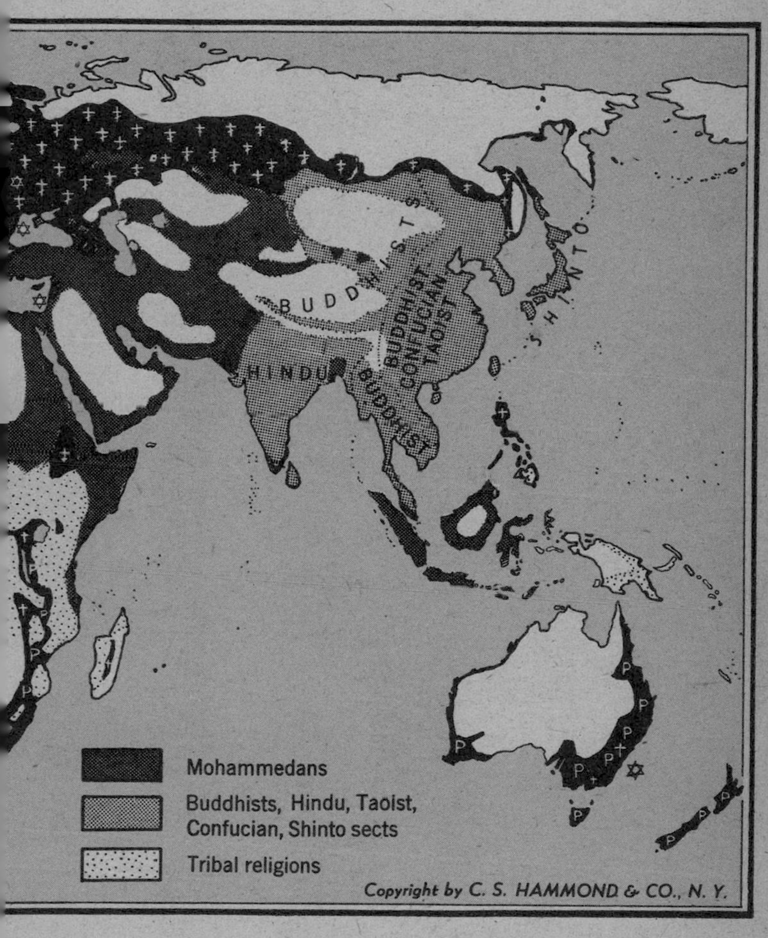

differs a great deal from Buddhism in Burma and Thailand. In the religion of China the teachings of Buddha, Confucius and Tao are mixed, while in Shinto a great deal of ancestor and emperor worship is added. About 11 million Hebrews live scattered over the globe, chiefly in cities and in the state of Israel.

INDEX OF THE WORLD

This alphabetical list gives statistics of population based on the latest official reports. Each line begins with the name of a place or feature, followed by the name of the country or state, the population, the index reference and the page number.

Capitals are designated by asterisks* † Including suburbs

Aachen, Germany, 242,453 ....B 3 35
Abadan, Iran, 340,000 ....F 5 62
Abeokuta, Nigeria, 253,000 ....E 7 79
Aberdeen, Scotland, 210,362 ....F 2 30
Abidjan,* Ivory Coast, 408,000............C 7 79
Abilene, Tex., 89,653 ....C 3 198
Abruzzi (reg.), Italy ....D 3 42
Abu Dhabi,* United Arab Emirates, 85,000 ....G 5 57
Acapulco, Mexico, 352,673 ....E 5 103
Accra,* Ghana, 564,194 ....D 7 79
Aconcagua (mt.), Argentina ....B 4 98
Adalia (gulf), Turkey ....D 4 58
Adana, Turkey, 383,046 ....F 4 58
Adapazari, Turkey, 101,283 ....D 2 58
Addis Ababa,* Ethiopia, 1,161,267..E 6 81
Adelaide, Australia, †809,482 ....D 7 87
Aden,* P.D.R. Yemen, 264,326 ....F 7 56
Adirondack (mts.), N.Y. ....F 1 177
Adriatic (sea), Europe ....D 2 42
Aegean (sea) ....D 4 49
Agra, India, 591,917 ....C 2 64
Aguascalientes, Mexico, 213,428....D 4 102
Ahmadabad, India, 1,585,544 ....C 2 64
Ahmadnagar, India, 118,236 ....C 3 64
Ahwaz, Iran, 350,000 ....F 5 62
Aix-en-Provence, France, 74,948....F 6 39
Ajaccio, France, 38,776 ....B 7 39
Ajmer, India, 262,851 ....C 2 64
Akita, Japan, 261,242 ....E 3 71
Akola, India, 168,438 ....C 2 64
Akron, Ohio, 275,425 ....D 2 182
Aktyubinsk, U.S.S.R., 150,000............B 4 52
Alameda, Calif., 70,968 ....E 2 122
Alaska (gulf), Alaska ....E 4 117
Albacete, Spain, 93,233 ....F 3 41
Albany, Ga., 72,623 ....B 5 133
Albany,* N.Y., 115,781 ....G 3 177
Albert (lake), Africa ....C 7 81
Ålborg, Denmark, 154,582 ....B 3 33
Albuquerque, N. Mex., 243,751......B 2 174
Alcoy, Spain, 61,371 ....F 3 41
Aleppo, Syria, 710,636 ....G 4 59
Alessandria, Italy, 78,644 ....B 2 42
Aleutian (isls.), Alaska ....B 5 116
Alexandretta (Iskenderun), Turkey, 79,297 ....G 4 59
Alexandria, Egypt, 2,259,000 ....C 2 80
Alexandria, Va., 110,938 ....G 1 205
Algeciras, Spain, 81,662 ....D 4 40
Algiers,* Algeria, 943,551 ....E 1 78
Alhambra, Calif., 62,125 ....E 3 122
Alicante, Spain, 184,716 ....F 3 41
Aligarh, India, 252,314 ....C 2 64
Alkmaar, Netherlands, 65,199 ....E 2 36
Al Kuwait,* Kuwait, 108,211 ....F 4 56
Allahabad, India, 490,622 ....D 2 64
Allentown, Pa., 109,527 ....F 3 189
Alleppey, India, 160,166 ....C 4 64
Alma-Ata, U.S.S.R., 851,000 ....D 5 52
Almería, Spain, 114,510............E 4 41
Alor Setar, Malaysia, 66,260 ....B 6 67
Alps (mts.), Europe ....F 4 28
Altay (mts.), Asia ....B 1 68
Altoona, Pa., 63,115 ....C 3 188
Amagasaki, Japan, 545,762 ....E 4 71
Amarillo, Tex., 127,010 ....A 2 198
Amazon (river), S. America ....A 4 96
Ambato, Ecuador, 77,052 ....B 4 94
Amboina, Indonesia, 79,636 ....H 6 75
Amersfoort, Neth., 88,440............F 3 36
Amiens, France, 116,107 ....D 3 38
Amman,* Jordan, 615,000 ....D 4 61
Amoy, China, 400,000 ....E 3 69
Amravati, India, 193,800 ....C 2 64
Amritsar, India, 407,628 ....C 1 64
Amsterdam,* Neth., 751,156 ....E 3 36
Amu-Dar'ya (river), Asia ....J 1 57
Amur (river), Asia ....G 1 69
Anaheim, Calif., 166,408 ....F 3 122
Anchorage, Alaska, 48,081 ....E 3 117
Ancona, Italy, 88,427 ....D 3 42
Andaman (isls.), India ....E 3 65
Anderlecht, Belgium, 103,796 ....D 6 37
Anderson, Ind., 70,787 ....D 2 140
Andes (mts.), S. America ....B 4 98
Andizhan, U.S.S.R., 188,000 ....D 5 52
Andorra la Vella, *Andorra, 12,000 ....G 1 41
Angel (fall), Venezuela ....E 2 94
Angers, France, 127,415 ....C 4 38
Ankara,* Turkey, 1,461,345 ....E 3 58
Annaba, Algeria, 152,423 ....F 1 78
An Najaf, Iraq, 134,027 ....C 4 62
Annapolis,* Md., 30,095 ....D 3 153
An Nasiriya, Iraq, 143,471............D 5 62
Anshan, China, 1,050,000 ....F 1 69
Antâkya, Turkey, 66,520 ....G 4 59
Antalya, Turkey, 95,616 ....D 4 58
Antananarivo,* Madagascar, 351,262 ....K 5 83
Antofagasta, Chile, 149,720 ....B 2 98
Antung, China, 450,000 ....F 1 69
Antwerp, Belgium, 224,543 ....D 5 37
Anyang, China, 225,000 ....E 2 69
Anzhero-Sudzhensk, U.S.S.R., 106,000 ....E 4 52
Aomori, Japan, 264,187 ....F 2 71
Apeldoorn, Netherlands, 134,055....G 3 36
Apia,* W. Samoa, †32,616............J 7 85

Ellice (isls.), Tuvalu ....H 6 84
El Minya, Egypt, 112,800 ....C 2 80
El Obeid, Sudán, 90,000 ....D 5 81
El Paso, Tex., 322,261 ....A 6 199
Eluru, India, 127,023 ....D 3 64
Elyria, Ohio, 53,427 ....C 2 182
Enfield, England, 260,900 ....B 5 31
Engels, U.S.S.R., 130,000 ....G 4 51
English (channel), Europe ....E 6 31
Enschede, Neth., 141,597 ....J 3 36
Ensenada, Mexico, 77,687 ....A 1 102
Enugu, Nigeria, 187,000 ....F 7 79
Epsom and Ewell, Eng., 70,700 ....B 6 31
Erfurt, Germany, 202,979 ....D 3 35
Erie (lake), N. Amer. ....K 2 111
Erie, Pa., 129,231 ....A 1 188
Eritrea (reg.), Ethiopia ....E 4 81
Erivan, U.S.S.R., 956,000 ....F 6 51
Erlangen, Germany, 100,671 ....D 4 35
Erzurum, Turkey, 133,444 ....J 3 59
Esbjerg, Denmark, 68,097 ....B 3 33
Esçuintla, Guat., 64,851 ....B 2 104
Esher, England, 63,970 ....B 6 31
Eskilstuna, Sweden, 92,628 ....C 3 33
Eskişehir, Turkey, 243,328 ....D 3 58
Esmeraldas, Ecuador, 60,132 ....A 3 94
Essen, Ger., 677,568 ....B 3 34
Essequibo (river), Guyana ....A 1 96
Esslingen, Germany, 95,298 ....C 4 35
Estremadura (reg.), Spain ....C 3 40
Etna (mt.), Italy ....E 6 43
Euclid, Ohio, 71,552 ....D 2 182
Eugene, Oreg., 78,389 ....B 2 186
Euphrates (river), Asia ....B 3 62
Evanston, Ill., 79,808 ....E 1 138
Evansville, Ind., 138,764 ....B 4 141
Everest (mt.), Asia ....D 2 65
Everett, Wash., 53,622 ....D 2 206
Everglades (swamp), Fla. ....C 3 131
Exeter, England, 93,300 ....E 5 31
Eyüp, Turkey, 86,384 ....D 6 58
Ez Zarqa', Jordan, 226,000 ....E 3 61

Faizabad, India, 102,835 ....D 2 64
Fall River, Mass., 96,898 ....E 3 155
Fargo, N. Dak., 53,365 .... H 3 181
Fear (cape), N.C. ....E 4 179
Ferrara, Italy, 97,507 ....C 2 42
Fez, Morocco, 321,460 ....D 2 78
Finland (gulf), Europe ....E 3 33
Flensburg, Germany, 93,213 ....C 1 34
Flint, Mich., 193,317 ....D 3 157
Florence, Italy, 465,823 ....C 3 42
Florianópolis, Brazil, 115,665 ....G 3 98
Florida (straits) ....K 6 111
Foggia, Italy, 136,436 ....E 4 43
Foochow, China, 680,000 ....E 3 69
Forlì, Italy, 83,303 ....C 2 42
Formosa (Taiwan) (isl.), Asia.........F 3 69
Fortaleza, Brazil, 520,175 ....E 3 96
Fort-de-France,* Martinique,
99,000 ....F 4 107
Forth (firth), Scotland ....E 2 30
Fort Lauderdale, Fla., 139,590 ......A 2 130
Fort Wayne, Ind., 178,021 ....D 1 140
Fort Worth, Tex., 393,476 ....D 2 198
Franca, Brazil, 86,852 ....C 7 97
Frankfort,* Ky., 21,902 ....D 2 147
Frankfurt am Main, Germany,
636,157 ....C 3 35
Frankfurt an der Oder, Germany,
70,817 ....F 2 34
Franz Josef Land (isls.),
U.S.S.R. ....L 1 27
Fraser (river), B.C. ....D 2 232
Fredericton,* New Brunswick,
45,248 ....D 3 218
Freetown,* S. Leone, 214,443 ....B 7 79
Freiburg, Germany, 175,371 ....B 5 35
Fremont, Calif., 100,869 ....E 2 122
Fresno, Calif., 165,972 ....D 4 123
Friesland (reg.), Neth. ....G 1 36
Frio (cape), Brazil ....D 7 97
Frunze, U.S.S.R., 430,600 ....D 5 52
Fuji (mt.), Japan ....E 4 71
Fukui, Japan, 231,365 ....D 3 70
Fukuoka, Japan, 1,002,214 ....C 4 70
Fukushima, Japan, 246,531 ....F 3 71
Fukuyama, Japan, 329,779 ....D 4 70
Fullerton, Calif., 85,987 ....F 3 122
Funabashi, Japan, 423,106 ....H 2 71
Funchal,* Madeira, 38,340 ....A 2 40
Fundy (bay), Canada ....B 2 216
Fürth, Germany, 101,639 ....D 4 35
Fushun, China, 1,080,000 ....F 1 69
Fusin, China, 350,000 ....F 1 69

Gaborone,* Botswana, 21,000 ....E 6 82
Gadag, India, 95,426 ....C 3 64
Gadsden, Ala., 53,928 ....D 1 114
Galápagos (isls.), Ecuador ....F 6 27
Galaţi, Rumania, 197,853 ....D 2 48
Galicia (reg.), Spain ....B 1 40
Galilee (reg.), Israel ....C 2 60
Galle, Sri Lanka, 72,720 ....C 4 64
Galveston, Tex., 61,809 ....E 7 199
Galway, Ireland, 27,726 ....B 4 31
Gander, Newfoundland, 9,301 ....C 4 215
Ganges (river), Asia ....D 2 64
Garden Grove, Calif., 121,371 ....E 3 122
Garland, Texas, 81,437 ....E 1 198
Garmisch-Partenkirchen, Germany,
26,831 ....D 5 35
Gary, Ind., 175,415 ....B 1 140
Gaspé (pen.), Quebec ....G 4 223
Gateshead, England, 222,000 ....E 3 30
Gauhati, India, 123,783 ....E 2 65
Gävle, Sweden, 86,934 ....C 2 33
Gaya, India, 179,884 ....D 2 64
Gaza, Egypt, 118,272 ....A 5 61
Gaziantep, Turkey, 275,000 ....G 4 59
Gdańsk, Poland, 402,200 ....D 1 44
Gdynia, Poland, 211,900 ....D 1 44
Gela, Italy, 66,845 ....E 6 43

Gelsenkirchen, Germany, 322,584....B 3 34
Geneva, Switz., 159,200 ....A 1 42
Genoa, Italy, 805,855 ....B 2 42
Georgetown,* Guyana, 100,855 ......A 1 96
Georgian (bay), Ont. ....B 2 224
Gera, Germany, 113,108 ....E 3 35
Germiston, S. Africa, 221,972 ....F 7 83
Ghent, Belgium, 148,860 ....C 5 37
Gibraltar,* Gibraltar, 26,833 ....D 5 40
Gibraltar (strait) ....D 5 40
Giessen, Germany, 75,481 ....C 3 35
Gifu, Japan, 408,699 ....E 4 70
Gijón, Spain, 187,612 ....D 1 40
Glacier Nat'l Park, Mont. ....B 1 164
Glasgow, Scotland, 880,617 ....B 1 30
Glendale, Calif., 132,752 ....E 3 122
Gliwice, Poland, 171,800 ....A 4 44
Gloucester, England, 90,550 ....F 5 31
Gobi (desert), Asia ....D 1 68
Godavari (river), India ....C 3 64
Goiânia, Brazil, 362,152 ....C 6 97
Gomel, U.S.S.R., 324,000 ....D 4 51
Gómez Palacio, Mexico, 79,650 ....D 2 102
Gorakhpur, India, 230,911 ....D 2 64
Gorki, U.S.S.R., 1,305,000 ....F 3 50
Görlitz, Germany, 84,658 ....F 3 35
Gorlovka, U.S.S.R., 341,000 ....E 5 51
Gorontalo, Indonesia, 82,320 ....G 5 75
Gorzów Wielkopolski, Poland,
76,200 ....B 2 44
Gosport, England, 82,830 ....F 5 31
Göteborg, Sweden, 442,236 ....B 3 33
Gotha, Germany, 59,243 ....D 3 35
Göttingen, Germany, 123,797 ....D 3 34
Gottwaldov, Czech., 84,300 ....D 2 47
Granada, Spain, 190,429 ....E 5 41
Granby, Quebec, 37,132 ....D 4 222
Grand (canal), China ....E 2 69
Grand Canyon Nat'l Park, Ariz. ....B 1 118
Grande, Rio (river), N. Amer. ......F 5 110
Grand Rapids, Mich., 197,649 ....B 3 157
Graz, Australia, 251,900 ....C 3 46
Great Barrier (reef), Austral. ......H 3 88
Great Bear (lake), N.W.T. ....B 3 236
Great Falls, Mont., 60,091 ....C 2 164
Great Salt (lake), Utah ....B 2 200
Great Slave (lake), N.W.T. ....C 3 236
Great Smoky (mts.), U.S. ....B 4 178
Great Wall, China ....D 2 68
Green (mts.), Vt. ....B 4 203
Green Bay, Wis., 87,809 ....F 4 211
Greenock, Scotland, 67,275 ....A 1 30
Greensboro, N.C., 144,076 ....C 1 178
Greenville, S.C., 61,436 ....B 2 192
Greenwich, Eng., 209,800 ....C 5 31
Greifswald, Germany, 53,940 ....E 1 34
Grenoble, France, 161,616 ....F 5 39
Grimsby, England, 94,400..................G 4 31
Grodno, U.S.S.R., 161,000 ....B 4 51
Groningen, Neth., 163,357 ....H 1 36
Grozny, U.S.S.R., 369,000 ....G 6 51
Grudziądz, Poland, 76,600 ....D 2 44
Guadalajara, Mexico, 1,560,805 ....D 4 102
Guadalcanal (isl.), Sol. Is. ....F 6 84
Guadalquivir (river), Spain............C 4 40
Guadiana (river), Europe ....D 3 40
Guantánamo, Cuba, 145,000 ....C 2 106
Guatemala,* Guat., 700,538 ....B 2 104
Guayaquil, Ecuador, 814,064 ....A 4 94
Guelph, Ontario, 67,538 ....B 3 224
Guernsey (isl.), Channel Is. ....E 6 31
Guinea (gulf), Africa ....E 8 79
Gulbarga, India, 145,588 ....C 3 64
Guntur, India, 269,991 ....C 3 64
Gur'yev, U.S.S.R., 114,000 ....B 5 52
Gütersloh, Germany, 77,128 ....C 3 34
Gwalior, India, 384,772 ....C 2 64
Győr, Hungary, 114,709 ....D 3 47

Haarlem, Netherlands, 165,861 ......D 3 36
Hachinohe, Japan, 224,213 ....F 2 71
Hachioji, Japan, 322,558 ....H 2 71
Haeju, N. Korea, 140,000 ....B 3 70
Hagen, Germany, 229,224 ....B 3 34
Hague, The ('s Gravenhage),
*Netherlands, 482,879 ....D 3 36
Haifa, Israel, 225,000 ....B 2 60
Hainan (isl.), China ....D 4 68
Haiphong, Vietnam, 276,300 ....D 2 66
Hakodate, Japan, 307,447 ....F 2 71
Halifax, England, 88,580 ....G 1 30
Halifax,* Nova Scotia, 117,882 ......D 3 216
Halle, Germany, 241,425 ....E 3 34
Halmstad, Sweden, 74,296 ....C 3 33
Hälsingborg, Sweden, 101,136 ....C 3 33
Hama, Syria, 137,421 ....G 5 59
Hamadan, Iran, 150,000 ....F 3 62
Hamamatsu, Japan, 468,886 ....E 4 71
Hamburg, Germany, 1,717,383 ......D 2 34
Hameln, Germany, 61,066 ....C 2 34
Hamhung, N. Korea, 484,000 ....B 3 70
Hamilton, N. Zealand, 90,100 ....K 6 89
Hamilton, Ohio, 67,865 ....A 4 183
Hamilton, Ontario, 312,000 ....C 3 224
Hamm, Germany, 172,210 ....C 3 34
Hammond, Ind., 107,888 ....B 1 140
Hampton, Va., 120,779 ....F 3 205
Hanau, Germany, 88,676 ....C 3 35
Hangchow, China, 960,000 ....E 2 69
Hannover, Germany, 552,955 ....C 2 34
Hanoi,* Vietnam, 643,000 ....D 2 66
Harbin, China, 2,750,000 ....F 1 69
Haringey, Eng., 228,200 ....B 5 31
Harrisburg,* Pa., 68,061 ....D 3 189
Harrogate, England, 64,620 ....F 3 31
Harrow, England, 200,200 ....B 5 31
Hartford,* Conn., 158,017 ....C 1 127
Harz (mts.), Germany ....D 3 34
Hastings, England, 74,600 ....G 5 31
Hatteras (cape), N.C. ....G 2 179
Havana,* Cuba, 1,084,540 ....A 2 106
Havering, Eng., 239,200 ....C 5 31
Hawaii (isl.), Hawaii ....F 4 135
Hayward, Calif., 93,058 ....E 2 122
Hebrides (isls.), Scot. ....B 2 30

Ismailia, Egypt, 189,700 .... D 1 80
Istanbul, Turkey, 2,376,296 .... D 6 58
Itasca (lake), Minn. .... B 2 158
Ivano-Frankovsk, U.S.S.R., 128,000 .... B 5 51
Iviza (isl.), Spain .... G 3 41
Ivry-sur-Seine, France, 60,342 .... B 2 38
Iwo Jima (isl.), Japan .... D 3 84
Izhevsk, U.S.S.R., 489,000 .... H 3 50
Izmir, Turkey, 590,997 .... B 3 58
Izmit, Turkey, 141,681 .... C 2 58

Jabalpur, India, 426,224 .... D 2 64
Jackson,* Miss., 153,968 .... B 3 161
Jacksonville, Fla., 528,865 .... C 1 130
Jaén, Spain, 78,156 .... D 4 40
Jaffa-Tel Aviv, Israel, 357,600 .... B 3 61
Jaffna, Sri Lanka, 112,000 .... D 4 64
Jaipur, India, 615,258 .... C 2 64
Jalapa, Mexico, 161,352 .... H 1 103
Jalgaon, India, 106,711 .... C 2 64
James (bay), Canada .... E 2 225
James (river), Va. .... E 3 205
Jammu, India, 155,338 .... C 1 64
Jamnagar, India, 214,816 .... B 2 64
Jamshedpur, India, 341,576 .... D 2 64
Jan Mayen (isl.), Norway .... J 2 27
Japan (sea), Asia .... G 1 69
Java (isl.), Indonesia .... K 1 75
Jefferson City,* Mo., 32,407 .... D 3 162
Jena, Germany, 99,431 .... D 3 35
Jerez de la Frontera, Spain, 149,867 .... D 4 40
Jersey (isl.), Channel Is. .... E 6 31
Jersey City, N.J., 260,545 .... A 2 172
Jerusalem,* Israel, 344,200 .... C 4 61
Jhansi, India, 173,292 .... C 2 64
Jidda, Saudi Arabia, 561,104 .... D 5 56
João Pessoa, Brazil, 197,398 .... F 4 97
Jodhpur, India, 317,612 .... C 2 64
Johannesburg, South Africa, †1,417,818 .... F 7 82
Johor Baharu, Malaysia, 136,229 .... D 6 67
Joliet, Ill., 78,887 .... E 2 138
Jönköping, Sweden, 108,405 .... C 3 33
Jonquière, Quebec, 60,691 .... E 1 223
Jordan (river), Asia .... D 3 60
Juàzeiro do Norte, Brazil, 79,796 .... E 4 96
Judaea (reg.), Asia .... B 5 61
Juiz de Fora, Brazil, 218,832 .... D 7 97
Jujuy, Argentina, 82,637 .... C 2 98
Jullundur, India, 296,106 .... C 1 64
Junagadh, India, 95,485 .... B 2 64
Juneau,* Alaska, 13,556 .... G 2 117
Jura (mts.), Europe .... F 4 39

Kabul,* Afghanistan, 318,094 .... B 1 64
Kabwe, Zambia, †83,100 .... F 4 83
Kadiköy, Turkey, 81,945 .... D 6 58
Kadievka, U.S.S.R., 140,000 .... E 5 51
Kaduna, Nigeria, 202,000 .... F 6 79
Kaesong, N. Korea, 175,000 .... B 3 70
Kagoshima, Japan, 456,818 .... C 5 70
Kaifeng, China, 330,000 .... E 2 69
Kaiserslautern, Ger., 100,886 .... B 4 35
Kakinada, India, 164,200 .... D 3 64
Kalahari (desert), Africa .... E 6 82
Kalamazoo, Mich., 85,555 .... C 3 157
Kalgan, China, 480,000 .... E 1 69
Kalinin, U.S.S.R., 383,000 .... D 3 50
Kaliningrad, U.S.S.R., 331,000 .... B 4 51
Kalisz, Poland, 82,400 .... D 3 44
Kaluga, U.S.S.R., 240,000 .... E 4 51
Kalyan, India, 99,547 .... C 3 64
Kamakura, Japan, 165,548 .... H 2 71
Kamchatka (pen.), U.S.S.R. .... N 4 53
Kampala,* Uganda, 380,000 .... D 7 81
Kamyshin, U.S.S.R., 103,000 .... F 4 51
Kanazawa, Japan, 395,262 .... D 3 70
Kanchipuram, India, 110,657 .... D 3 64
Kanchow, China, 135,000 .... E 3 69
Kandahar, Afghan., 160,684 .... B 1 64
Kandy, Sri Lanka, 93,602 .... D 4 64
Kano, Nigeria, 399,000 .... F 6 79
Kanpur, India, 1,154,388 .... C 2 64
Kansas City, Kans., 168,213 .... H 2 145
Kansas City, Mo., 507,330 .... H 3 163
Kaohsiung, China, 1,000,000 .... F 3 69
Kaolack, Senegal, 106,899 .... A 6 79
Kara (sea), U.S.S.R. .... C 2 52
Karachi, Pakistan, 3,498,634 .... B 2 64
Karaganda, U.S.S.R., 523,000 .... D 5 52
Karakoram (mts.), Asia .... C 1 64
Karbala, Iraq, 211,214 .... C 4 62
Kariba (lake), Africa .... F 5 83
Karl-Marx-Stadt, Ger., 302,000 .... E 3 35
Karlskrona, Sweden, 60,009 .... C 3 33
Karlsruhe, Germany, 280,488 .... C 4 35
Karlstad, Sweden, 72,401 .... C 3 33
Karviná, Czech., 79,100 .... D 2 47
Kashan, Iran, 80,000 .... G 3 63
Kashgar, China, 175,000 .... A 2 68
Kassala, Sudan, 99,000 .... E 4 81
Kassel, Germany, 205,534 .... C 3 34
Katanga (reg.), Zaire .... M 3 82
Kathmandu,* Nepal, 150,402 .... D 2 64
Katowice, Poland, 320,400 .... B 4 44
Katsina, Nigeria, 120,000 .... F 6 79
Kaunas, U.S.S.R., 337,000 .... B 3 50
Kawaguchi, Japan, 345,547 .... H 2 71
Kawasaki, Japan, 1,015,022 .... H 2 71
Kayseri, Turkey, 183,128 .... F 3 59
Kazan, U.S.S.R., 970,000 .... G 3 50
Kazvin, Iran, 110,000 .... G 2 62
Kecskemét, Hungary, 77,963 .... E 3 47
Kediri, Indonesia, 178,865 .... K 2 75
Kelang, Malaysia, 113,607 .... C 7 67
Kemerovo, U.S.S.R., 385,000 .... E 4 52
Kénitra, Morocco, 135,960 .... B 2 78
Kenosha, Wis., 78,815 .... F 6 211
Kenya (mt.), Kenya .... E 8 81
Kerch, U.S.S.R., 145,000 .... E 5 51
Kerguélen (isl.), ... M 8 27
Kerman, Iran, 110,000 .... K 5 63